HORIZONS IN NEUROSCIENCE RESEARCH

VOLUME 15

Horizons in Neuroscience Research

Additional books in this series can be found on Nova's website
under the Series tab.

Additional e-books in this series can be found on Nova's website
under the e-book tab.

HORIZONS IN NEUROSCIENCE RESEARCH

VOLUME 15

ANDRES COSTA
AND
EUGENIO VILLALBA
EDITORS

New York

For permission to use material from this book please contact us:
Telephone 631-231-7269; Fax 631-231-8175
Web Site: http://www.novapublishers.com

Library of Congress Cataloging-in-Publication Data

ISSN: 2159-113X

ISBN: 978-1-63321-523-8

Published by Nova Science Publishers, Inc. † New York

Contents

Preface

This compilation presents original study results on the leading edge of neuroscience research. Each article has been carefully selected in an attempt to present substantial research results across a broad spectrum. Included in this volume are molecular mechanisms integrating adenylyl cyclase responsiveness to metabolic control on long term emotional memory and associated disorders; homeostatic synaptic plasticity; reorganization of elementary functions (ref) after brain injury and in the intact brain; the effects of acoustic stimulation on comatose patients; the splitting of the brain; the role of mineralocorticoid receptors in major depressive disorder; betaine in the brain; clinical outcome prediction using Bayesian regression and artificial neural networks; pathological changes of event-related brain oscillations in Parkinson's disease; the impact of sub thalamic stimulation on autonomic dysfunction in Parkinson's disease.

Chapter 1 - The individual and separate effects of Mg^{2+}, Mn^{2+}, Ca^{2+}, ATP^{4-} and their complexes on the kinetics of rat's brain adenylyl cyclase (AC), were jointly investigated with AC of fat cells and the insulin receptor tyrosine kinase (IRTK) of liver. The purpose was to evidence the integration of neurotransmitter responsiveness of AC, with the regulatory feedbacks of the metabolic network, supporting neuronal aerobic glycolysis. $MgCl_2$ saturation curves of AC show cooperative interactions at three sites: i) An obligatory free Mg^{2+} requirement for the enzyme responsiveness to noradrenaline (NA), with a cooperative value n=4 and 300% increase of Vmax. The NA effect may mediate an emotionally activated long-term memory, which requires a sudden increase of glucose and oxygen consumption. ii) Memory affirmation may require a restless basal AC-activity: Km(MgATP)=1.05 mM. The active site requires an excess of free Mg^{2+} over substrate, to prevent dead end inhibition by ATP^{4-} and/or CaATP. Basal AC-activity shows that the substrate MnATP increases affinity Km(MnATP)=0.06 mM and 200% the Vmax. AC-dependent memory may have Mn^{2+}-activated switches, for differentiable gating of nerve impulses, along specific synapses networks. iii) Basal activity shows a cooperativity value, n=2.6, surging from the interaction between active and regulatory sites. The latter, affinity values: K0.5(E-Mg^{2+})=4.7 mM, K0.5(E-Mn^{2+})=3.8 mM and K0.5(Ca^{2+})=0.02 mM. The AC kinetics could be integrated in a random rapid equilibrium (RARE BiBi), with dead end inhibition Ki(ATP^{4-})=0.27 mM and Ki(CaATP)=0.01 mM. An increase in chelating metabolites increases both dead end inhibitors. Alternatively, a decrease in chelating metabolites decreases CaATP, strongly activating AC-mediated and cAMP-dependent activation of pathways for memory affirmation. The insulin-independent uptake of glucose, supports anaerobic glycolysis in the erythrocyte, allowing maximal activity of 2,3-DPG mutase, at the cerebral spinal fluid (CSF) pH=7.4. The control of the erythrocyte 2,3-DPG level fulfills, the role of a glucose level sensorial capability of the Hb-red cell system. The molecular dynamics proposed allows that the sharing of some of the Hb R-groups allow binding of O_2 and the activatory ligands Mg^{2+} or Zn^{2+}, with exclusion of 2,3-DPG or viceversa. Hence, allowing matching of the rates of glucose and O_2 consumption during brain aerobic glycolysis. The overall metabolic feedback by the "in common" dependence on free Mg^{2+} or Mn^{2+} of AC of brain, fat cells and the insulin receptor tyrosine kinase (IRTK) in liver, allows homoestatic network integration. The Na^+/K^+-pump by breaking ATP^{4-} into ADP^{3-} and phosphate (Pi^{2-}) increases the concentration of ionic Mg^{2+}. The latter, by activating AC links the nerve impulse with the cAMP activation of protein kinases and the cAMP response element binding proteins. Hence, the dominant role of ATP within brain energy consuming processes, allows metabolic feedbacks which operate as a unificatory backbone of the processes leading to long-term memory consolidation.

Chapter 2 - Homeostatic regulation is a negative feedback response that is similar to other processes in the body that regulate things such as body temperature, hydration levels, pH and osmolarity of fluids, or respiration rates. Each homeostatic mechanism returns the system to an intrinsic physiological "set point". The brain is a dynamic organ that constantly evolves as we perceive and adapt to the world around us. This plasticity allows the brain to learn and form new memories, refine movements, recover after injuries, and to predict and obtain rewards. With billions of neurons making up to 100,000 synaptic connections each, the mammalian brain is a very complex system constantly undergoing modifications as it responds to these changes in the environment. A fundamental question is how the nervous system can constantly undergo modifications without losing its stability or the content encoded within neural circuits. To maintain stability in brain function, neurons must adopt a mechanism of homeostatic regulation by which they can assess the activity levels and make constant adjustment so as to maintain their activity within a normal physiological range.

At the neuronal level, homeostatic plasticity aims to maintain a stable rate in firing action potentials. This can be achieved through adjustments in the strength of synaptic inputs, neuronal excitability, neuronal connectivity or the balance between excitation and inhibition. Among these possibilities, the regulation of synaptic strength has been the most extensively studied and believed the most crucial measure in homeostatic regulation, which is known as homeostatic synaptic plasticity, or synaptic scaling.

Chapter 3 - Integration between psyche and soma was analyzed at the molecular level. Signaling by sensor neurons activates the fight-or-flight response through adenylyl cyclases (AC) "In common" signaling resulting from depletion of chelating metabolites by the tendencies of ionic equilibrium to increase free Mg2+ vs the decrease of inhibitors CaATP, ATP4-, and Ca2+. Mg2+-dependent responsiveness of AC to noradrenaline in brain and adrenaline on the tissue integrates hormone-dependent circadian rhythms. The response to stress of the hypothalamic-pituitary-adrenal (HTPA) axis is mediated by hypothalamic glucose sensor neurons, which signal increasing the adrenal glands secretion of adrenaline and cortisol. These on the pancreas stimulate secretion of glucagon and inhibit the release of insulin, distorting the pulsatile rhythm of glucose metabolism. Thus, flattens the normal circadian modulation of glucose availability to brain. At the tissue network, supporting brain, the rate of ATP generation and consumption could exceed the modulatory limits of Energy Charge. Thus, opposes the ATP-requiring regeneration of glycogen and fat reserves. The lipolytic hormones: adrenaline, ACTH, glucagon, etc. releases free fatty acids from fat tissue to support liver's gluconeogenesis supporting the glucose needs of brain. The basal AC activity of fat cell shows a Mg2+ saturation curve of cooperativity value "n" of about two because saturation by Mg2+ an activator sums-up with desaturation of ATP4- an inhibitor of the enzyme. The effect of hormones increases "n" to about 4 implicate activation of additional Mg2+ saturation sites, mediating in the membrane for the bonding of AC and the associated G-protein receptor. Cortisol that crosses the blood-brain barrier (BBB) exerts a negative feedback on the HTPA axis, but adrenaline does not, allowing persistence of its stimulatory effect on the HTPA axis to favor an uncontrolled secretion of cortisol. Persistent adrenaline-dependent inhibition of insulin secretion interferes with its capability to maintain a circadian rhythm, controlling availability of glucose to muscle, and brain. Parkinson and schizophrenia share on muscle-tremor symptomatology, which relates to depletion of the ATP required for muscular sliding. The overexposure to adrenaline has an inactivatory effect on the adrenaline-AC complex, which is considerably less stable, than the complexes of AC with either glucagon or ACTH. The lack of an adrenaline feedback could have been imposed by an evolutionary tendency of mind/brain dominance over soma/body itself. Thus, fight-or-flight response became a selection path for the development of an alternatively responsiveness adapted to environmental changes formative of long-term memory. Hence, the confluence of the sense of danger and metabolic mobilization resulted in a NA in brain vs adrenaline in blood conforming and axis leading to the corpus coeruleus neurons with their long axons to reach many cerebral regions to expand harmonize the range of perception and response. The presynaptically discharged NA acts postsynaptically on the Mg2+ sympathetic receptor or on the Ca2+-dependent alpha-2-receptor for a separated AC responsiveness to an activator free Mg2+ or inhibitor Ca2+. The Mg2+ incremented cAMP became a ligand of the cellular transcription factor, cAMP response element-binding protein (CREB). This one is generally accepted to participate in the plasticity of the neuronal circuits involved in long-term memory. Thus, allows an integrative role to the brain adrenergic system, synchronizing the glutamate-mediated neurotransmission. Additionally, AC variable responsiveness to NA also fits with the effect of Ca2+ vs Mg2+ in locally adjusting the glucose permeability of each capillary-neuronal BBB. The evolutionary increment of brain capabilities were supported by the role of astrocytes to keep a homeostatic temperature. This cell drives a rapid flux of H-bond depleted water cluster, out of the system, adjusting turnover to 3.7 times of the 160 ml cerebrospinal fluid per day. This interdependent network potentiates the adrenergic receptors to endow the brain with a potent working memory, but extremely sensitive to stress. Acute-panic could be associated to a lack of a promptly reuptake of NA, allowing a symptomatology of anxiety followed by depression, due to the exposure of the NA-AC complex, to the time-dependent inactivatory effects of NA on the AC enzyme.

Chapter 4 - The REF (Reorganization of Elementary Functions) model describes the dynamic neural and cognitive reorganizations mediating behavioural and conscious flexibility. These mechanisms have phylogenetically evolved to enable such flexibility in intact individuals. It is, however, the same mechanisms, which mediate the functional recovery after brain injury. Thus, the REF-model provides a theoretical framework for the understanding of normally occurring cognitive flexibility as well as the posttraumatic reorganization of the injured brain and for the development of improved therapeutic procedures in the cognitive rehabilitation of brain injured patients. A key feature of the model is that the "reprogramming" of the injured brain depends crucially on the feedback to the patient in the specific training situation. The resultant cognitive strategies may lead to proficient task solutions in the training situation (e.g. in the clinical setting) while not being able to generalize to the daily life of the patient. It is therefore important to develop more "ecologically valid" training methods in the therapeutic settings and to continue the rehabilitative training into the daily life of the patient. The use of advanced technologies is one of the promising ways to "bring reality into the institution" and to "bring training into the real world". Frequently, however, traditional methods of training and testing cannot easily be translated into advanced technology applications. Such efforts have to be undertaken as part of a thorough research process. This research process, on the other hand, provides a deeper insight into previously overlooked neural and cognitive mechanisms of the normal and injured brain. It is also possible to achieve an improved generalization from clinical training to reallife situations by specifically aiming the training at the development of what is in the REF-model called Algorithmic Modules (AMs). An AM participates in many different cognitive programmes (each of which is related to a specific task). Rehabilitative training aimed at improving AMs can therefore give the patient basic mechanisms, which can be applied broadly. The REF-model stresses that the prognosis regarding posttraumatic cognitive rehabilitation depends significantly on the pretraumatic experience of the patient. And that the rehabilitative training has to be tailored to the "cognitive profile" of the patient. The REF-model reconciles the concepts of "functional localization" and "functional recovery" by describing a connectionist network (with a high flexibility and potential for reprogramming) within which the basic units are strictly localized basic information processing units. This theoretical reconceptualisation highlights important challenges to neurorehabilitation and points the way towards an improved clinical practise. The model also represents an alternative to "modular" as well as "connectionist" models of the neurocognitive organization of the brain.

Chapter 5 - One of the most important clinical challenges in patients with severe brain damage is to estimate different aspects of their actual state of consciousness and, in particular, their residual conscious perception of the environment. Different methods have been developed to quantify the preservation of responses in the ascending sensory systems and the projection areas, e.g., bedside behavioural assessment and analyses of short-latency event-related potentials.

This study includes a detailed, objective and systematic analysis of the effects of acoustic stimulation on comatose patients using a standardised stimulus message. The curves of electrocardiography, blood pressure and pulse plethysmography and the pulse frequency, systolic, diastolic and mean arterial pressure, oxygen saturation level and peripheral body temperature, were monitored before, during and after the auditory stimulation of comatose patients with an unknown voice and a familiar voice. This study contributes new evidence to improve the quality of care and help in establishing the prognosis.

Chapter 6 - Mnemonic processing involves large scale networks, where several fiber bundles play significant roles, as they enable adequate and smooth flow of information between and within networks. The damage of these bundles might lead to various degrees of memory impairments, including amnesia. Cases of amnesia after the damage of white matter tracts have been reported for a long time, especially after frank alterations of white matter pathways due to strokes. The recent wider employment of structural brain imaging methodologies, such as diffusion tensor imaging or magnetization transfer ratio measurements has however improved the ability to estimate the integrity of the brain white matter and resulted in an increase in findings of fiber bundles' damage in association with memory impairments, including amnesia. This led to a rejuvenated interest in the mechanism of disconnection underlying amnesia in the scientific literature. Several medical, neurological and substance use- related conditions might be accompanied by both white matter damage and memory impairments. Furthermore, recent research data suggest that subtle micro-structural white matter tracts' changes may account for memory impairments in several psychiatric conditions, including dissociative amnesia. In the present chapter, after a brief introduction of the historical conceptualization of amnesia as a disconnection syndrome, an overview is provided of the fiber bundles that are important for mnemonic processing and causes and conditions that might alter their integrity and lead to amnesic conditions are discussed.

Chapter 7 - Corticosteroids regulate the Hypothalamic-Pituitary-Adrenal (HPA) axis via two types of receptors: the glucocorticoid receptor (GR) and the mineralocorticoid receptor (MR). The dysfunction of both receptors and a disturbed balance in the MR/GR ratio is thought to be implicated in stress related illness such as Major Depressive Disorder (MDD). Until recently there has been a lack of investigation into MR function in MDD. This review presents findings that reveal the vital role of MR in MDD. MR function is overactive in MDD, possibly compensating for a disturbed GR function, also dysfunction of MR is associated with treatment resistant depression and could also be an indicator of poor outcome following treatment. MR gene variants MRI180V and 215G/C could increase vulnerability to MDD. Future therapies for MDD should perhaps target the MR/GR ratio, MR function and expression and the interactions MR have with serotonin's 5HT1A receptors and with corticotrophin releasing hormone, in order to accelerate and improve treatment response in MDD. The exact mechanisms that underlie the role of MR in MDD are still enigmatic, thus future *in vitro* and *in vivo* investigations are required.

Chapter 8 - Elevated levels of plasma homocysteine have been negatively linked to the development of cardiovascular disease, impaired neurological function and most recently to neurodegenerative diseases. Homocystinuria, an in born error of metabolism results in significantly high levels of plasma homocysteine. Choline, along with its' metabolite betaine can lower levels of plasma homocysteine by remethylating it back to methionine with the aid of the enzyme betaine homocysteine *S*-methyltransferase (BHMT). Folates can also methylate homocysteine to methionine. Mice with genetic deficiencies in methylenetratrahydrofolate reductase (MTHFR) and methionine synthase reductase (MTRR) enzymes involved in folate and homocysteine metabolism have reduced levels of betaine in brain tissue. Interestingly, when patients suffering from homocystinuria are treated with betaine there is a reversal of most negative symptoms including reducing plasma homocysteine levels, however neurological symptoms still remain. This chapter will discuss interactions of betaine with folate metabolism, implications of reduced betaine levels in brain function and importance of this nutrient in normal and disease brain function.

Chapter 9 - The goal of this chapter is to use a combination of traditional and Bayesian logistic regression, as well as artificial neural networks for clinical outcome prediction. Using the example of ruptured brain aneurysms, this chapter provides novel insights on specific brain-body interactions that are essential in influencing outcome in patients with ruptured brain aneurysms. Thus far, there is scarce existing literature on such important relationships. The multicenter Tirilazad database (3550 patients) was used to create this clinical outcome prediction model in order to elucidate significant brain-body interactions. Traditional logistic regression main effects model included five statistically significant single prognostic variables, namely, neurological grade, age, diastolic blood pressure, time to surgery and stroke. Logistic regression models found the significance of liver disease and hypertension in development of brain swelling, and the negative consequences of seizures in those with history of myocardial infarction and post admission fever worsening neurological outcome. Bayesian logistic regression using informative prior probability likelihoods detected a number of prognostic factors that may play a role in multiplicative interaction effects, as well as possible heterogeneity in patient subgroups. An artificial neural networks model, with similar model discrimination values as traditional logistic regression model, was created to explore the influence of non-linear relationships on the importance of prognostic factors.

Chapter 10 - Brain event-related oscillations provide temporal and spatial codes for information processing and give a useful tool for detecting subtle abnormalities in cognitive processes. They are one of the promising candidates explaining the neural mechanisms at central nervous system in Parkinson's patients (PP) and healthy subjects (HS). This study reports auditorily elicited brain oscillatory responses in association with sensorimotor and cognitive processes. Brain oscillations at specific frequencies and times provide a window into the organization and transfer of information underlying such processes in the human brain. Brain event-related oscillatory responses evoked from given auditory stimuli (low-800, high tone type-1000 Hz) were quantified by means of event-related desynchronization / synchronization (ERD/ERS) technique for PP and HS. The differences between the ERD/ERS responses at different frequency bands (delta 1.5–4, theta 4–7, alpha 7–14, beta1 13-20, beta2 20-32, gamma 32-50 Hz) for PP and HS during auditory discrimination tasks are of relevance to the understanding of these cortical functions. Widespread delta- and theta- ERS, and alpha-ERD were significantly pronounced in HS in response to either tone during sensory information processing (0-250 ms after the stimulus onset). More prominent fronto-parietal delta-ERS was observed in response to a low than to a high tone, with predominantly theta- and alpha-ERD in HS following either tone during the cognitive processing (250-600 ms after the stimulus onset). Beta1-, beta2- and gamma-ERD were revealed everywhere in HS after either tone within both intervals except for frontal beta1- and gamma-ERS after a high tone during the cognitive processing. When comparing the PP and HS during the sensory processing, PP delta- and theta- ERS were less pronounced after either tone with exception of frontal higher delta-ERS and theta-ERD at the left motor area in response to a high tone. Widespread alpha-ERS was found only in PP during the sensory processing in response to either tone. PP delta-ERD after a low tone and delta-ERS after a high tone, theta- , alpha-ERD and parietal theta-ERS following either tone were revealed during cognitive processing. Only frontal theta- and alpha-ERD were more prominent for PP than for HS. Beta1-ERD was less pronounced in PP within both intervals after either tone. Beta2-ERS was observed only in PP, but the most prominent beta2-ERS followed the high tone during cognitive processing. Gamma-ERD was observed in PP during sensory processing except for fronto-parietal ERS after a low tone. During cognitive processing, either tone, prominent gamma-ERS was revealed everywhere for PP. Deviations of the ERD/ERS at specific frequencies for the PP compared with HS during sensorimotor and cognitive processing are clear evidence for disturbances in the temporal and regional integration of these frequency components and the relationships between cortical and basal ganglia circuits in parkinsonism.

Chapter 11 - Parkinson's disease (PD) is a degenerative movement disorder, with rigidity, bradykinesia and tremor being its cardinal symptoms. Even in the early stages of PD, Autonomic Nervous System (ANS) involvement is established with cardiovascular, urogenital, gastrointestinal, sudomotor and thermoregulatory manifestations. Degeneration changes in hypothalamus, intermediolateral nucleus of the thoracic cord, sympathetic ganglia and in sacral parasympathetic nuclei might be responsible for the dysautonomia in PD. There is substantial evidence that Deep Brain Stimulation of the Subthalamic Nucleus (STN-DBS) is particularly effective in improving motor impairment and reducing the medication needs in patients with PD. However, the influence of this modern therapy on the autonomic regulation is not yet clarified. There are several standard tests to estimate the integrity of each autonomic function and therefore the STN-DBS influence upon it, such as: Sympathetic Skin Response, Quantitative Sudomotor Axon Reflex Test, Thermoregulatory Sweat Test, salivation test, plasma catecholamine levels and urodynamic studies; specifically for the cardiovascular dysautonomia the tests are divided into those estimating the sympathetic (blood pressure changes to standing, to head-up Tilt testing and during Valsalva maneuver) and those estimating the parasympathetic part of the regulation (recording of heart rate changes to standing, to head-up Tilt-table test and during Valsalva maneuver, heart rate variation to deep breathing) or both (spectral analysis of RR interval variation). The DBS mechanism of action in the basal ganglia is complex and not completely understood, but since the basal ganglia project to a number of nuclei involved in the control of the ANS, as well as to thalamus, a possible direct effect of STN-DBS on ANS functions might be expected. It is worth mentioning that STN-DBS has been found to increase blood flow in the anterior cingulated cortex and the insular cortex, which are primary centers of autonomic control. On the other hand, ANS manifestations could subside due to post-DBS medication reduction and general improvement of motor function seen in the majority of patients. Nevertheless, the results of the - up to date- reports on whether STN-DBS could influence the dysautonomia in PD are controversial. Some study groups have reported in direct effect of STN-DBS on autonomic cardiac function, whereas others have found improvement in several cardiovascular variables and in Sympathetic Skin Response, as well as an amelioration of constipation, dyshidrosis and urological dysfunction. In this chapter the authors present the current status of the investigations on DBS effect on ANS function in humans.

In: Horizons in Neuroscience Research. Volume 15
Editors: Andres Costa and Eugenio Villalba
ISBN: 978-1-63321-523-8

Chapter 1

Molecular Mechanisms Integrating Adenylyl Cyclase Responsiveness to Metabolic Control on Long-Term Emotional Memory and Associated Disorders

A. Bennun
Full Professor- Graduate School Rutgers University.
Joint Graduate Program Rutgers and New Jersey College Medicine and Dentistry (NJCMD), Newark, NJ, US
CONICET, AR

Abstract

The individual and separate effects of Mg^{2+}, Mn^{2+}, Ca^{2+}, ATP^{4-} and their complexes on the kinetics of rat's brain adenylyl cyclase (AC), were jointly investigated with AC of fat cells and the insulin receptor tyrosine kinase (IRTK) of liver. The purpose was to evidence the integration of neurotransmitter responsiveness of AC, with the regulatory feedbacks of the metabolic network, supporting neuronal aerobic glycolysis. $MgCl_2$ saturation curves of AC show cooperative interactions at three sites: i) An obligatory free Mg^{2+} requirement for the enzyme responsiveness to noradrenaline (NA), with a cooperative value n=4 and 300% increase of Vmax. The NA effect may mediate an emotionally activated long-term memory, which requires a sudden increase of glucose and oxygen consumption. ii) Memory affirmation may require a restless basal AC-activity: Km(MgATP)=1.05 mM. The active site requires an excess of free Mg^{2+} over substrate, to prevent dead end inhibition by ATP^{4-} and/or CaATP. Basal AC-activity shows that the substrate MnATP increases affinity Km(MnATP)=0.06 mM and 200% the Vmax. AC-dependent memory may have Mn^{2+}-activated switches, for differentiable gating of nerve impulses, along specific synapses networks. iii) Basal activity shows a cooperativity value, n=2.6, surging from the interaction between active and regulatory sites. The latter, affinity values: $K_{0.5}(E\text{-}Mg^{2+})$=4.7 mM, $K_{0.5}(E\text{-}Mn^{2+})$=3.8 mM and

$K_{0.5}(Ca^{2+})$=0.02 mM. The AC kinetics could be integrated in a random rapid equilibrium (RARE BiBi), with dead end inhibition $K_i(ATP^{4-})$=0.27 mM and K_i(CaATP)=0.01 mM. An increase in chelating metabolites increases both dead end inhibitors. Alternatively, a decrease in chelating metabolites decreases CaATP, strongly activating AC-mediated and cAMP-dependent activation of pathways for memory affirmation. The insulin-independent uptake of glucose, supports anaerobic glycolysis in the erythrocyte, allowing maximal activity of 2,3-DPG mutase, at the cerebral spinal fluid (CSF) pH=7.4. The control of the erythrocyte 2,3-DPG level fulfills, the role of a glucose level sensorial capability of the Hb-red cell system. The molecular dynamics proposed allows that the sharing of some of the Hb R-groups allow binding of O_2 and the activatory ligands Mg^{2+} or Zn^{2+}, with exclusion of 2,3-DPG or viceversa. Hence, allowing matching of the rates of glucose and O_2 consumption during brain aerobic glycolysis. The overall metabolic feedback by the "in common" dependence on free Mg^{2+} or Mn^{2+} of AC of brain, fat cells and the insulin receptor tyrosine kinase (IRTK) in liver, allows homoestatic network integration. The Na^+/K^+-pump by breaking ATP^{4-} into ADP^{3-} and phosphate (Pi^{2-}) increases the concentration of ionic Mg^{2+}. The latter, by activating AC links the nerve impulse with the cAMP activation of protein kinases and the cAMP response element binding proteins. Hence, the dominant role of ATP within brain energy consuming processes, allows metabolic feedbacks which operate as a unificatory backbone of the processes leading to long-term memory consolidation.

Introduction

Long term memory requires the construction of new synapses [1] [2] Changes in synaptic strength participate in learning and memory. The process involves neuro-transmitters, receptors and new synapses generating pathways which reinforce the strength of connectivity between neurons. Noradrenalin (NA) binding activates various subtypes of adrenergic receptors, for a differential role of signaling pathways, which functionally operate by inducing synthesis of second messengers.

The *striatum* and the hippocampus have been shown by Functional Magnetic Resonance Imaging (FMRI), to interact cooperatively to support episodic memory formation [3]. New long-lasting details of specific events could be lost to people that show amnesia, a dysfunction that can be related to hippocampus damage [4].

The *cerebral cortex* is capable of semantic memory, the ability to learn facts and relationships mediated by changes in the connections between cells, related by their conducting of specific types of information [5].

The ventromedial *hypothalamus* (VMH) has been shown to respond to catecholamine released into the glucose-sensing region, through a counter-regulatory response, to hypoglycemia by stimulation of the beta-2-adrenergic-receptor (B2AR) [6].

Emotional charged events have been related to activation of the locus coeruleus release of noradrenalin [7], activating a second messenger system [8]. Noradrenalin activates AC in membrane preparations obtained from rat's cerebral cortex, corpus striatum and hypothalamus [8]. However, over-exposure to noradrenalin of AC during pre-incubation, an assay previous to the one measuring the activatory effects of the neurotransmitter [9], results in inactivation of the enzyme.

Hence, it could be inferred that noradrenalin-dependent psychosomatic damage, lead in the course of an evolutionary selection process to a costly protection mechanism. This one, turns-off the enzyme rather than allowing a too perdurable permanence of the organism into the fight or flight state.

The results of the experimentally obtained prolonged association of NA with its receptor may correspond to a stressing of the quaternary structure, during the activatory state [9]. This catecholamine-dependent inactivatory event was prevented by a pre-incubation condition, in which the presence of MgATP protects from NA, by the formation of an enzyme substrate complex. Hence, the NA AC complex manifests thermic instability, whereas the NA-AC-MgATP allows to protect the enzyme activity, but to the cost of decreasing its responsiveness to a catecholamine activatory effect [8] [9] [10].

Stress mediated by increasing adrenalin secretion, usually shows a pattern of hyperfunction follow by hypofunction, which when the stressed organ is brain it could be associated with a symptomatology of persistent anxiety followed by depression [9].

Under prolonged stress, the synthesis of new AC, may not match the rate of noradrenalin-dependent inactivation. The viability of functional synapses may be controlled by the speed of synthesis of the proteins, which make-up the hormonal receptor sites and/or AC-itself, versus the rate of inactivation or destruction of AC. This generalized molecular modeling for the evolution of psychosomatic diseases [9] could be tested, with regard to more specific physiological conditions and permit prognosis of dysfunction, allowing preventing treatments.

Already medical applications have been developed, based in the use of Beta blockers to interfere with a stressful condition, attenuating the memory of a traumatic event.

Characterization of the NA activation of brain AC indicated that could participate in the formation of molecular memory pathways [8] [11]. This proposition was extended to show its role in cAMP-dependent pathways, involved in the development of behavior patterns [12] [13] [14] [15] [16]. In absence of noradrenalin, AC was shown to have a basal activity subject to up-regulation by Mg^{2+} and Mn^{2+} ions and their ATP complexes: MgATP and MnATP as substrates [11]. Na^+-pump activity changes ionic equilibrium by the neuronal transfer Na^+-out coupled to K^+ intake. Additional changes, relate to ATPase activity splitting ATP^{4-} into ADP^{3-} and P_i^{2-}, with an ionic equilibriums allowing the endogenous release of free Mg^{2+} ions. The breakdown of ATP also decreases the concentration ATP^{4-} and CaATP, which were shown to act as strong inhibitors of AC activity [11]. More recent studies obtained from the molecular, behavioral cognitive, signal transduction, physiologic and pathological research, has shown the relevance of cAMP-mediated long-term memory [17] [18] [19] [20] [21] [22] [23] [24] [25].

It could be expected that a large number of the pathways associated with neuronal activity, could be integrated into the complex process of memory formation and consolidation, by an “in common” coupling by the interdependence of signal and ATP-energy transduction mechanisms [26] [27] [28] [29].

Up-regulation of AC by ionic equilibrium in dependence of changes in ATP concentrations, led to investigate possible metabolic-ionic feedbacks, related to the energy requirement of the NA-AC system. Stimulated neurons require O_2 to maintain the supply of ATP for the Na^+-pump. O_2 increases turnover of both electron transport chain and Krebs cycle, which occurs through the increase in dicarboxylic and tricarboxylic acids concentrations in mitochondria. Effect which increases Ca^{2+} retention in mitochondria and up-regulate the Na^+/K^+-ATPase-AC system. The relationship between the rate of aerobic glycolysis and neuronal work was therefore examined, in the context of signal transduction within tissue networking. The latter, allows a large increment in the rate glucose and O_2 uptake by stimulated brain areas, even when sustaining homeostatically the glucose levels in blood.

Methods

Preparations

Female Sprague-Dawley rats weighing 210-300g were ovariectomized to decrease interference by estrus cycles [8] and decapitated at 4°C two weeks after surgery. The brains were dissected in the cold (4°C), to isolate and separately collect the hypothalamus, cortex and striatum from groups of 10 to 15 animals [8] [30]. The total wet weight of each pool of tissues was colected in 20 ml of ice-cold 2 mM Tris-HCI containing 10 mM NaCl, 10 mM KCl and 5μM EDTA, buffer at pH=7.4. These were rinsed twice with 15 ml/g wet weight of tissue and homogenized using a motor-driven (Inframo, Wayne, New Jersey) Teflon homogenizer at 400 rpm for 2 min. Each homogenate was centrifuged at 4°C and 6,000 g_{av} in a RCA Sorvall, head SS-34 for 20 min. The resulting supernatants were discarded and the pellet washed, resuspended and centrifuged again once. Aliquots stored at -70°C. Protein concentration for each membrane preparation was determined by the method of Lowry (Lowry et al., 1951) using bovine serum albumin as the standard. Particle suspensions and reaction mixtures were tested, with a calcium ion electrode for Ca^{2+} contamination with negative results.

Assay of Adenylyl Cyclase Activity

Adenylyl cyclase (AC), systematic name: ATP diphosphate-lyase (cyclizing, 3´,5´-cyclic-AMP-forming), EC 4.6.1.1, in the membrane preparations of cortex, corpus striatum and hypothalamus were kinetically characterized. The tissues were selected in expectation of finding differential kinetic properties, which could be related to specific areas of rat´s brain [8].

Defrosted particulate preparations (from 1.7 to 6 μg protein/assay) were re-suspended, in a standard reaction mixture, containing 20 mM Tris-HCl buffer (pH 7.4), 5 mM theophylline, $MgCl_2$, plus ATP disodium salt, from equine muscle, with GTP content of less than 2% and traces of other nucleotides from Sigma. This ATP/GTP mixture was used when neurotransmitter dependent activation, was desirable as in the experiments reported in figure 1 a) and b) and table 1. The ATP disodium salt from P-L Biochemicals, Inc (Milwaukee, Wisconsin) was used for the assay of AC basal activity. Other additions indicated in the figures legends.

After incubating in a shaking bath at 37°C for 15 min, the reaction was terminated by heating in a boiling water bath for 5 min. Time-dependent studies using the membrane preparations, showed linear responses for no less than 30 min.

Cyclic AMP formed by AC activity was assayed by the Gilman method [31] modified by Brown et al. [32] according to publish specifications [33] but with cyclic [^{3}H]AMP increased to 20,000 c.p.m./assay. These membrane preparations show an ATPase activity [34] of only 0.1 nmol P_i/h/µg-protein [8].

Results were calculated by formula, (C- b)/(T- b), C: c.p.m. from the assay of the enzyme activity, T: c.p.m. from total binding (zero-time blanks) and b: c.p.m. in the millipore filters (moisture blanks). Calibration curves are linear for % of C- b/T- b versus log cAMP without interference by ATP or other reagents. Results mean values of triplicates or more: nanomoles of cAMP formed per hour per milligram of protein. The standard error of the mean (S.E.M.) did not exceed 6% [35].

Results

Previous kinetic studies of the noradrenalin-activated adenylyl cyclase (AC), evidenced that the neurotransmitter response could be differentiated from a basal response, according to their different Vmax, which could explain memory at two differentiable levels. The noradrenalin dependent activity of AC, could be induced by events with strong emotional connections, related to the fight or flight reaction. The activity of the neurotransmitter-enzyme complex correlates to a reinforcement over the basal activity elicited by free Mg^{2+}- or Mn^{2+}-dependent activation of the basal state of the enzyme [9] [10]. However, prolonged effects may elicit a CNS excited or anxious state, which prolonged results in a depressive state [9] [36]. The latter, could be related to memory deficiencies so emotional potentiation of memory capability became limited by collapsing under stress conditions.

The present studies were directed: I) To elucidate the kinetic properties of AC relevant to modeling mechanism: a) for NA-dependent emotional eliciting and reinforcement of AC-dependent long-term memory, b) for continuous memory processing by activation of the basal AC activity by free divalent metals; II) Characterization of hemoglobin red cell-system as a sensor of glucose levels in blood, and responsiveness for O_2-release matching glucose uptake III) To detect a metabolic feedbacks of a tissue network, capable to shift the rates of O_2 and glucose consumption, between hormonal activated and the basal memory processing stage. The matching of the rates of brain glucose consumption and tissue oxygenation requires "in common" molecular signaling. This one should operate within a network, by a healthy physiologycal response through the sensing role of the red cell-Hb system, matching the brain metabolism with the homeostatic need of balancing the metabolic flows, within the liver fat-cell subsystems.

Relevant Brain AC Kinetic Properties for Emotional Reinforcement and Basal Memory Processes

Purified enzymes preparations from rat's tissues were obtained for adenylyl cyclase (AC) of fat cells, brain and insulin receptor tyrosine kinase (IRTK) from liver. This preserved their allotopically associated receptors since did not extend purification to an extraction procedure separating the enzyme from the membrane [34]. These enzyme-membrane systems were shown to preserve modulatory properties, required for their physiological role.

The preparations of AC [ATP pyrophosphate-lyase (cyclizing) (EC 4.6.1.1)] obtained from corpus striatum, hypothalamus and cortex of the brains of rats show that noradrenalin (nor-epinephrine) activation is dependent of the concentration of magnesium ion (Mg^{2+}).

Impulse transmission and/or noradrenaline binding the receptor induce the depolarization of the plasma membrane. Neuronal transmission is terminated by the neurotransmitter reuptake linked to Na^{+} entry. The β-receptor is present in the hormone-binding subunits of the enzyme AC.

Table 1. Kinetic characterization of adenylyl cyclase obtained from specifics brain areas

Particulate Membrane preparations	Km(MgATP) [mM]	Cooperativity (n)
Cortex	2.2	1.4
Striatum	2.0	1.1
Hypothalamus	2.0	1.1

The values obtained do not allow inferring differences between the responsiveness to the substrate of the basal adenylyl cyclase activity.

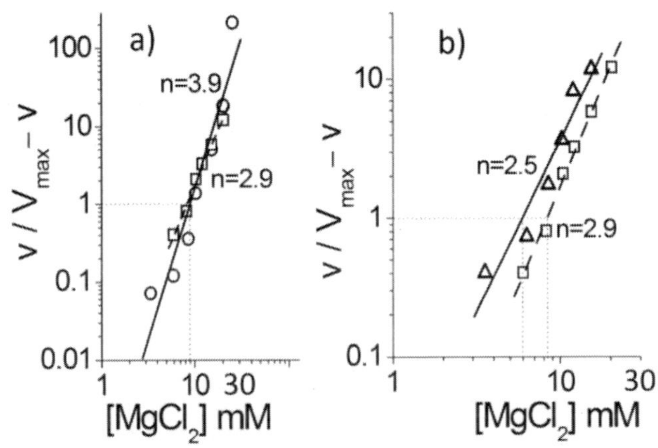

Figure 1. a) and b). Control by ΔMg_T (total Mg or $MgCl_2$ added) at constant 1mM-ATP of the responsiveness of noradrenaline activated AC. Samples of an EDTA-particles preparation from cortex containing 100μg of protein were incubated at 37oC for 1hour, in 40mM-Tricine/Tris buffer, pH=7.4, 6.67mM-caffeine, constant 1mM-ATP (Sigma purified from horse muscle), in a final 0.5ml with the indicated concentration of $MgCl_2$ (Mg_T). Vmax for each curve was determined at the indicated $MgCl_2$ concentration. Activity reported: nano-moles of cAMP formed/h per mg of membrane protein. Figure 1.a). $MgCl_2$ saturation curves: o) +1mM-adrenaline, at 30mM [$MgCl_2$] were determined Vmax values, 212 ±14(4);Δ) basal, 78±7(4). Figure 1.b) $MgCl_2$ saturation curves: basal, at 15 mM [$MgCl_2$], 78±7(4); □) +1mM-adrenaline +0.3 mM $CaCl_2$ at 25mM [$MgCl_2$], 27±2(4).

When noradrenaline binds to its receptor, this one becomes activated to bind G-protein, which is activated by GTP. As has been reported elsewhere, the neurotransmitter activated enzyme, increases Vmax. The activated enzyme generates endogenous cAMP, which rather than acting as a messenger between one neuron and another, has the capability to induce changes in the function of ion channels within the membrane. The cAMP has effect on gene expression in the nucleus of the cell and in the processes of growth and development.

Changes on the genetic material of cells may lead to long-term alterations of behavior. Second messengers play a role in the manufacture and release of neurotransmitters, intracellular movements, and a carbohydrate metabolism in the brain.

Hence, Table 1 shows that allotopic enzyme-membrane properties are similar for the brain regions tested.

Figure 1.a) shows the activatory effect of noradrenaline versus basal on the AC of cortex increasing by 300% basal Vmax. The cortex enzyme shows cooperativity for total-magnesium in the presence of 0.1mM-noradrenaline with a value $n(Mg_T)=3.9$.

Figure 1.b) shows that the AC basal cooperativity decreases from $n(Mg_T)=2.9$ and by addition of noradrenaline plus $CaCl_2$ decreasing to $n(Mg_T)=2.5$ and decreased Vmax to 1/3.

Comparative results on the effect of noradrenaline in AC of corpus striatum are reported in table 2.

In table 2, the AC of corpus striatum at basal conditions and for 0.1mM-noradrenaline and for +0.1mM-noradrenaline+0.3mM $CaCl_2$, show similar affinity $S_{0.5}(Mg_T)$ and $S_{0.5}(Mg^{2+})$ of about 9mM. The affinity for the inhibitory specie $S_{0.5}$ $(ATP^{4-})=0.1mM$. Hence, noradrenalin mainly increases Vmax but does not change by much the enzyme affinity for either Mg^{2+} or ATP^{4-}. The AC of corpus striatum cooperativity for 0.1mM-noradrenaline n=4.4. Cooperativity for basal n=2.7, since for basal $n(Mg^{2+})$, $n(ATP^{4-})$ and n(MgATP) show values close to 1.

Hence, the values obtained for basal activity $n(Mg^{2+})$, are influenced by the summation to the activatory effect of Mg^{2+}, and the disappearance of the inhibitory effect of ATP^{4-}.

The enzyme saturation system, shows a n(adrenaline) values of about 4, because the neurotransmitter potentiates Mg^{2+} mediated binding to the membrane. The hormonal receptor for the expression of properties dependent of its position in the membrane, allotopic, may require Mg^{2+} [34] for the activatory binding of G-protein.

The enzyme shows a single active site, which requires Mg^{2+} and MgATP for basal activity, and a separate for the binding of an adrenaline receptor.

The applicability of the random of the RARE BiBi model is confirmed by the ability of the model to predict a change of Hill cooperativity from a basal value of n=2.6 to 4.0 in the presence of noradrenaline (Figure 1 and Table 2). This indicates that excess ions Mg^{2+} over that required for the saturation of sites, for substrate formation and for the Mg^{2+} regulatory activation. Hence, an additional saturation by ionic Mg^{2+} is an obligatory requirement for the manifestation of third site linked to the responsiveness to hormonal dependent activation [8] [33] [35]. Lineweaver-Burk plots of AC activity as a function of substrate concentration at different concentrations of Mg_T show a noncompetitive behavior, indicating that Mg^{2+} interacts with a distinct regulatory site differentiable from the active site [11]. If the data of Figure 5 is plotted according to Lineweaver-Burk, the addition of CaATP shows Ki=0.015mM. However, Mg^{2+} (Mg_T) can easily displace Ca^{2+} from ATP. The equilibrium favors that Mg^{2+} can easily displace Ca^{2+} from ATP, and reduce the concentration of CaATP. Since $Ka(MgATP)=-20\times10^3M^{-1}$ versus $Ka(CaATP)=-9\times10^3M^{-1}$.

Evaluation of the responsiveness of the enzyme indicates that there is a requirement that the concentration of free Mg^{2+} exceeds that of MgATP to activate the enzyme. The AC response appear coupled to the operation of the Na^+ gates augmenting free ionic Mg^{2+}. The gates associated ATPase reaction releases free Mg^{2+} because since ADP^{3-} has a much weaker association constant for the divalent metal than ATP^{4-}. The Hill Plot in figure 3 was obtained from saturation curves with a Vmax(MnATP): 19.2nmol cAMP/h/mg, a considerable higher activity than the corresponding Vmax(MgATP): 8.2nmol cAMP/h/mg. Comparing both Km, it could be shown a marked increase for the basal activity of adenylyl cyclase in the presence of MnATP, with an affinity increment of over 17 times than the Km MgATP. Moreover, Vmax was increased for the substrate MnATP 2.3 times over that of Vmax MgATP. Equilibrium conditions for MnATP (log Ka=4.6) allowed to increase the concentration of MnATP at negligible concentration of free ionic Mn^{2+}. Hence, it could be infered that if both substrates are separately compartmentalized, the adenylyl cyclase consuming MnATP would be able to produce cAMP at much higher rates than in the presence of MgATP alone.

Table 2. Effects of Mg^{2+} and ATP^{4-} on the response to noradrenaline and calcium of adenylyl cyclase of corpus striatum

Kinetic Constants	Basal	0.1mM-Noradrenaline	0.1mM-noradrenaline +0.3mM $CaCl_2$
Mg_T	n=2.7	n=4.4	n=2.3
Mg_T	$S_{0.5}$=8mM	$S_{0.5}$=9mM	$S_{0.5}$=9.4mM
Mg^{2+}	n= 1.1	n=1.9	n=1.3
Mg^{2+}	$S_{0.5}$=4.7mM	$S_{0.5}$=4.6mM	$S_{0.5}$=4.6mM
ATP^{4-}	n=1.1	n=1.9	n=1.3
ATP^{4-}	$S_{0.5}$=0.1mM	$S_{0.5}$=0.1mM	$S_{0.5}$=0.1mM

To investigate the individual and separate effects of free Mg^{2+} and of ATP^{4-}, their concentration were determined by multiple-equilibrium equations.

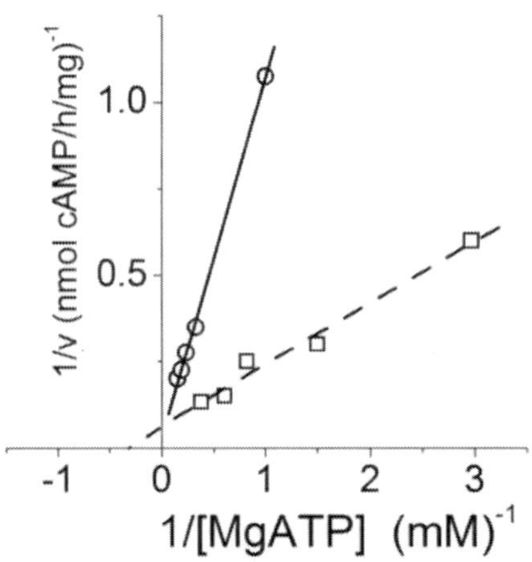

Figure 2. Lineweaver-Burke plot. Effect of ATP^{4-} on the substrate saturation curve of AC particulate preparation from corpus striatum. Portions of an AC particulate preparation containing 2.5 µg of protein were incubated. Basal, MgATP saturation curve, at 5mM MgATP, Vmax: 9nmol cAMP/h/per mg of protein; o) basal+2.5 mM ATP^{4-} at 9mM MgATP, $Vmax_{app}$: 7nmol cAMP/h/per mg of protein constant equilibrium concentration adjusted according to table 3.

Table 3. Equilibrium concentrations of MgATP, ATP^{4-} and Mg^{2+} under experimental conditions where ATP^{4-} is kept constant and MgATP is increased. Equilibrium concentrations, adjusted from the calculated ionic equilibrium [37] [38]

Mg^{2+} (mM)	ATP^{4-} (mM)	MgATP (mM)	$[MgATP]/[Mg^{2+}]$
0.0058	2.20	0.244	42.1
0.0229	2.23	0.976	42.6
0.0455	2.25	1.950	43.4
0.0896	2.29	3.910	43.4
0.1110	2.31	4.880	43.9
0.1320	2.33	5.860	44.0
0.2000	2.33	8.900	44.0

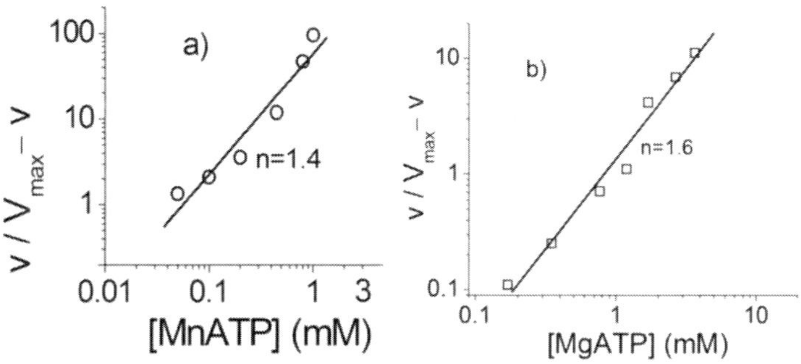

Figure 3. a) MnATP substrate saturation curve and 3.b) MgATP substrate saturation curve. Portions of AC particulate preparation from corpus striatum containing 6μg proteins were incubated in the conditions for assay of basal activity described for figure 1. Activity reported: nmol cAMP formed/h per mg of membrane protein. Figure 3.a), MnATP saturation curve: o) At 1mM MnATP reached Vmax value: 19.2nmol cAMP, cooperativity value n(MnATP)=1.4, Km(MnATP)=0.05mM. Figure 3.b), MgATP saturation curve: At 3.6mM MgATP reached Vmax: 8.2nmol cAMP, cooperativity value n(MgATP)=1.6, Km(MgATP)=0.84mM.

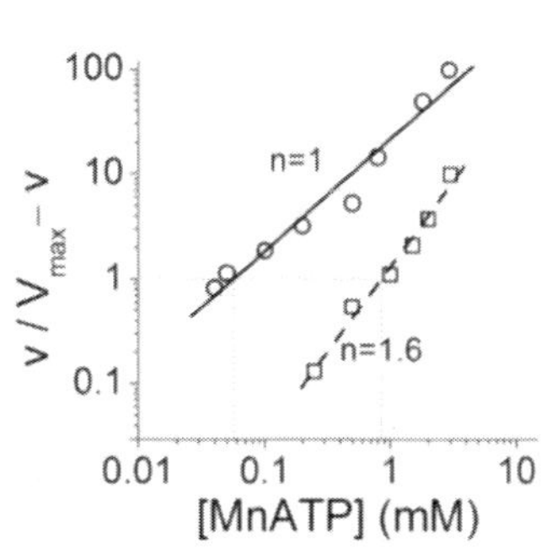

Figure 4. Effect of inhibitor ATP^{4-} on the substrate MnATP saturation curve of corpus striatum adenylyl cyclase. 6μg protein portions of an adenylyl cyclase-membrane preparation were incubated at 37°C as indicated in the legend to figure 1 o) Basal; with the indicated increasing equilibrium concentration of MnATP at 1mM MnATP reached Vmax: 19.2nmol cAMP/h/per mg; basal+inhibitor 5.5mM ATP^{4-} at 4mM reached Vmax: 15.2nmol cAMP/h/per mg.

Additionally, the adenylyl cyclase in the presence of substrate MnATP manifests generation of cAMP independently of the already discussed free Mg^{2+} activatory requirement (figure 1), controlling responsiveness at the neurotransmitter and basal levels. Hence, it could be infer the existence of neurons or synaptic structures able to orchestrate the response of adenylyl cyclase utilizing MgATP, with a control role in the generation of neuronal memory nets.

The increasing concentration of $[MnATP]/[Mn^{2+}][ATP^{4-}]=4\times10^4$, implies that equilibrium maintains negligible concentration of free Mn^{2+}. Hence, the addition to the basal curve of ATP^{4-} maintains its equilibrium concentration well above that in the basal curve. However, the ATP^{4-} inhibitory-curve approaches the basal at the higher maximal concentration, showing unchanged Vmax.

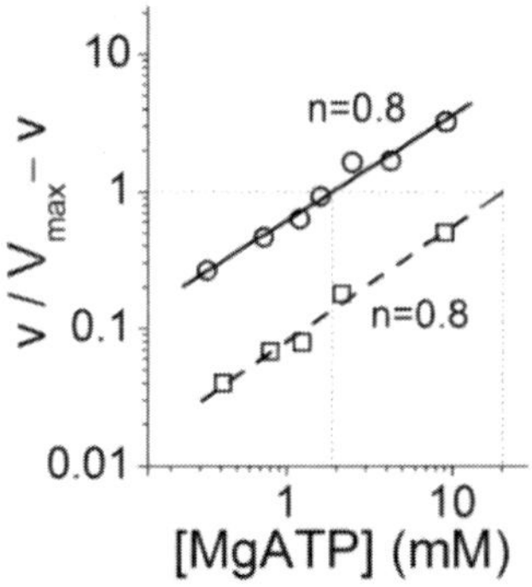

Figure 5. Effect of inhibitor CaATP on the substrate MgATP saturation curve of hypothalamic adenylyl cyclase. 3.3μg protein portions of a hypothalamic-membrane preparation were incubated at 37°C as described in the legend to figure 1 with increasing equilibrium concentrations of 0.41 to 9.3 mM MgATP: MgATP saturation curve: o) Basal at 9.3mM MgATP Vmax: 13.2nmol cAMP/h/mg; □) basal+inhibitor 0.37±0.05mM CaATP at 9.3mM MgATP Vmax: 5.8nmol cAMP/h/mg. Cooperativity value: n, computational obtained were included in the figure.

Table 4. Kinetic values from Hill and Lineweaver and thermodynamics constant from Figure 2, 4 and 5 for the indicated E-Complexes. Kd dissociation constant; Kf formation constant for enzyme complexes; ΔG°, Gibbs free energy. Kd values [37] [38] change by only 9% when ionic strength changes 100%. Hence, Kd values where affected by considerably lower deviations

Substrate Saturation	Additions	Enzyme Complex	Km (mM)	Kd (mM)	Ki (mM)	Kf×10^4 (M^{-1})	Cooperativity (n)	ΔG° (Kcal)
MnATP	---	E-MnATP	0.065	---	---	1.43	0.8	5.6
MgATP	---	E-MgATP	1.05	---	---	0.10	1.1	4.1
MgATP	+0.5mM CaATP	E-CaATP	---	---	0.015	6.67	0.9	6.5
---	---	E-Ca	---	0.02	---	5.0	---	6.4
MgATP	+2.26mM ATP^{4-}	E-ATP^{4-}	---	---	0.27	0.37	1.49	4.8
MgATP	+2.2mM Mg^{2+}	E-Mg	---	4.7	---	0.021	1.1	3.2
MnATP	+5.5mM ATP^{4-}	E-Mn	---	3.8	---	0.026	---	3.3

K_{mapp}(MnATP) changes from 0.075 to 1.5 mM (Table I). The calculated $K_i(ATP^{4-})$=0.25mM similar to the value obtained for substrate MgATP, $K_i(ATP^{4-})$=0.27mM.

Hence, the presence of MnATP does not alter the affinity for adenylyl cyclase for ATP^{4-}, ruling out, indirect modulation by Mn^{2+} on the affinity of the enzyme for ATP^{4-}.

Figure 5 shows that addition of CaATP produce competitive inhibition decreasing affinity from 2mM to 20mM K_{mapp}(MgATP) [11].

Energy charge [39] is described as the intracellular relationship of [ATP]+1/2[ADP]/([ATP]+[ADP]+[AMP]), can have a value ranging from 0 (all AMP) to 1(all ATP).

This index could be given a role in ionic equilibrium by its relationship to the control of the availability of free divalent metals. The latter, related to the stronger binding constants of ATP^{4-} for Mg^{2+} and Ca^{2+} than for ADP^{3-} and AMP^{2-}.

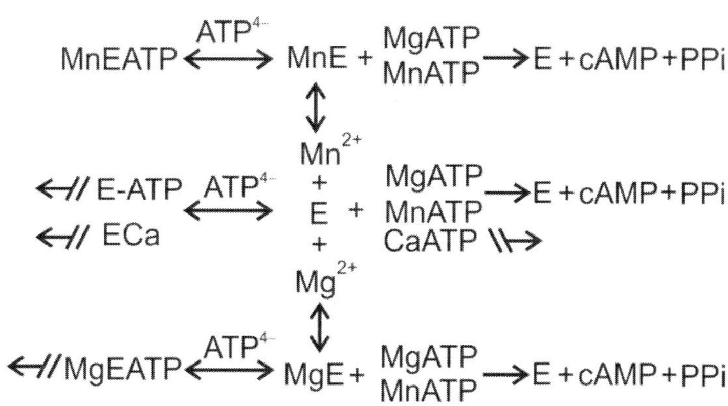

Scheme 1. Equilibriums for the RARE BiBi kinetic mechanism of the basal adenylyl cyclase. Study of initial velocity patterns gave evidence, that at the active site the substrates MgATP and MnATP by group transfer exchange PPi between ATP^{4-} and a pyrophosphate (PPi) accepting group, releasing cAMP and PPi. Mn^{2+} substitutes for Mg^{2+} for substrate formation and as an enzyme modulator. The catalytic rates and dissociation constants for Mn^{2+} and MnATP differ from Mg^{2+} and MgATP (Table 4), with active site binding by inhibitor CaATP and ATP^{4-} resulting in dead-end inhibition (⇸).

In table 4 is reported that K_m(MnATP)=0.065mM versus K_m(MgATP)=1.05mM, correspond to 16 times lower affinity of the basal enzyme for the substrate MgATP than for MnATP. Table 4 dissociation constants, show an ionic equilibrium order $K_d(Mg^{2+})$=4.7mM>$K_d(Mn^{2+})$=3.8, indicating that Mg^{2+} can only at higher concentrations than Mn^{2+} displace the latter from the divalent metal activatory site.

The activity order MnATP>MgATP and both ATP^{4-} and CaATP act as inhibitors. Since $K_i(ATP^{4-})$=0.27 and K_i(CaATP)=0.015mM the competitive affinity order CaATP>MnATP>ATP^{4-}>MgATP.

Table 4, indicates that at lower concentration of excess ATP^{4-} the greater Ka of the nucleotide for either Mn^{2+} or Mg^{2+} than for Ca^{2+} could maintain the enzyme active.

However, at higher concentration of excess ATP^{4-} the greater affinity (K_i=0.01) of the enzyme for CaATP, could predominate reducing the availability of cAMP for sustainable memory persistence. Hence, eradication of particular memory, confer to the system plasticity to re-conform synaptic connectivity for acquisition of a new memory or reinterpretation of the preceding one. However, Mg^{2+} can easily displace Ca^{2+} from ATP and reduce the concentration of CaATP, because the association constant: K_a(MgATP)=20×10^3 M^{-1} is greater than K_a (CaATP)=9×10^3 M^{-1} [37] [38].

The mutual competitive inhibition of ATP^{4-} and CaATP on the substrates, MgATP and MnATP, and a constant cooperative value of about one indicate that the ionic and divalent metal complexes of ATP, bind to the same site on basal adenylyl cyclase.

A requirement of cAMP leading to increment neuronal connectivity implies that the level of AC activity may play a partial role of neuronal plasticity [40] [41]. This process results from a perdurable lack of cAMP to drive the creation and/or reinforcement required to maintain synaptic connectivity. The requirement to control cAMP levels indicates the applicability of AC kinetics, for modeling the versatility of AC responsiveness. Enzyme plasticity is implied in the scheme 1, showing a random BiBi sequential group transfer mechanism (RARE BiBi) with dead-end inhibition, since the binding at the active site of CaATP or ATP^{4-} results in dead-end inhibitions.

In the intracellular fluid at 1mM [Mg^{2+}] is considerable higher than 1μM Ca^{2+} [42]. The ratio is practically constant, except for certain pathological states [43]. However, the equilibrium dynamics of divalent metal existing as Mg^{2+} and MgATP, as well as Mn^{2+} and MnATP fluctuate. The latter, due to accumulation of endogenous metabolite intermediates like ATP^{4-}, amino acids, dicarboxylic acids, and organic phosphates, which can complex Mg^{2+} or Mn^{2+}. The cytoplasmic concentration of Ca^{2+} is in a dynamic state by the same metabolic variables. Additionally, 1mM extracellular and 10mM mitochondrial concentration of Ca^{2+} allow import and export of Ca^{2+} across structural compartmentalization [44] [45].

A strong restriction to the adenylyl cyclase activation, became evident from $S_{0.5}(ATP^{4-})$ of about 0.1mM (table 4). Hence, any decrease in the rate of ATP^{4-} consumption by the Na^+ gate activity and others functions, should turn off the adenylyl cyclase at both the neurotransmitter and basal levels. It is suggested that a matching the rate of ATP production and consumption should be optimal because the brain lacks storage capability. If the rate of ATP production overcomes consumption and endogenous accumulation occurs, ATP^{4-} may compete for the divalent metal structurally present for binding receptors and enzyme to the membrane [34].

This requirement may have important implications, since neurons may need to adjust endogenous aerobic glycolysis to prevent ATP^{4-} accumulation, overcoming the stability of receptors in the membrane.

Characterization of the Dynamics of the Quaternary Structure of Hemoglobin as an Erythrocyte Sensor Role Function of Glucose Levels in Blood

The insulin-regulated transporter GLUT4 is found in adipose tissue and striated muscle. GLUT1 is present in erythrocytes and endothelial cells of barrier tissues like the Blood brain barrier (BBB), is also expressed in liver cells which require bidirectionality to uptake glucose and by having glucose 6-phosphatase to release it from glucose-6-phosphate (G6P). GLUT3 is expressed mostly in neurons. Hence, brain and liver use transporters which do not require insulin for efficient uptake of glucose.

Cerebrum cells, lack of storage pathways are nutritionally dependent of a continuous uptake of glucose and oxygenation, to support aerobic glycolysis. Under starvation ketone bodies can only partially substitute for glucose. Hence, the need for signal transduction between the tissues involved, required for matching the rates of O_2 and glucose consumption. This capability allows cerebral areas to increase by 10 times the rate of aerobic glycolysis, without overcoming the homeostatic controls of glucose level in blood.

Heterotropic allosteric modulation was shown for: 2,3-diphosphoglycerate (2,3-DPG) [46]. 2,3-DPG cross-links subunits β_1 with β_2 of deoxyHb, at the tissues low pH a central cavity of the Hb tetramer became structured as an binding site for 2,3-DPG [47]. The latter, became surrounded by the terminal amino acid, Lys 82, and the protonated imidazole rings of histidines His 143 and His 2, in the β_1 and β_2 chains, which bear two NH bonds and show overall positive charge.

The 2,3-DPG-Hb complex of much lower affinity for O_2 than isolated Hb could be regarded as the physiological T structure [48] [49].

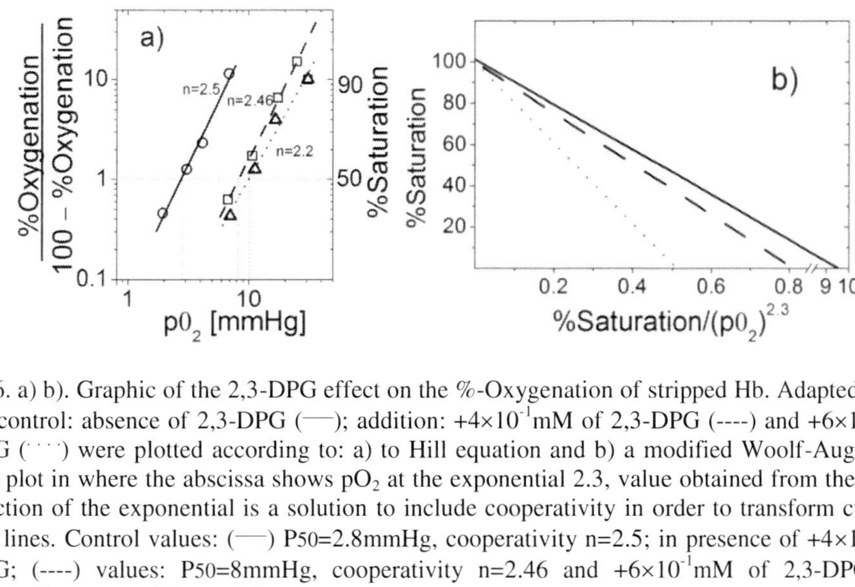

Figure 6. a) b). Graphic of the 2,3-DPG effect on the %-Oxygenation of stripped Hb. Adapted data [53] for the control: absence of 2,3-DPG (—); addition: $+4\times10^{-1}$mM of 2,3-DPG (----) and $+6\times10^{-1}$mM of 2,3-DPG (· · ·) were plotted according to: a) to Hill equation and b) a modified Woolf-Augustinsson-Hofstee plot in where the abscissa shows pO_2 at the exponential 2.3, value obtained from the Hill Plot. Introduction of the exponential is a solution to include cooperativity in order to transform curves into straight lines. Control values: (—) P_{50}=2.8mmHg, cooperativity n=2.5; in presence of $+4\times10^{-1}$mM of 2,3-DPG; (----) values: P_{50}=8mmHg, cooperativity n=2.46 and $+6\times10^{-1}$mM of 2,3-DPG (· · ·) P_{50}=10mmHg cooperativity n=2.2.

Upon deoxygenation increase the concentration of free Mg^{2+} ion [50] showing that deoxyHb and Mg^{2+} were competitors binding reciprocally with organic phosphates [50]. At the pH of lungs, the protons dissociate from R-histidines and the imidazole ring changes from overall positive to manifest the reactivity of negatively charged nitrogen atoms. Thus, now negatively charged groups like His 143 [51] repels the similarly charged 2,3-DPG.

The R-groups participants in H^+ release or Bohr effect have been widly reported and allows to molecular characterization of proton disociated Hb as the R configuration within oxyHb. Interconvertion of deoxyHb pKa=8±0.2 to oxyHb pKa=6.4±0.2 releases Bohr protons from R-groups, within the two β chains: from Cysβ_1 and $\beta_2$93, plus that from Aspβ_1 and $\beta_2$94 [52] [53] [54].

Treatment to obtain human Hb stripped-out of 2,3-DPG and other ligands increases Hb affinity for O_2 decreasing $P_{50}(O_2)$ 2.8mmHg [53]. The figure 6.a) illustrates a Hill plot showing that addition of 2,3-DPG [55] decreases Hb affinity for O_2 by increasing P_{50}, but without affecting Hb cooperativity for O_2.

A modified Woolf–Augustinsson–Hofstee (WAH) plot of %saturation on the ordinate versus %saturation/$(pO_2)^{2.3}$ on abscissa, figure 6.b) allowed to obtain an straight lines intersection at the theoretical 100%saturation on the ordinate.

The plot indicates that under the ideal condition of increasing pO_2 until reaching a theoretical maximum O_2-saturation, the presence of a constant concentration of 2,3-DPG may not prevent the homotropic Hb response of forming oxyHb.

Hence, a mutual exclusion between pO_2 and 2,3-DPG could lead to predict, that a theoretical increase in pO_2, could reduce an inhibitory effect of a constant 2,3-DPG concentration [55].

However, under physiological conditions Hb binds 2,3-DPG into a specific site and forms a complex in which the Hb quaternary structure, became stabilized as the deoxyHb form by the release of O_2.

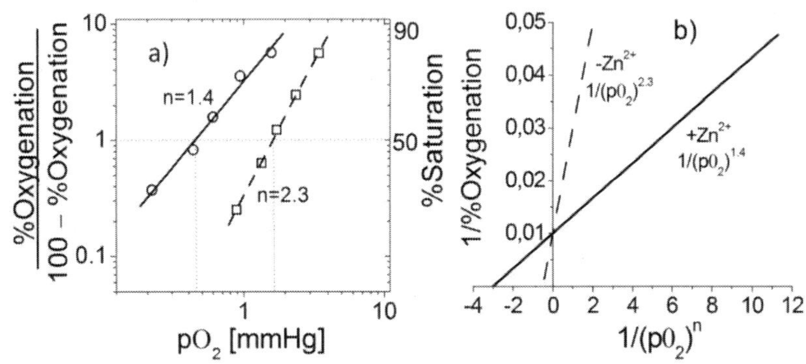

Figure 7. Graphic of the effect of Zn^{2+} on the %-Oxygenation of stripped Hb. Adapted data [53] for the control: absence of Zn^{2+} (----), values: P50=1.6mmHg; and addition of 4.47×10^{-2}mM Zn^{2+} (—), values: P50=0.5mmHg was plotted according to a) Hill equation and b) a modified doble reciprocal Lineweaver-Burk plot in were the abscissa shows pO_2 at the exponentials values obtained for the individual curves plotted according to Hill. Thus, the $-Zn^{2+}$: $1/(pO2)^{2.3}$ and $+Zn^{2+}$: 1/(pO2)1.4.

Lineweaver-Burk a double reciprocal plot adapted to subtract cooperativity, by introducing the cooperativity n-value for each curve in the abscissa $1/(pO_2)^n$. The plot becomes 1/%Oxygenation versus $1/(pO_2)^n$ and allowed to obtain intersection of straight lines for a theoretical maximal $1/(pO_2)^{1.5}$ and $1/(pO_2)^{2.3}$. Hence, at a theoretical maximal pO_2 ($1/pO_2=1/\infty\rightarrow0$) saturation of Hb by O_2 would be reached even in the absence of Zn^{2+}. Hence, the activatory effect was to decrease the control value $S_{0.5}$=1.64mmHg by addition of Zn^{2+} to $S_{0.5}$=0.5mmHg [53] [55]. The strip method to free of ligands the Hb molecule appears to reduce steric hindrance. Even so Zn^{2+} becomes a strong enhancer of striped Hb affinity for O_2. The effects of 2,3-DPG decreasing affinity (figure 6) and Zn^{2+} increasing affinity (figure 7) of Hb for O_2, suggest by allosteric considerations alone, that a single site for both actions, should be discarded. However, even if solving by classical competitive interaction could not be applied, other treatments were tested. The nullification of 2,3-DPG inhibitory (figure 6.b) and of the activatory Zn^{2+} actions (figure 7.b), became evident when the plottings allow to assume a tendency for pO_2 to reach the equivalent of infinite concentration. The figures 6 and

7 illustrate that because these ligands were at an invariant concentration, could be displaced from Hb by a competitive mechanism involving pO_2 increment.

Therefore, competition between O_2 with inhibitor 2,3-DPG (figure 6.b) do not occur at the same site but may result from this heterotropic ligand tendency to bind and stabilize one or more R-groups restricting O_2-saturation of the Heme site of Hb. Accordingly, increasing pO_2 would eventually release the hindrance preventing O_2-binding to the Heme. The activatory effect by Zn^{2+} (figure 7.b) implicates that the effect of divalent metal is to decrease the concentration of O_2 required to remove the hindrance effect. Hence, a model (figure 8 and 9) was developed that shows that O_2 *per se* has full homotropic capability for the interconversion of deoxy- to oxyHb. Hence, competition for a modulatory site, which control hindrance, could explain Hb heterotropic reactivity with ligands decreasing (figure 6.b) or increasing (figure 7.b) affinity for O_2. These dynamics imply that the modulatory sites could not be fully covalently structured, but are rather structures supported by coordinative bonding. The affinity of Hb for the ligands Mg^{2+} and/or Zn^{2+} supports the tendency to conform two coordinative structures of Hb, which fit the theoretical and physiological findings. The plasticity of this divalent metal or sugar phosphates modulated sites became evident in term of a glucose sensor role for the Hb-erythrocyte system. This one, adding the physiological regulation by the endogenous metabolic state of the red cell, expressed by endogenous 2,3-DPG concentration, to the physiological ΔpO_2 between lung and CSF. Hence, when Hb becomes characterized in the sole presence of O_2 shows the quaternary changes in the structural links between T- to R- forms, which characterize homotropic interconversion. However, the Hb potential to acquire metal chelating sites allows the molecular quaternary structures T- and R-, to participate in 2,3-DPG-Hb and divalent metal-Hb complexes stabilizing changes in the O_2 affinity. These ones induced on Hb by endogenous heterotropic ligands within the red cells, allowing the adaptive response to the O_2 needs of tissues and fulfilling in brain the matching of the rates of glucose and O_2 uptake.

The homotropic tendency of isolated Hb shows that during oxygenation a link between Hisβ146 and Aspβ94 breaks up and the imidazole ring at above pH 6 takes up a proton (alkaline Bohr Effect), which may be able to trigger the changes from T to R as a sole function of O_2 through a ΔG=3Kcal [56] [57].

The erythrocyte concentrations: 65μg/g for Mg^{2+} and 11μg/g for Zn^{2+}, lead to physiological significant forms of the coordinative complexes: $(Mg^{2+})_2$Hb, $(Zn^{2+})_2$Hb and (Mg^{2+}or Zn^{2+}) Hb stabilizing complexes of higher affinity for O_2 than the R-form of Hb [53] [55] [58] [59].

In figure 6 b) and figure 7 b), were reported plots [55] [59], both directed to use the cooperativity value n as exponential of the pressure of oxygen: $(pO_2)^n$. This treatment allows obtaining straight curves. These predicts that the extrapolation to a maximal $(pO_2)^n$, renders an homotropic dependence on the transition between T and R forms of Hb. However, it could be predicted that the physiological ligand-Hb equilibriums, within the red cell, would lead to an heterotropic mutual inclusion of O_2 and divalent metals Zn^{2+} or Mg^{2+}, with a mutual exclusion of 2,3-DPG and H^+. The exergonic association per atom of Zn^{2+} to Hb: ΔG=−9.7Kcal [53], the tetramer has two binding sites, for either Zn^{2+} or Mg^{2+}. In the direction of mutual inclusion of divalent metal and oxygen the exergonic binding of Zn^{2+} of 19.4Kcal plus pH dependent neutralization of Bohr protons, exceeds the endergonic dissociation of 2,3-DPG from Hb of only: ΔG=5.7Kcal. Hence, in the lungs this thermodynamic coupling integrates pO_2 with other contributions for a decrease in free energy ($\Delta G<0$). At lower pH than 7, the increasing stability order: $(Mg^{2+})_2$-oxyHb < (Zn^{2+},Mg^{2+})-oxyHb < $(Zn^{2+})_2$-oxyHb, indicates pH-differentiable complexes with O_2 dissociation all along the body pH-axis. Hence, the multiplicity of complexes allows that deoxygenation could occur over an extended pH-range, than could be possible for any particular divalent metal complex of Hb. The pH of blood remains 7.4 but surroundings tissues allow increasingly lower pH from the trunk to the limbs. Hence, deoxygenation could follow a pattern of gradual release of O_2, first from the Mg^{2+}-Hb and thereafter from the Zn^{2+}-Hb species. However, the augment of glucose level in blood, in untreated diabetes, induces in the erythrocyte an increase of sugar phosphates.

These ones by releasing the divalent metal from liganded state of Hb decrease the molecule affinity for O_2. Hence, a premature release of O_2, allows O_2-depleted Hb, to reach the body limbs and subsequence tissue anoxia results in the ulcerations characteristics of untreated diabetes. In the red cell, at the dimerization interphases $\beta_2\alpha_1$ and $\beta_1\alpha_2$ of Hb, pO_2 is mutually inclusive with the heterotropic ligands Mg^{2+} and Zn^{2+}, and could changes the ions from "bi-" to "tetra", and "hexa"-dentate coordinative states.

The more stable geometric configuration could be an octahedron, with the β_2 R-groups: His 143, Cys 93, Asp 94, His146 and the α R-group His 87, at the α_1-Heme. A symmetric structuring of the R-groups at the interphase $\beta_1\alpha_2$ will result in the formation of an oxyHb$(Me^{2+})_2$ complex, in which the quaternary form of Hb would show a lower tendency for O_2 dissociation. Figure 8 illustrates the dynamics of divalent metals association with exclusion of 2,3-DPG in Hb and vice versa. As scheme 2 illustrates low pH inhibits 2,3-DPG mutase, allowing anaerobic glycolysis without generation of 2,3-DPG. The 2,3-DPG-depleated erythrocyte, arrives at the lung level.

These events facilitate up to an 180° turn-around over their α-carbon of the R-groups of Hb. This effect adds to the directional position of histidines another 0.5nm length, to reach the 1nm coordinative distance to the metal binding site. The chelating structure became conformationaly stabilized by the forming of the divalent metal-Hb complex. Simultaneously, at the lung pH=7.4 neutralization of H^+ released from the histidines of Hb, induce a same charge rejection, favoring the release of 2,3-DPG from its state as ligand of Hb and increasing O_2-affinity allowing that the higher pO_2 by decreasing free energy favors the direction of oxygenation versus that of deoxygenation. At tissue level the change in free energy became associate to deoxygenation by the increase in the erythrocyte 2,3-DPG concentration (scheme 2) and the much lower pO_2 (figure 8).

At tissue low pH, the tendency of divalent metals to became surrounded by hydration spheres, became an exergonic contribution to their release as ligands from Hb. Effect, facilitating that β His 143 and other histidines H^+-release, to became positively charged and by rotating along their α-carbon, to participate in the conformation of a binding site for 2,3-DPG (figure 8). Along the blood path to the tissues, progressive accumulation of 2,3-DPG (scheme 2), would provide a constant decrease in the ratio of oxy- to deoxyHb.

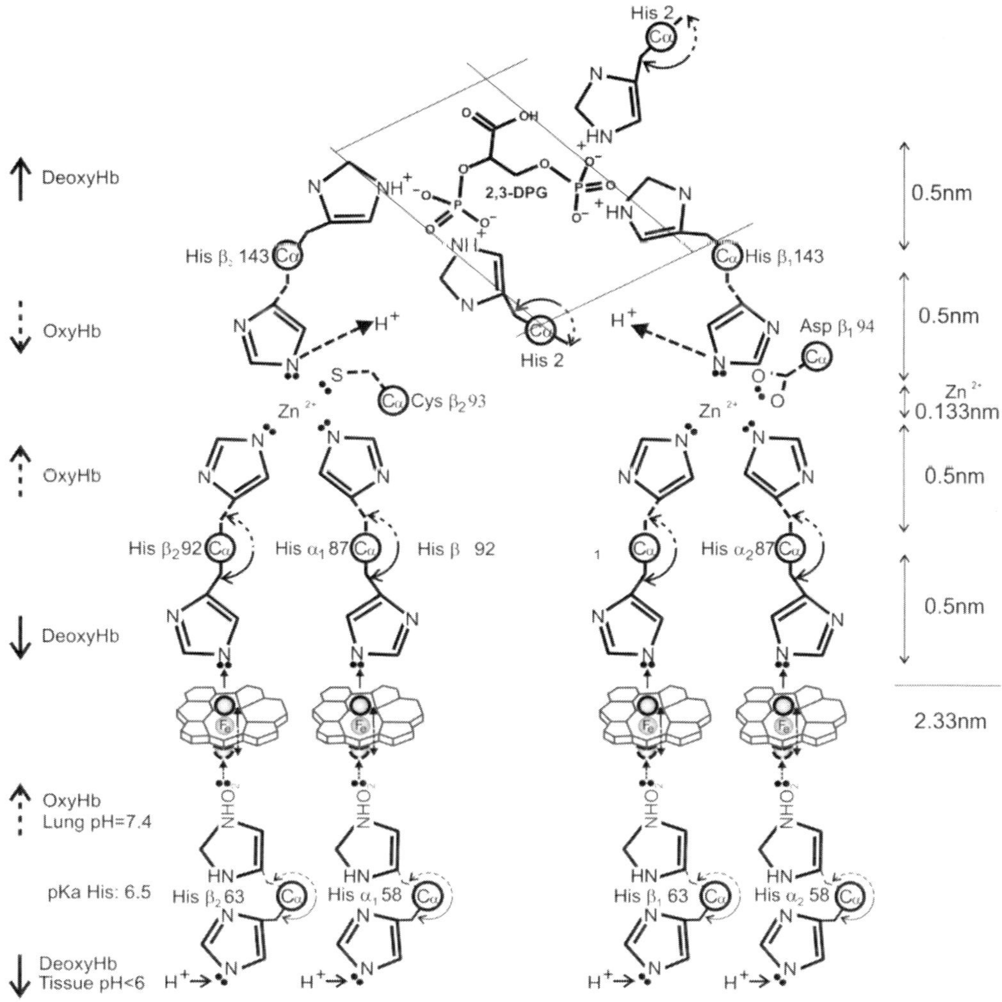

Figure 8. Model of the R-groups dynamics of Hb for mutually inclusive interactions of divalent metals and pO2 allowing simultaneous exclusion of 2,3-DPG and H^+. According to saturation ranges of free (dissociated) divalent metals (Zn^{2+} or Mg^{2+}) could induce a coordinative-fit of ligand R-groups of Hb conforming two chelating sites. Hence, at the subunit interface β_2, α_1 and β_1, α_2 by displacement of His $\beta_2$92, His $\alpha_1$87, His $\beta_1$92 and His $\alpha_2$87 the Fe^{2+} atoms move from their position at the heme planes in the deoxyHb state ( ) to allow their reactive position for oxygenation ( ) and form with the 4 distal histidyl residues, $\alpha_1$58His, $\alpha_2$58His, $\beta_1$63His and $\beta_2$63His, to form H-bonds with the O_2 ligands. The oxygenation pathway (shown in broken lines: - - - -) predicts a physiological cooperative number n=2.8 for an oxygen saturation curve of Hb. This cooperativity number sums up the increase in affinity for O_2 and divalent metals Me^{2+} to a decrease in affinity for 2,3 DPG and H^+. Deoxygenation pathway (shown in solid lines: ——) the R-groups shift-out of the chelation complex and His $\beta_2$143 and His $\beta_1$143 to capture protons and the resultant imino groups attract 2,3-DPG.

Figure 8 shows that α-carbon rotation allows to couple the endergonic dissociation of 2,3-DPG with the exergonic association of Mg^{2+} or Zn^{2+}. Oxygenation in the red cell promote Mg^{2+} coordination to Cys-$\beta_2$93 inducing fit of His-$\beta_2$92, and facilitate the movement of the F-helix toward the H-helix [60] [58] [59] [61]. Thermodynamics considerations indicate that modeling of Hb has to balance oxygenation vs. deoxygenation for an energy efficiency coupling between the endergonic dissociation of 2,3-DPG from deoxyHb: ΔG=5.7KCal [53] and the binding of Zn^{2+} with a ΔG^o=–80KJ per 2 Zn^{2+} atoms [62]. The solvation or hydration interaction between a solute like Hb and the solvent media of the cytoplasm, allows an Hb hydration sphere. The tendency of a divalent metal to acquire a hydration sphere could facilitate Me^{2+} and O_2 dissociation from oxy to deoxy. The model graphics patterns a solution involving the symmetric turnarounds of both His β 92 and both His α 87, for dynamic transfer of the structural changes expanding the 2.33nm required for Heme to Heme interaction. Hence, figure 8 illustrates a model of mutual inclusion of divalent metals and O_2, which could lead from mono to the tetra oxygenated molecule: $HbO_2 \rightarrow Hb(O_2)_2 \rightarrow Hb(O_2)_3 \rightarrow Hb(O_2)_4$. Each state of oxygenation, coupled by the concerted interaction between the two divalent metals ligand sites, in which the coordinatively sharing of His β_2 143 and His β_1 143 are being subtracted step by step. This effect jointly with the displacement of the other ligands, that were conforming the binding site of single 2,3-DPG, allow for the gradual disappearance of an structure capable to bind 2,3-DPG. Hence, the binding of each O_2, decreasing the free energy requirements for the binding of the subsequent oxygens.

The Role of BBB in Modulation of Red Cell Control of Hb Deoxygenation

Figure 9 illustrates brain as "bathing" in the cerebrospinal fluid (CSF) within the subarachnoid space and the ventricular system, around and inside the brain and spinal cord.

The blood-brain barrier (BBB), show capillaries without fenestrations or pinocytotic (transportation) vesicles, and have tight junctions that almost fuse adjacent cells. Efficient mechanisms guard the constancy of the CSF pH 7.30–7.36 units, even in non-respiratory and, to some extent, in respiratory acid-base disturbances.

The pH of blood is maintained around 7.4, and buffering is required to maintain homeostasis. The blood contains Na^+, K^+, Ca^{2+} and Mg^{2+} cations, and Cl^-, PO_4^{3-} and SO_4^{2-} anions, which play a role in buffering. In surrounding peripheral tissues pCO_2 is about 50mm Hg, which allows diffusion into plasma and erythrocytes at about 40 mm Hg. Into the erythrocyte, the H^+ dissociated from HCO_3^- ion is buffered by hemoglobin. The production of H^+ within erythrocytes and their subsequent buffering by hemoglobin, results in a reduced affinity of hemoglobin for oxygen. This leads to a release of O_2 to the peripheral tissues, a phenomenon is termed the Bohr Effect, which is potentiated by the change in the endogenous 2,3-DPG concentration in the erythrocyte (scheme 2).

The presence of chlorides ions: 120-130 mEq/l may not only play an isotonic role, because across the CSF barrier may help dissociation of the divalent metals from oxyHb and facilitate deoxygenation. Cerebral blood flow may be constricted at high CSF pressure. Disorders of CSF flow may also affect the intracranial blood flow, with subsequent neuronal and glial vulnerabilities.

CSF contains < 10-12 U/ml of adenosine deaminase that breaks down adenosine into Inosine and ammonia. Adenosine, has inhibitory effects over neuronal activity, sedates and increments the non-rapid eye movement sleep (NREM) in stage IV. Caffeine blocking the adenosine binding to the receptor decreases sleeping tendency.

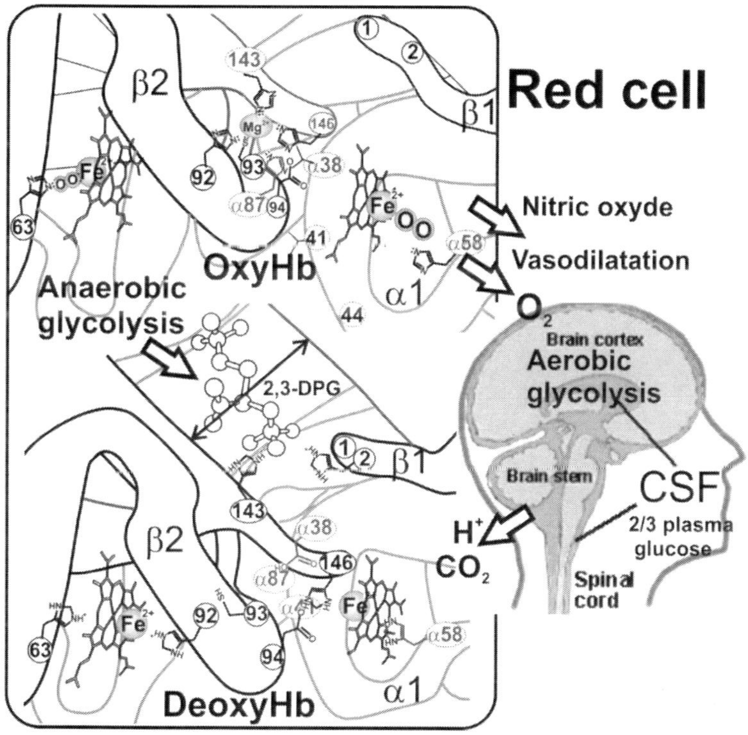

Figure 9. In the transition from oxyHb to deoxyHb, inclusion of 2,3-DPG and H^+ couples with mutual deoxygenation and dissociation of Mg^{2+} or Zn^{2+}. The illustration has been scaled for Hb projected with a diameter of 5.5nm [61]. Freedom for α-carbon rotation of the side chains implicated for their participation on the coupling model described chelated structures, were tested with the generous assistance of Fermi, G. and Milstein, C. in the large tri-dimensional model of Hb, located in the laboratory of M.J. Perutz, Cambridge, England. The figures show only the $\beta_2\alpha_1$ interface, since structural symmetry allows an equivalent set of changes at the $\beta_1\alpha_2$ interface. Peptide bonds are rigid and fixed in a plane in where two α-carbons, 0.36nm apart, rotate 180° over angles ϕ (fi) and ψ (psi). The longitudinal lengths from α-carbon to their coordinating negative atoms: His=0.5nm, Cys=0.23nm and Asp=0.39nm; Magnesium Radius: atomic=0.150 nm and ionic (2+ ion)=0.086 nm; Zinc radius: atomic=0.135 nm and ionic (2+ ion)=0.088 nm; Iron radius: atomic=0.140 nm and ionic (2+ ion)=0.077 nm.

A Mechanism for Hb-Deoxygenation at Neutral pH

By incubating cells in media containing acetate, pyruvate, etc., the total concentrations of acid-soluble organic phosphate accumulation fractions depends on the extracellular phosphate level [63] [64] [65] [66]. In the human erythrocyte incubation with pyruvate and inosine, the total concentration of acid-soluble organic phosphates in the red cells rises from 14 to 55 μmol/g red cells. The organic phosphate other than 2,3-DPG can be elevated in vitro to 24 μmol/g [66]. The CSF pH=7.4 prevents pH-changes at the brain level, but the red cell heightens the concentration of the sugar phosphate intermediates of anaerobic glycolysis.

The concentration of 2,3-DPG became maximal at pH=7.4 because 2,3-DPG mutase became inhibited by low pH (scheme 2) allowing the shift the from oxy- to deoxy-Hb.

The energy necessities of the red cell, devoid of mitochondria, indicate the important role of anaerobic glycolysis. The regulation of oxygen-hemoglobin erythrocyte-system have requirement for both NADPH produced by the hexose monophosphate pathway (HMP) and 2,3-DPG the most concentrated organophosphate which varies inversely with pH [67].

cAMP [68] and cGMP [69] are transported into human erythrocyte (Human-RBC) against significantly large differentials of the intracellular-extracellular concentrations [70] [71]. 0.1mM Dibutyryl-cAMP and dibutyryl-cGMP, compounds were found to diffuse freely into the erythrocyte.

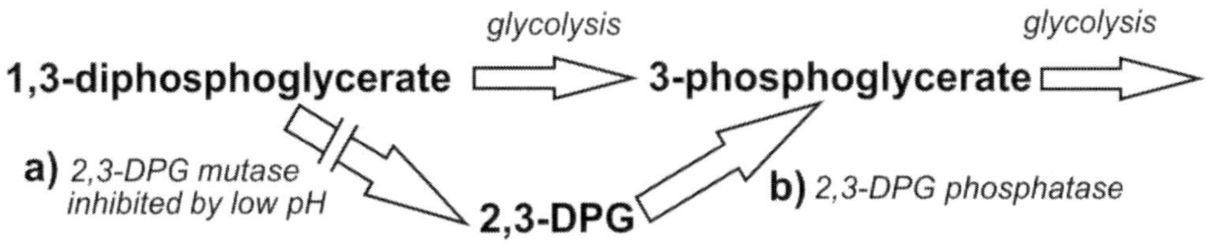

Scheme 2. Rapapport-Luebering scheme. a) diphosphoglycerate mutase transfers from a carbon one (C1) phosphoryl group of 1,3-DPG to C2 of 2,3-DPG that by action of b) diphosphoglycerate phosphatase forms 3-PG. In the erythrocyte pH is an inhibitor of 2,3-diphosphoglyceromutase. The insulin-independent glucose uptake of erythrocyte allows its endogenous sugar concentrations to be directly related to that in blood. Anaerobic glycolysis within the erythrocyte could generate 2,3-DPG, to reach is highest endogenous concentration, because the pH inside the blood capillaries surrounding the CSF, would be close to the latter homeostatic-pH 7.4. The lack of a pH differential across the BBB indicates that the displacement of the equilibrium from the oxy to deoxyHb for the continuous release of O_2 into CSF, becomes only dependent in the increment in 2,3-DPG. The exergonic binding of 2,3-DPG of $\Delta G = -5.7$Kcal favoring the release of O_2 to match oxygenation for a sustainable aerobic glycolysis of the activated brain areas.

However, even if cAMP and cGMP catabolism did occur, were not able to elevate endogenous lactate and 2,3-DPG. Absence a guanylyl or adenylyl cyclase activity in the erythrocyte or the presence of cyclic nucleotides signaling was shown [71] [72]. However, at 0.1μM dibutyryl-cGMP inhibits lactate production, which could be a mechanism contributing to pH homeostasis at the BBB.

Control of metabolic shifts on the erythrocyte appears to rest mainly in the interconnection of glycolytic pathways with the catabolism of cyclic nucleotides. The HMP-pathway production of CO_2 is decreased by cAMP and cGMP. Effect prevented by theophylline-dependent inhibition of phosphodiesterase, the enzyme producing AMP and GMP. Inhibition of glycolysis occurs as a feedback by the mass action of ribose or ribose phosphate, as a product buildup from the oxidative branch of the HMP pathway. Hence, increase in the non-oxidative branch might increase 2,3-DPG levels (scheme 2). Inosine, could also provide relatively small increases in the pathway of 2,3-DPG [71] [73] [74] [75] [76].

Human-RBC, hemolysates show an increment of $^{14}CO_2$ released from the consumption of [1-^{14}C]-glucose, within the HMP pathway by the Mn^{2+} stimulation of the redox recycling of NADPH, which contribute to pH of homeostasis because otherwise could favor lactic acid formation [77].

Structure and Function of Neuronal Regions

The structure of neuronal regions [78] allows functional differentiation of the hippocampus [79] and the striatum, which have distinct roles, but functional magnetic resonance imaging (FMRI) have revealed that successful memory require greater activity of both, for their jointly support of episodic memory formation [80] [81] [82] [83]. Long-term potentiation (LTP) and long-term depression (LTD) of synaptic transmission, are mechanisms which vary according to the synapses and circuits involved [80] [84]. These mechanisms can contribute to experience-dependent modifications of brain function [85] [34] [86] [87] [88]. Figure 10 illustrates the relationship of noradrenaline activated AC catalyzing cAMP formation and its relationship to other second messenger-dependent systems [11] [8] [9] [86] [87] [88].

The level of cAMP is controlled by phosphodiesterase which breaks down cAMP into AMP. The entry of calcium into the neuronal cell inhibits AC and allows completing the cycle of a neuronal transmission pulse.

The beta-1 adrenergic receptors interact with stimulatory Gs [86] [88] by coupling mediated through a cAMP-induced phosphorylation of protein kinase A (PKA) of the receptor [89] [90] [91] [92] [93] [94] [95]. The B2AR-channel complex by activating adenylyl cyclase forms cAMP, activating PKA to counter balancing the protein phosphatase 2 (PP2A). Sensitivity to subsequent stimulus could condition its answer, which may be functional for long-term memory formation [96] [97].

In the nervous system, based on their structural and functional characteristics, neurotransmitter receptors can be classified into two broad categories: i) Ionotropic receptors activated by opening a pore channel that allows ions flows such as Na^+, K^+, or Cl^- [98] [94] [99] ii) Metabotropic, subtype of membrane receptors at the surface or in cell vesicles, which are indirectly linked to activated ion-channels.

These are associated to a second messenger signal transduction mechanisms. G protein-coupled receptors are encoded in the human genome in large numbers [96] has been detected about 350 of which 150 do not have known functions but the others has been related to hormones, growth factors and other endogenous ligands.

G-protein-coupled-receptor (GPCR) ligand binding induces a conformational modification into guanine nucleotide exchange factor (GEF). This activates an associated G-protein by exchanging its bound GDP for a GTP. The α-subunit of G-protein and bound GTP dissociate from the β and γ subunits. This allows interaction with endogenous signaling or target proteins directly depending on the α subunit type ($G_{\alpha s}$, $G_{\alpha i/o}$, $G_{\alpha q/11}$, $G_{\alpha 12/13}$) [88] [96].

The glutamate agonist N-methyl D-aspartate conforms a receptor system (NMDAR) multiple isoforms, to distinct distribution and functional properties. The ion channel in NMDAR is blocked by extracellular Mg^{2+}, the flow of Na^+- and Ca^{2+}-in and K^+-out of the neuron, is activated by voltage and gating by ligands glutamate (Glu.) and glycine (Gly.) [25] [100] [101] [102].

The Na^+/K^+ pump binds 3 intracellular Na^+ ions. ATP becomes hydrolyzed by phosphorylation of an aspartate residue and subsequent release of ADP. cAMP up-regulates the Na^+/K^+-ATPase. The process includes the ligands of the Gs- protein-coupled receptor (GPCRs). A decrease in cAMP and the effect of Gi-GPCRs down-regulates the enzyme [100] [101] [102] [103].

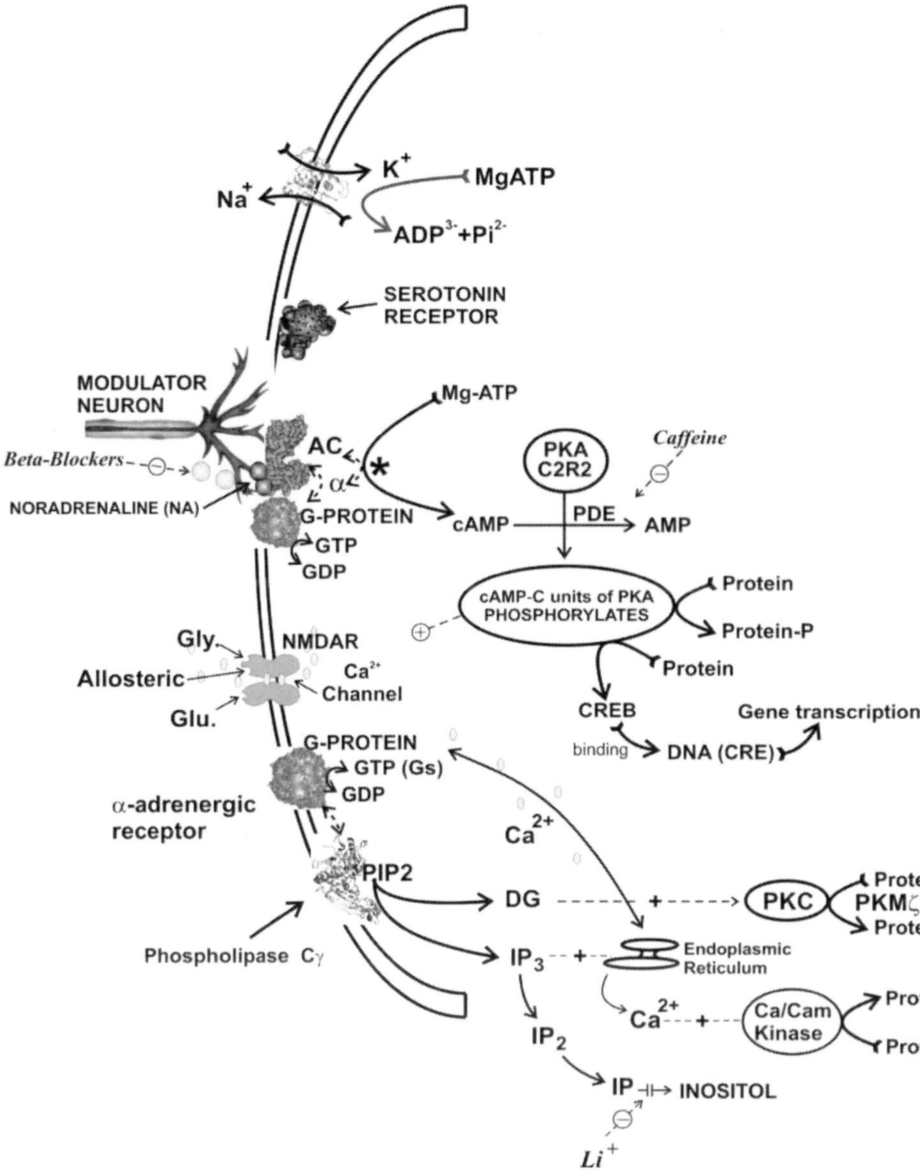

Figure 10. Integrative ionic-metabolic signaling to second messenger-dependent systems. G protein-coupled receptor (GPCRs) exchange it's binding to GDP inactive for GTP active (GDP/GTP-exchange) for noradrenaline (NA) receptor coupling to AC. The GPCRs α, β, γ subunits dissociate into separate α and βγ activated complexes [*]. Once the α-subunit hydrolyzes bound GTP to GDP, it re-associates with the βγ complex for ending receptor activation. The system has a requirement for free Mg^{2+}, for AC conversion of MgATP or MnATP into endogenous cAMP. The latter, could be converted by phosphodiesterase (PDE) in AMP or could activate protein kinase-A (PKA, with subunits C2R2), by dissociating regulatory (R) from catalytic (C) units. cAMP-C acts by phosphorylating proteins and enzymes, changing these from inactive to active forms or vice versa. The WM relates to phospholipase C [17] and endogenous intracellular IP_3-mediated Ca^{2+} level [23] [24] [99].

The phosphorylated form of the pump has a low affinity for Na^+ ions a conformational change in the pump transfers the three Na^+ ions, out of the cell for every two extracellular K^+ ions. The dephosphorylated likeness of the pump, has a higher affinity for Na^+ ions than K^+

ions, so the two bound K^+ ions, are released endogenously. The electrogenic transmembrane Na^+/K^+-ATPase-pump-system consumes ATP^{4-} generating ADP^{3-} plus inorganic phosphate (Pi^{2-}). The reaction modifies ionic equilibrium, by releasing free Mg^{2+} which activates basal AC and responsiveness to l-noradreanline [9] [11]. Up-regulation of the ATPase by the cAMP formed are therefore synchronized with a nerve impulse. The depolarization phase by increasing Ca^{2+}, within the cytoplasmic side of the membrane, could form CaATP which could turn-off AC activity [11] [33].

The carrier, Na^+-Ca^{2+} exchanger (NCX) uses the Na^+ gradient generated by the Na^+-K^+ pump, to transfer out the endogenous Ca^{2+}. A slowing down of the Na^+-K^+ pump increases intracellular Ca^{2+}. The site level at which $[Ca^{2+}]$I regulates NCX activity is different from the one required for Ca^{2+} transport [103] [104]. Sub-micromolar endogenous concentrations 0.1–0.3 μM Ca^{2+} are needed to activate the anti-porter and removal of Ca^{2+} ions completely blocks NCX activity [103] [104] [105]. ATP participates in the NCX molecule phosphorylation process by protein kinase A (PKA) and protein kinase C (PKC) [106]. The phosphorylation of NCX increases affinity for both endogenous Ca^{2+} and exogenous Na^+, and decreases its inhibition by endogenous Na^+. Nitrogen oxide (NO) is able to stimulate NCX in the reverse mode of operation in neuronal preparations and astrocytes through a cGMP-dependent mechanism [107]. A 0.4 decrease in the intracellular pH can induce a 90% inhibition of NCX activity [107] [108], increasing intracellular Ca^{2+} and inhibiting AC. Acting as a phosphoryl donor molecule. ATP may increase the activity of the exchanger in a number of ways [109]. ATP cellular depletion inhibits NCX1 and NCX2 isoforms but does not affect NCX3 activity [110]. A dual-store memory model predicts that a short-term memory allows capturing events that can be recalled easily for about 30s. Thus, should differ structurally and functionally from long term memory. A single-store retrieved model predicts biophysics contexts, which do not differentiate between the initial events of short versus long-term memory. Working memory (WM) behavior is a process involving retention of information for a short period of time. GPCRs also couple to G-protein activation of phospholipase C-γ (PLCγ). The activated form hydrolyzes membrane phospholipids into fatty acids and other lipophilic substances, there are four major classes A, B, C and D. PLCγ attack of substrate phosphatidylinositol 4, 5-bisphosphate (PIP2) increases the levels of inositol 1,4,5-triphosphate (IP3), the primary mechanism for Ca^{2+} release, and diacylglycerol (DAG) (figure 10). Downstream signaling proteins are phosphorylated on serine and threonine by PKA and DAG-activated protein kinase C (PKC) leading to alterations in their activities. Membrane-associated and intracellular tyrosine kinases phosphorylate specific tyrosine residues on target enzymes and other regulatory proteins. Synaptic plasticity and neuroprotection involves DNA sequences with a CRE motif, which interact with a cAMP response element binding protein (CREB) to increase or decrease transcription. The binding activity of cAMP on PKA is rapidly suppressed and the activation of CREB-mediated transcription of genes rapidly diminishes [111]. The system constitutes a convergence point for many signaling cascades. These are related in structure and function to CREM and activating transcription factor-1 (ATF-1), involved in long-term memory formation and neuronal plasticity [111] [112] [113] [114] [115]. Expressed CREB proteins include Brain-derived neurotrophic factor (BDNF), tyrosine hydroxylase, and many neuropeptides, like somatostatin, enkephalin, VGF and corticotropin-releasing hormone [114] [115]. Dopamine regulates hormonal responses, directing the hypothalamus to manufacture hormones and store then in the pituitary gland [106] [107]. Dopamine activates in brain a per-durance phase, with a control role of the persistence process, remembrance required for long-term memory storage [70]. It also controls muscle movements, patients with Parkinson's disease have practically no dopamine. Several psychiatric disorders are related, example an excess of dopamine in the frontal lobules of the brain has been implicated in schizophrenia, and abnormalities on the cerebral dopaminergic centers of psychopaths.

The noradrenaline-dependent acceleration of the destruction-inactivation rate of AC is tempered by glucose metabolite [116]. Induction of deficiency of adrenalin in blood by Bilateral Electro-coagulation of the Adrenal Medulla (BEAM), has been reported as a restorative treatment, effective for Disappearance of Parkinson Symptoms [116]. The relevance of the energy charge as the intracellular mechanism supporting memory formation has to be evaluated, within the more inclusive control by extracellular hormones and synaptic neurotransmitters.

Free Mg^{2+} and Mn^{2+} metabolic feedbacks control an "*in common*" tissue network, supporting the bioenergetics of brain processes in fat cells [33] [36], cAMP protein kinase activation of hormone-sensitive triglyceride lipase releasing free fatty acid and in liver by controlling the kinetic response IRTK [58] [76]. MnATP stimulates the insulin response of IRTK phosphorylating [Val^5] angiotensin II, which potentiates the kinetic response of insulin-activated IRTK to MnATP [73] [117] [118] [119].

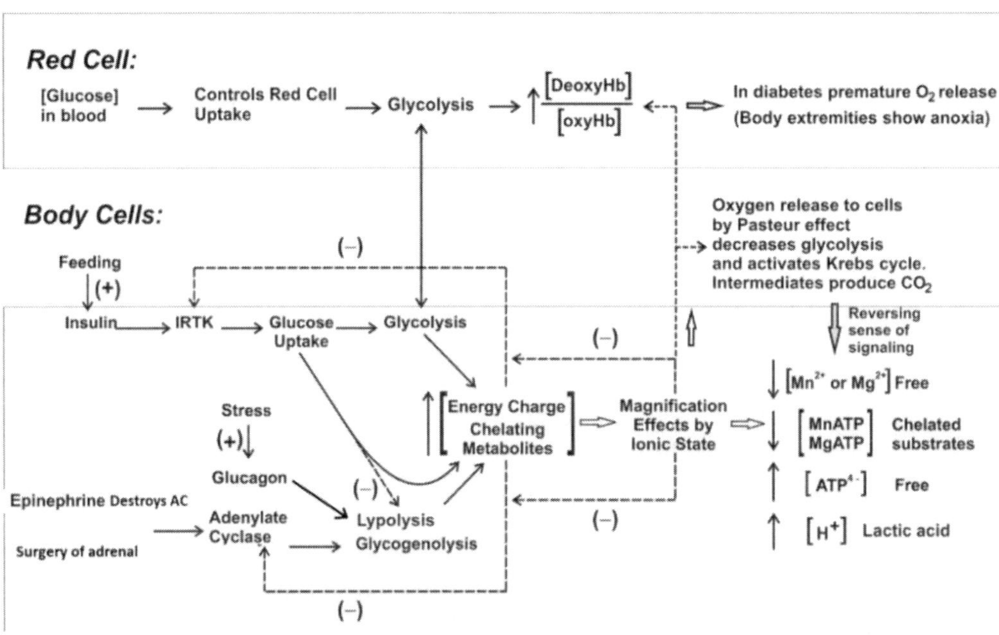

Figure 11. Integration of hormonal responsiveness of AC and IRTK to metabolic-ionic feedback. Network enhancement may include receptor kinases like liver IRTK, which by binding of insulin, could phosphorylate downstream proteins inducing changes in cell responsiveness. These ones may modulate metabolic relationships of fat cells, releasing free fatty acids and liver, glucose and ketone bodies. The network was idealized by an "in common" feedback controlling of hormone activity, by the metabolite modulated availability of free Mg^{2+} or Mn^{2+}.

Conclusion

- *Long-term memory:* is involved into the affirmative process associated with emotional experiences. Is sensitive to a beta-blocking of the noradrenalin receptor of AC. Persistence of an activate state of AC allows activation by cAMP of kinase-A, which by phosphorylation modifies the responses of enzymes and proteins within neuronal pathways.
- Additionally, free Mg^{2+} is required for binding of oxytocin [120] [121] to if receptor. AC dependent emotional intelligence, could also involves, oxytocin favoring attachments which involves a response to reward and the desire to obtain it.
- *Short-term memory:* the release of Ca^{2+} inhibits AC, allowing completion of a neuronal transmission impulse,
- The pulse of endogenous generated cAMP controls Ca^{2+} fluxes.
- The nerve impulse releases Ca^{2+} for Ca^{2+}-actin formation regulating *Microtubule* Self-Assembly. ATP^{4-} binds Ca^{2+} releasing actin.
- cAMP activate calmodulin binding free Ca^{2+} and allows its return to the endoplasmic reticulum.
- cAMP is consumed by phosphodiesterase enzymes, sensitive to inhibition by caffeine, prolonging cAMP action.
- In short term memory a high uptake of glucose versus a low oxygenation may affect the turnover of the Na^{+}-pump dependent cyclic fluctuations of ATP^{4-} level.
- Noradrenaline dependent activation of AC of brain is fully dependent on free Mg^{2+}, in excess of MgATP.
- In brain, the Na^{+} pump consumes ATP^{4-} for the propagation of nerve impulses, releasing free Mg^{2+}. The event increases AC responsiveness because ATP^{4-} has a much higher binding constant for Mg^{2+}, than ADP^{3-} and AMP^{2-}.
- AC of brain, liver, and fat cells as well than insulin receptor tyrosine-kinase (IRTK) show 3 differentiable sites as a function of free $[Mg^{2+}]$.
- An active site for the formation of the enzymes substrate complex: MgATP.
- A basal site(s) activatory: for free $[Mg^{2+}]$ saturation and by decreasing saturation with $[ATP^{4-}]$.
- A free $[Mg^{2+}]$ requirement for the hormonal-receptor sites.
- The computation separable saturation curves, of the 3 sites, yields a joint saturation response with a cooperative value of about 2.6 resulting of the sum of activatory sites for MgATP and for Mg^{2+}, plus desaturation of one inhibitory site for $[ATP^{4-}]$.
- The affinity for O_2 of the Heme Prosthetic Group of Hb, is mutually inclusive for the binding of free Mg^{2+} or Zn^{2+}, at the R-groups of interphase $\beta_2\alpha_1$: His$\beta_2$143, His$\beta_2$92, Cys$\beta_2$93, His$\alpha_1$87 and for the respective R-groups in $\beta_1\alpha_2$. Moreover, these bindings are mutually exclusive of the H^{+} and 2,3-DPG binding. Hence, the pO_2 saturation curve of Hb by inclusion of the activators Mg^{2+} plus O_2 binding and exclusion of inhibitors shows a cooperative value of about to 2.6.
- Not only noradrenalin, but numerous neurotransmitter are released into the extracellular space in the brain and their biological actions are mediated by specific receptors, which interconnect or link with the AC 2nd messenger system.

- At the organismal level, the fight or flight reaction releases adrenalin in blood activating AC of several tissues. Adrenalin in liver increases glucose level in blood and adrenalin plus other hormones in fat cells releases free fatty acids to produce in liver ketone bodies.
- An action or hormone dependent increase of glucose in blood results in an increase in sugar phosphates within the red cell decreasing free Mg^{2+} and Hb releasing O_2 that diffuses into the cerebrospinal fluid (CSF).
- The glia cells support minimal aerobic glycolysis. The activation effect on the membrane receptors of brain, by increasing free Mg^{2+}, consumes MgATP by AC and Na^+-pump. This requires a sudden increase of aerobic glycolysis to regenerate MgATP. The Mg^{2+}-dependent cooperative increase in neuronal activity, needs to be match by the promptness of the cooperative responsiveness of Hb deoxygenation. This mechanism adjusts the supply of O_2 to that of G in CSF. This allows the sudden 10 times increment of the rate of aerobic glycolysis, at the activated neuronal circuits.
- Hence, a metabolic-ionic feedback for metabolic integration rests in a hemoglobin sensor role.

The aforementioned results were examined to obtain a reductionist modeling, circumscribed to the participation of a core energy transduction molecular mechanism, within a generalized perspective of neuronal function. However, a wider more specific perspective of long-term memory depends of very significant studies interrelating molecular, physiological, learning, behavioral and pathological levels, briefly treated under discussion.

Modeled: kinetics of metabolic relationships between the nerve impulse and AC.

The action potential opens the K^+-out Na^+-in gates and activates the Na^+/K^+ ATPase pump.

The breakdown of MgATP by the Na^+/K^+ ATPase, releases ADP^{3-} and Pi^{2-} modifing ionic equilibrium, by releasing free Mg^{2+}.

The free $[Mg^{2+}]$ up-regulates responsiveness to l-noradreanline stimulated-AC and basal-AC activities.

During the depolarization phase, the inside of the neuron becoming more positive. The AC-generated-cAMP up-regulates the Na^+/K^+ ATPase. Activation of the pump expels Na^+ out of the cytoplasm.

Modeled: A metabolic network between erythrocyte and neuron. A Pi^{2-} increment in cerebral spinal fluid (CSF), allows augmenting the uptake of Pi^{2-} and glucose by the erythrocyte incrementing anaerobic glycolysis and sugar phosphates. At pH=7.4 maximal production of 2,3-DPG in the Hb-red cell system results in O_2-release, which in the mitochondria of the neuron, enhance turnover of the electron transport chain generating ATP.

Repolarization synchronize within the cytoplasmic increases in the$[ATP^{4-}]$ and $[Ca^{2+}]$ generating [CaATP].

Ca^{2+}-dependent inhibition of AC.

Stimulated neurons up-regulate Krebs cycle, stepping-up mitochondrial concentrations of $citrate^{3-}$ and $dicarboxylic^{2-}$ acids. These ones chelate Ca^{2+}, and allow its influx into mitochondria, limiting a Ca^{2+}-dependent inhibition of basal AC activity.

Modeled: Coding and deconding

Pathways loops allow propagation of the nerve impulse and at the required stages leading to time-consuming metabolic processes. Basal-AC supports a flow of cAMP, its binding to PKA, C2R2 units, releases the inhibitory R and forms cAMP-C units (activated PKA). Activated PKA phosphorylates the adrenergic receptor coupled with the stimulatory G-protein (Gs) and cAMP-response element-binding proteins (CREB). Its binding to a CRE region of a DNA results in transcription of coding proteins or polipeptides.

The nature of the decoding mechanism, may relate to the coupling of energy transduction processes, involving modification of the quaternary structures of proteins [26][28][29]. As an original proposition, the mechanism could be based in the fluctuation between two states of hydration/solvation of coding proteins [122][123][124][125][126].

An ionic kosmotropes Mg^{2+} and Na^{+} have stronger structuring interactions with water, than the latter with itself. This order-maker effect breaks hydrogen bonds between water molecules. An ionic chaotropes K^{+} exhibits weaker interactions with water and lower interference on the hydrogen bonding of surrounding water. These interactions allow thermodynamically metastable phases on internal chemical bonding of molecules surrounded by water [127][128].

The mechanism may involve interaction of Mg^{2+} with proteins, which trough Na^{+}-influx dependent modification, fluctuate from a hydration state inactive to one active. The influx of K^{+} restoring solvation to the initial resting state.

Involvement of a β-adrenergic/cAMP mechanism in long-term memory may be explained by "*in common*" dependence on free Mg^{2+}. The integrative mechanism: norepinephrine and nitric oxide, increasing access of O_2 and glucose to a brain area. The mitochondrial generation of ATP becoming depleted during Na^{+} pump operation, activating AC. The forming of cAMP-response element-binding proteins (CREB) activating the CRE region for DNA transcription. The generated polypeptide molecules acting as receptors bound by Mg^{2+} to the membrane, may respond to Na^{+}/K^{+}-exchanges by signaling through changes on the hydration shells the timing for neuronal firings. These could become integrated within a neuronal circuit of "in common" vibrational resonance.

A separate mechanism for working memory may not require a totally independent pathway, if resulting from a process of accumulation of long-term memory molecules. Natural selection may adjudicate lasting neuronal patterns out of experiences conditioning integrative responsiveness. These ones like oxytocin bonding which require a receptor bound into the membrane by Mg^{2+} is a long-term process. Bonding could be transformed into sensorial patterns of recognition, creating a working memory out of conditioning reflexes, by nurturing process along the life learning period, characterizing each animal species.

Acknowledgments

The contributions of my PhD. Students: S. Brydon-Golz, H. Ohanian, R.H. Harris, V.A. De Bari, P.P. Vicario, R. Cruz and K. Borhanian, from special projects of the College of Medicine and Dentistry of New Jersey (CMDNJ). S. De Farias on leave from the Instituto Butantan Sao Paulo, Brazil and under graduate project N. Seidler, plus the assistances of E.M. De Pahn, N.A. Novak and N. Bennun, are all gratefully recognized.

References

[1] L. Cahill and J.L. McGaugh, "Mechanisms of emotional arousal and lasting declarative memory", *Trends Neurosci.*, 21, 294-9 (1998).

[2] M. Costa-Mattioli, N. Sonenberg, "Translational control of gene expression: a molecular switch for memory storage", *Prog. Brain Res.*, 169, 81-95 (2008).

[3] T. Sadeh, D. Shohamy, D.R. Levy, N. Reggev and A. Maril, "Cooperation between the hippocampus and the striatum during episodic encoding", *J. Cogn. Neurosci.*, 23 (7), 1597-608 (2010).

[4] E. Tulving and H.J. Markowitsch, "Episodic and declarative memory: role of the hippocampus", *Hippocampus*, 8 (3), 198-204 (1998).

[5] Martin and L.L. Chao, "Semantic memory and the brain: structures and processes", *Current Opinion in Neurobiology*, 11 (2), 194-201(2001).

[6] Szepietowska, W. Zhu, O. Chan, A. Horblitt, J. Dziura and R.S. Sherwin, "Modulation of β-Adrenergic Receptors in the Ventromedial Hypothalamus Influences Counterregulatory Responses to Hypoglycemia", *Diabetes.*, 60 (12), 3154-8 (2011).

[7] S.S. Kety, "The biogenic amines in the central nervous system: their possible roles in arousal, emotion and learning", *In the neurosciences: second study program*, 324-335 (1970).

[8] S. Brydon-Golz, H. Ohanian and A. Bennun, "Effects of noradrenaline on the activation and the stability of brain adenylate cyclase", *Biochem. J.*, 166, 473-483 (1977).

[9] Bennun, "Characterization of the norepinephrine-activation of adenylate cyclase suggests a role in memory affirmation pathways. Overexposure to epinephrine inactivates adenylate cyclase, a casual pathway for stress-pathologies", *BioSystems*, 100, 87-93(2010).

[10] H. Ohanian, K. Borhanian and A. Bennun, "The effect of manganese on the regulation of brain adenylate cyclase by magnesium and adenosine triphosphate", *Biochem Soc Trans.*, 6(6), 1179-82 (1978).

[11] H. Ohanian, K. Borhanian, S. De Farias and A. Bennun, "A model for the regulation of brain adenylate cyclase by ionic equilibria", *Journal of Bienergetics and biomembranes*, 13, Nos. 5/6 (1981).

[12] K. Tully, V.Y. Bolshakov, "Emotional enhancement of memory: how norepinephrine enables synaptic plasticity", *Mol Brain*, 13(3), 15 (2010).

[13] J.R. McReynolds, K. Donowho, A. Abdi, J.L. McGaugh, B. Roozendaal, C.K. McIntyre, "Memory-enhancing corticosterone treatment increases amygdala norepinephrine and Arc protein expression in hippocampal synaptic fractions", *Neurobiol Learn Mem.*, 93(3), 312-21 (2010).

[14] D. Mueller and S.P., "Cahill Noradrenergic modulation of extinction learning and exposure therapy", *Behav Brain Res.*, 208(1), 1-11 (2010).

[15] D.L. Korol and P.E. Gold, "Epinephrine converts long-term potentiation from transient to durable form in awake rats", *Hippocampus.*, 18(1), 81-91 (2008).

[16] H. Mitoma and S. Konishi, "Monoaminergic long-term facilitation of GABA-mediated inhibitory transmission at cerebellar synapses", *Neuroscience.*, 88(3), 871-83 (1999).

[17] J.F. López-Téllez, M.F. López-Aranda, I. Navarro-Lobato, M. Masmudi-Martín, E.M. Montañez, E.B. Calvo and Z.U. Khan. "Prefrontal inositol triphosphate is molecular

correlate of working memory in nonhuman primates", *J. Neurosci.*, 30(8), 3067-71 (2010).

[18] T.J. Gould and P.C. Bickford. "Age-related deficits in the cerebellar beta adrenergic signal transduction cascade in Fischer 344 rats", *J. Pharmacol Exp. Ther.*, 281(2), 965-71 (1997).

[19] P.J. Scarpace and I.B. Abrass, "Alpha- and beta-adrenergic receptor function in the brain during senescence", *Neurobiology of Aging*, 9, 53-58 (1988).

[20] P. Valverius, S. Borg, M.R. Valverius, P.L. Hoffman and B. Tabakoff. "Beta-adrenergic receptor binding in brain of alcoholics", *Experimental Neurology*, 105, Issue: 3, Pages: 280-286 (1989).

[21] G.N. Pandey. and J.M. Davis. "Treatment with antidepressants and down regulation of beta-adrenergic receptors", *Drug Development Research, 3, Issue 5*, 393-406 (1983).

[22] D. León, C.A. Castillo, M.A. Ruiz, J.L. Albasanz and M. Martín., "Metabotropic glutamate receptor/phospholipase C pathway is increased in rat brain at the end of pregnancy", *Neurochem Int.*, 50(5), 681-8 (2007).

[23] J.F. López-Téllez, M.F. López-Aranda, I. Navarro-Lobato, M. Masmudi-Martín, E. Martín Montañez, E. Blanco Calvo and Z.U. Khan. "Prefrontal Inositol Triphosphate Is Molecular Correlate of Working Memory in Nonhuman Primates", *The Journal of Neuroscience*, 30(8), 3067-3071 (2010).

[24] R. Bourtchuladze, B. Frenguelli, J. Blendy, D. Cioffi, G. Schutz and A.J. Silva. "Deficient long-term memory in mice with a targeted mutation of the cAMP-responsive element-binding protein", *Cell,* 79(1), 59-68 (1994).

[25] Y. Liu and J. Zhang., "Recent development in NMDA receptors", *Chin. Med. J.*, 113 (10), 948-56 (2000).

[26] Bennun, "The unitary hypothesis on the coupling of energy transduction and its relevance to the modeling of mechanisms", *Annals of the New York Academy of Sciences*, 227, 116-145 (1974).

[27] S. Brydon-Golz and A. Bennun, "Postsynthetic stabilized modification of adenylate cyclase by metabolites", *Biochemical Society Transactions*, 3, 721-724 (1975).

[28] Bennun, "Hypothesis for coupling energy transduction with ATP synthesis or ATP hydrolysis", *Nature New Biology*, 233 (35), 5-8 (1971).

[29] Bennun, "Hypothesis on the role of liganded states of proteins in energy transducing systems", *BioSytems, 7*, 230-244 (1975).

[30] J. Glowinski and L.C. Iverson, *J. Neurochem.*, 13, 655-669 (1966).

[31] A.G. Gilman, *Proc. Natl. Acad. Sci. USA*, 67, 305-312 (1970).

[32] B.L. Brown, R.P. Ekis and J.D.M. Albano, "In advances in Cyclic Nucleotide", *Research*, 2, 25-40 (1972).

[33] R.H. Harris, R. Cruz, A. Bennun, "The effect of hormones on metal and metal-ATP interactions with fat cell adenylate cyclase", *BioSystems*, 11, 29-46 (1979).

[34] Bennun and E. Racker, "Partial resolution of the enzymes catalyzing photophospharylation IV. Interaction of coupling factor I from chloroplast with components of the chloroplast membrane", *J. Biol. Chem.*, 244, 1325-1331 (1969).

[35] R. Harris and A. Bennun, "Hormonal control of fat cells adenylate cyclase", *Mol. Cell. Biochem.*, 13, 141-146 (1976).

[36] G.L. Collingridge, S. Peineau, J.G. Howland and Y.T. Wang, "Long-term depression in the CNS", *Nat Rev Neurosci.*, 11(7), 459-73 (2010).

[37] M.M. Taqui-Khan, A.E. Martell, *J. Phys. Chem.*, 66, 10-15 (1962).
[38] M.M. Taqui-Khan, A.E. Martell, *J. Am. Chem. Soc.*, 88, 668-□□□ (1966).
[39] D.E. Atkinson, *Biochemistry*, 7, 4030-4034 (1968).
[40] C.H. Bailey, E.R. Kandel, "Synaptic remodeling, synaptic growth and the storage of long-term memory in Aplysia", *Prog Brain Res.*, 169, 179-98 (2008).
[41] G.F. Reis, M.B. Lee, A.S. Huang, K.D. Parfitt, "Adenylate cyclase-mediated forms of neuronal plasticity in hippocampal area CA1 are reduced with aging", *J. Neurophysiol.*, 93(6), 3381-9. (2005).
[42] D. Veloso, R.W. Guynn, M. Oskarsson and R.L. Veech, *J. Biol. Chem.*, 248, 4811-4819 (1973).
[43] M.S. Seeling, "In magnesium deficiency in the pathogenesis of disease", *Plenum Press*, New York (1980).
[44] H. Rasmussen, *Science*, 170, 404-412 (1970).
[45] H. Rasmussen, D.B.P. Goodman, *Ann. N.Y. Acad. Sci.*, 253, 789-796 (1975).
[46] Chanutin and R.R. Curnish, "Effect of organic and inorganic phosphates on the oxygen equilibrium of human erythrocytes", *Arch Biochem Biophys.*, 121(1), 96-102 (1967).
[47] Arnone, "X-ray diffraction study of binding of 2,3-diphosphoglycerate to human deoxyhaemoglobin", *Nature*, 237, 146-9 (1972).
[48] M.F. Perutz, "Stereochemistry of cooperative effects in hcamoglobin", *Nature*, 228, 726-739 (1970).
[49] M.F. Perutz, A.J. Wilkinson, M. Paoli, G.G. Dodson, "The stereochemical mechanism of the cooperative effects in hemoglobin revisited", *Annu. Rev. Biophys. Biomol. Struct.*, 27, 1-34 (1998).
[50] H.F. Bunn, B.J. Ransil, A. Chao, "The interaction between erythrocyte organic phosphates, magnesium ion, and hemoglobin", *J. Biol. Chem.*, 246(17), 5273-9 (1971).
[51] R. Benesch and R.E. Benesch, "The effect of organic phosphates from the human erythrocyte on the allosteric properties of hemoglobin", *Biochem. Biophys. Res. Commun.*, 26, 162-174 (1967).
[52] E. Antonini and M. Brunori, *J. Biol. Chem.*, 244, 3909-3912 (1969).
[53] R. Benesch and R.E. Benesch, *Nature (London)*, 221, 618-622 (1969).
[54] R.E. Benesch, "The stability of the heme-globin linkage: measurement of heme exchange", Methods Enzymol., 231, 496-502 (1994).
[55] Bennun, N. Seidler and V.A. De Bari, "A model for the regulation of haemoglobin affinity for oxygen", *Biochemical Society Transactions*, 13, 364-366 (1985).
[56] P.C. Churchil and M.C. Churchil, *Arch. Int. Pharmacodyn. Ther.*, 258, 300-312 (1982).
[57] M.F. Perutz, M. Paoli, A.M. Lesk, "Fix L, a haemoglobin that acts as an oxygen sensor: signalling mechanism and structural basis of its homology with PAS domains", *Chem. Biol.*, 6, 291-7 (1999).
[58] P.P. Vicario and A. Bennun, "Separate effects of Mg^{2+}, MgATP, and ATP^{4-} on the kinetic mechanism for insulin receptor tyrosine kinase", *Archives of Biochemistry and Biophysics*, 278, 99-105 (1990)
[59] Bennun, N. Seidler and V.A. De Bari, "Divalent Metals in the Regulation of Hemoglobin Affinity for Oxygen", *Ann. N.Y. Acad. Sci.*, 463, 76-79 (1986).
[60] G. Gerber, H. Berger, G.R. Janig and S.M. Rapoport, *Eur. J. Biochem.*, 38, 563-571 (1973).

[61] G. Fermi, M.F. Perutz, in: Atlas of molecular structures in Biology, 2 Haemoglobin and Myoglobin D.C., Phillips and Richards F.M. (Eds.) Clarendon Press, Oxford (1982).
[62] J.M. Rifkind, J.M. Heim, *Biochemistry*, 16, 4438-4443 (1977).
[63] A.O.M. Stoppani, A. Bennun and E.M. De Pahn, "Energy requirement for the anaerobic oxidation of acetate in baker's yeast", *Biochimica et Biophysica*, 92, 176-178 (1964).
[64] Bennun, E.M. De Pahn and A.O.M. Stoppani, "Some properties of particle-bound intracellular ATPase from baker's yeast", *Biochimica et Biophysica*, 89, 532-539 (1964).
[65] A.O.M. Stoppani, A. Bennun and E.M. De Pahn, "Effect of 2,4-dinitrophenol on krebs cycle and phosphate metabolism in baker's yeast", *Archives of Biochemistry and Biophysics*, 108(2), 258-265 (1964).
[66] Deuticke, J. Duhm and R. Dierkesmann, "Maximal elevation of 2,3-diphosphoglycerate concentrations in human erythrocytes: Influence on glycolytic metabolism and intracellular Ph", *Pflügers Archiv European Journal of Physiology*, 326 (1), 15-34 (1971).
[67] Brewer, G.J. General red cell metabolism. In: Surgenor D. ed. The red blood cell. New York: Academic Press, 387-433, (1974).
[68] E.L. Thomas, L.E. King, M. Morrison, "The uptake of cyclic AMP by human erythrocytes and its effect of membrane phosphorylation", *Archs. Biochem. Biophys.*, 196, 459-464 (1979).
[69] V.A. DeBari and A. Bennun, "Cyclic GMP in the human erythrocyte. Intracellular levels and transport in normal subjects and chronic hemodialysis patients", *Clin. Biochem.*, 15, 219-221 (1982).
[70] V.A. DeBari, N.A. Novak, A. Bennun, "Cyclic Nucleotide Metabolism in the Human Erythrocyte", *Clin. Physiol. Biochem.*, 2, 227-238 (1984).
[71] P. Novembre, J. Nicotra, V.A. DeBari, M.A. Needle and A. Bennun, "Erythrocyte transport of cyclic nucleotides", *Annals of the New York Academy of Sciences*, 435, 190-194 (1984).
[72] J. Marie, L. Tichonicky, J.C. Dreyfus, A. Kahn, "Endogenous, cyclic 3',5' AMP-dependent phosphorylation of Human red cell pyruvate kinase", *Bichem. Biophys. Res. Commun.*, 87, 862-867 (1979).
[73] P.P., Vicario and A. Bennun, "Interaction of MnATP and peptide substrate with insulin receptor tyrosine kinase", *Biochem. Soc. Trans.*, 17(6), 1108-9 (1989).
[74] V.A. De Bari and A. Bennun, "Cyclic GMP in the human erythrocyte. Intracellular levels and transport in normal subjects and chronic hemodialysis patients", *Clinical Biochemistry*, 15(4), 219-221 (1982).
[75] V.A. De Bari, N. Novak and A. Bennun, "Cyclic nucleotide metabolism in the human erythrocyte", *Clinical Physiology and Biochemistry*, 2, 227-238 (1984).
[76] P.P., Vicario and A. Bennun, "Regulation of insulin receptor tyrosine kinase by metabolic intermediates", *Biochem. Soc. Trans.*, 17(6), 1110-1 (1989).
[77] Bennun, M.A. Needle, V.A. De Bari, "Stimulation of the hexose monophosphate pathway in the human erythrocyte by Mn^{2+}: Evidence for a Mn^{2+}-dependent NADPH peroxidase activity", *Biochemical Medicine*, 33, 17-21 (1985).
[78] S. Ramón y Cajal, "The Croonian Lecture: La Fine Structure des Centres Nerveux", *Proceedings of the Royal Society of London*, 55, 444-468 (1894).

[79] Mørk, "Actions of Lithium on the Cyclic AMP Signalling System in Various Regions of the Brain - Possible Relations to its Psychotropic Actions -: A Study on the Adenylate Cyclase in Rat Cerebral Cortex, Corpus Striatum and Hippocampus", *Pharmacology and Toxicology*, 73 (1993).

[80] T. Lømo, "The discovery of long-term potentiation", *Phil. Trans. Roy. Soc. London B (Biological Sciences)*, 358(1432), 617-620 (2003).

[81] K. Doya, "Complementary roles of basal ganglia and cerebellum in learning and motor control", *Current Opinion in Neurobiology*, 10 (6), 732-739 (2000).

[82] J.H. Medina, J.I. Rossato, L.R.M. Bevilaqua, I. Izquierdo, M. Cammarota, "Dopamine controls persistence of long term memory storage", *Science*, 21, 1017-1020 (2009).

[83] Bennun, "A coupling mechanism to inter-relate regulatory with haem-haem interactions of haemoglobin", *Biomed. Biochim.*, 46, 314-319 (1987).

[84] R. Malenka and M. Bear, "LTP and LTD: an embarrassment of riches", *Neuron* 44 (1), 5-21 (2004).

[85] C.E. Curtis, M. D'Esposito, "Persistent activity in the prefrontal cortex during working memory", *Trends in Cognitive Sciences*, 7(9): 415-423 (2003).

[86] A.G. Gilman., "G Proteins: Transducers of Receptor-Generated Signals", *Annual Review of Biochemistry*, 56, 615-649 (1987).

[87] R.C. Malenka, M.F. Bear, "LTP and LTD: an embarrassment of riches", *Neuron*, 44, 5-21 (2004).

[88] N. Wettschureck, S. Offermanns, "Mammalian G proteins and their cell type specific functions", *Physiol. Rev.*, 85(4), 1159-204 (2005).

[89] S. Brydon-Golz and A. Bennun, "Postsynthetic stabilized modification of adenylate cyclase by metabolites", *Biochem. Soc. Trans.*, 3, 721-724 (1975).

[90] Von Hungen and S. Roberts, "Adenylate-cyclase receptors for adrenergic neurotransmitters in rat cerebral cortex", *Eur. J. Biochem.*, 36(2), 391-401 (1973).

[91] S. Kakiuchi, T.W. Rall, "Studies on adenosine 3,5-phosphate in rabbit cerebral cortex", *Mol. Pharmacol.*, 4(4), 379-388 (1968).

[92] K. Tanaka, "Alteration of second messengers during acute cerebral ischemia - adenylate cyclase, cyclic AMP-dependent protein kinase, and cyclic AMP response element binding protein", *Prog. Neurobiol.*, 65(2), 173-207 (2001).

[93] Karaçay, G. Li, N.J. Pantazis and D.J. Bonthius, "Stimulation of the cAMP pathway protects cultured cerebellar granule neurons against alcohol-induced cell death by activating the neuronal nitric oxide synthase (nNOS) gene", *Brain Res.*, 1143, 34-45 (2007).

[94] Mørk, "Calmodulin-dependent adenylate cyclase activity in rat cerebral cortex : effects of divalent cations, forskolin and isoprenaline", *Archives of Physiology and Biochemistry*, 97(3) (1989).

[95] Mørk, "Mode of Action of Lithium on the Catalytic Unit of Adenylate Cyclase from Rat Brain", *Pharmacology and Toxicology*, 60(4) (1987).

[96] D.K. Vassilatis, J.G. Hohmann, H. Zeng, F. Li and J.E. Ranchalis., "The G protein-coupled receptor repertoires of human and mouse", *Proc. Natl. Acad. Sci. USA*, 100 (8), 4903-4908 (2003).

[97] N. Del Campo, S.R. Chamberlain, B.J. Sahakian and T.W. Robbins, "The roles of dopamine and noradrenaline in the pathophysiology and treatment of attention-deficit/hyperactivity disorder", *Biol. Psychiatry.*, 69(12), 145-57 (2011).

[98] J. Prince, "Catecholamine dysfunction in attention-deficit/hyperactivity disorder: an update", *J. Clin. Psychopharmacol.*, 28, 39-45 (2008).

[99] Mørk, "Effects of Lithium on Calmodulin-stimulated Adenylate Cyclase Activity in Cortical Membranes from Rat Brain", *Pharmacology and Toxicology*, 60(1) (1987).

[100] S. Cull-Candy, S. Brickley and M. Farrant, "NMDA receptor subunits: diversity, development and disease", *Curr. Opin. Neurobiol.*, 11(3), 327-35 (2001).

[101] P. Paoletti, J. Neyton, "NMDA receptor subunits: function and pharmacology", *Curr. Opin. Pharmacol.*, 7(1), 39-47 (2007).

[102] R. Dingledine, K. Borges, D. Bowie and S.F. Traynelis, "The glutamate receptor ion channels", *Pharmacol. Rev.*, 51(1), 7-61 (1999).

[103] D.O. Levitsky, D.A. Nicoll and K.D., "Philipson Identification of the high affinity Ca^{2+}-binding domain of the cardiac Na^{+}-Ca^{2+} exchanger", *J. Biol. Chem.*, 269, 22847-22852 (1994).

[104] L. Annunziato, G. Pignataro and G.F. Di Renzo, "Pharmacology of brain Na+/Ca2+ exchanger: from molecular biology to therapeutic perspectives", *Pharmacol Rev.*, 56(4), 633-54 (2004).

[105] K.D. Philipson and D.A. Nicoll, "Sodium-calcium exchange: a molecular perspective", *Annu. Rev. Physiol.*, 62, 111-33 (2000).

[106] P. Caroni and E. Carafoli, "The regulation of the Na^{+}-Ca^{2+} exchanger of heart sarcolemma", *Eur. J. Biochem.*, 132, 451-460 (1983).

[107] S. Asano, T. Matsuda, K. Takuma, H.S. Kim, T. Sato, T. Nishikawa and A. Baba, "Nitroprusside and cyclic GMP stimulate Na^{+}/Ca^{2+} exchange activity in neuronal preparations and cultured rat astrocytes", *J Neurochem*, 64, 2437-2441 (1995).

[108] W. Harley, C. Floyd, T. Dunn, X.D. Zhang, T.Y. Chen, M. Hegde, H. Palandoken, M.H. Nantz, L. Leon, K.L. Carraway, B. Lyeth and F.A. Gorin, "Dual inhibition of sodium-mediated proton and calcium efflux triggers non-apoptotic cell death in malignant gliomas", *Brain Res.*, 2(1363), 159-69 (2010).

[109] M.P. Blaustein and E.M. Santiago, "Effects of internal and external cations and ATP on sodium-calcium exchange and calcium-calcium exchange in squid axons", *Biophys J.*, 20, 79-111 (1977).

[110] Linck, Z. Qiu, Z. He, Q. Tong, D.W. Hilgemann and K.D. Philipson, "Functional comparison of the three isoforms of the Na^{+}/Ca^{2+} exchanger (NCX1, NCX2, NCX3)", *Am. J. Physiol.*, 274, C415-C423 (1998).

[111] A.J. Silva, J.H. Kogan, P.W. Frankland and S. Kida, "Creb and memory", *Annu. Rev. Neurosci.*, 21, 127-48 (1998).

[112] Purves, J. A. George, D. Fitzpatrick, W.C. Hall, A. LaMantia, J.O. McNamara and L.E. White. *Neuroscience. 4th ed.*. 170-6 (2008).

[113] Y. Wang, A. Ghezzi, J.C. Yin and N.S. Atkinson, "CREB regulation of BK channel gene expression underlies rapid drug tolerance", *Genes, Brain, and Behavior*, 8(4), 369-76 (2009).

[114] Barco, C. Bailey and E. Kandel, "Common molecular mechanisms in explicit and implicit memory", *J. Neurochem.*, 97(6), 1520-33 (2006).

[115] Mayr and M. Montminy, "Transcriptional regulation by the phosphorylation-dependent factor CREB", *Nat. Rev. Mol. Cell Biol.*, 2(8), 599-609 (2001).

[116] J. Mackliff, "Bilateral Electro Coagulation of Adrenal Medulla a Pioneer Surgery for Disappearence of Parkinson Symptons", 2nd world annual congress of Neurotalk 2011, Dalian, China (2011) http://www.drjosemackliff.com.
[117] Delorenzi, M.E. Pedreira, A. Romano, S.I. Garcia, C.J. Pirola, V.E. Nahmod and H. Maldonado, "Angiotensin II enhances long-term memory in the crab Chasmagnathus", Brain Research Bulletin Volume: 41, Issue: 4, Pages: 211-220 (1996).
[118] J.M. Saavedra, "Brain angiotensin II: new developments, unanswered questions and therapeutic opportunities", *Cell Mol. Neurobiol.*, 25(3-4), 485-512 (2005).
[119] J.W. Wright and J.W. Harding; "Brain renin-angiotensin--a new look at an old system", *Prog. Neurobiol.*, 95(1), 49-67 (2011).
[120] L.C. Craig, E.J. Harfenist and A.C. Paladini, "Dialysis studies. VII. The behavior of angiotensin, oxytocin, vasopressin, and some of their analogs", *Biochemistry.*, 3, 764-9 (1964).
[121] R. Cipriani, P. Villa, G. Chece, C. Lauro, A. Paladini, E. Micotti, C. Perego, M.G. De Simoni, B.B. Fredholm, F. Eusebi and C. Limatola, "CX3CL1 Is Neuroprotective in Permanent Focal Cerebral Ischemia in Rodents", *J. Neurosci.*, 31(45), 16327-35 (2011).
[122] Bennun, "A model mechanism for coupled phosphorylation", Proceedings 3rd International Congress on Photosynthesis, Rehovoth (M. Avron, ed.), 2, 1107-1120, Elsevier Sciences Pub. Co., Amsterdam (1974).
[123] Bennun, "Interaction of the chloroplast coupling factor with protons and water, Congreso Argentino de Ciencias Biológicas - 1970, en "Recientes adelantos en Biología" (J.A. Moguilevsky and R. Mejía, eds.), 254-264, University of Buenos Aires Press (1971).
[124] Bennun, "Properties of chloroplast's coupling factor-1 and a hypothesis for a mechanism of energy transduction", Proceedings First European Biophysics Congress, Baden, Austria, 1971, in "Photosynthesis, bioenergetics, regulation, origin of life" (E. Broda, A. Locker and H. Sprínger-Lederer, eds.), IV, 85-91, Wiener Medizinischen Akademíe, Vienna (1971).
[125] Bennun and N. Bennun, "Hypothesis for a mechanism of energy transduction. Sigmoidal kinetics of chloroplast's heat-activated ATPase", Proceedings 2nd International Congress on Photosynthesis Research (G. Fortí, M. Avron and A. Melardri, eds.), 2, 1115-1124, Dr. W. Junk N.V. Pub., The Hague (1972).
[126] A.L. Lomize, I.D. Pogozheva and H.I. Mosberg: "Anisotropic solvent model of the lipid bilayer. 2. Energetics of insertion of small molecules, peptides and proteins in membranes". *J. Chem. Inf. Model.*, 51 (4), 930–46 (2011).
[127] C.L. Gibb and B.C. Gibb, "Anion binding to hydrophobic concavity is central to the salting-in effects of Hofmeister chaotropes", *J. Am. Chem. Soc.,* 133(19):7344-7 (2011).
[128] D.A. Schmidt, O. Birer, S. Funkner, B.P. Born, R. Gnanasekaran, G.W. Schwaab, D.M. Leitner and M. Havenith, "Rattling in the cage: ions as probes of sub-picosecond water network dynamics", *J. Am. Chem. Soc.,* 131(51), 18512-7 (2009).

In: Horizons in Neuroscience Research. Volume 15
Editors: Andres Costa and Eugenio Villalba
ISBN: 978-1-63321-523-8

Chapter 2

Homeostatic Synaptic Plasticity: Molecular Mechanisms and Implications in Neurological Disorders

James P. Gilbert, Guan Wang and Heng-Ye Man *
Department of Biology, Boston University, US

Introduction

Homeostatic regulation is a negative feedback response that is similar to other processes in the body that regulate things such as body temperature, hydration levels, pH and osmolarity of fluids, or respiration rates. Each homeostatic mechanism returns the system to an intrinsic physiological "set point". The brain is a dynamic organ that constantly evolves as we perceive and adapt to the world around us. This plasticity allows the brain to learn and form new memories, refine movements, recover after injuries, and to predict and obtain rewards. With billions of neurons making up to 100,000 synaptic connections each, the mammalian brain is a very complex system constantly undergoing modifications as it responds to these changes in the environment. A fundamental question is how the nervous system can constantly undergo modifications without losing its stability or the content encoded within neural circuits. To maintain stability in brain function, neurons must adopt a mechanism of homeostatic regulation by which they can assess the activity levels and make constant adjustment so as to maintain their activity within a normal physiological range (Burrone et al., 2002; Davis and Bezprozvanny, 2001; Turrigiano et al., 1998; Turrigiano and Nelson, 2004).

At the neuronal level, homeostatic plasticity aims to maintain a stable rate in firing action potentials. This can be achieved through adjustments in the strength of synaptic inputs, neuronal excitability, neuronal connectivity or the balance between excitation and inhibition. Among these possibilities, the regulation of synaptic strength has been the most extensively studied and believed the most crucial measure in homeostatic regulation, which is known as

*Correspondence should be addressed to Heng-Ye Man, Department of Biology, Boston University, 5 Cummington St, Boston MA 02215. E-mail: hman@bu.edu.

homeostatic synaptic plasticity, or synaptic scaling (Burrone et al., 2002; Davis and Bezprozvanny, 2001; Turrigiano et al., 1998).

Basic Paradigms in Homeostatic Synaptic Plasticity

Homeostatic synaptic plasticity has mostly been studied in cultured cortical and hippocampal neurons (Turrigiano et al., 1998; Burrone and Murthy, 2002; Wierenga et al., 2006). A typical paradigm for studying this phenomenon is to manipulate the global activity levels with a bath application of pharmacological reagents to suppress or enhance neuronal activity. The homeostatic responses are measured electrophysiologically by analyzing the miniature excitatory, or inhibitory, postsynaptic currents (mEPSCs or mIPSCs, respectively). In doing so, one can observe changes in the amplitude or frequency of these postsynaptic potentials after chronic drug treatments. A change in amplitude will reflect alterations in postsynaptic neurotransmitter receptor number, while a change in frequency indicates a presynaptic change in the probability of neurotransmitter release or number of synaptic connections. Following 1-2 days of incubation with tetrodotoxin (TTX) in cultured neurons, a sodium channel blocker that inhibits action potentials and silences network activity, a compensatory increase in mEPSC amplitudes is observed. This indicates that the inhibited neuron tries to recover its activity by enhancing synaptic efficacy (Turrigiano et al., 1998; Malinow and Malenka, 2002; Frank et al., 2006). Conversely, chronic bath application of bicuculline, an antagonist of the inhibitory gamma-aminobutyric acid A ($GABA_A$) receptor that triggers neuronal hyperactivity during the ealier phase of treatment, leading to a homeostatic reduction in mEPSC amplitude, indicating an activation-dependent homeostatic down-regulation in synaptic strength (Turrigiano et al., 1998; Malenka, 2003). These pharmacological treatments do not alter mEPSC frequencies, indicating a postsynaptic-dependent mechanism in the expression of homeostatic synaptic regulation.

Homeostatic Synaptic Plasticity versus Hebbian Plasticity

Hebbian synaptic plasticity, including long-term potentiation (LTP) and long-term depression (LTD), is believed to be the cellular substrate for learning and memory. Hebbian plasticity is positive-feedback in nature, and over time, an LTP- or LTD-dominant neuron could potentially reach functional saturation or complete silence, respectively. It is therefore necessary for neurons to have an intrinsic sensor of their own activity and the ability to restore its activity to a basal set point. Unlike Hebbian plasticity that occurs at single synapses, homeostatic synaptic plasticity is believed to regulate the entire synapse population of a neuron. Because global homeostatic regulation is believed to scale the activity of synapses up or down proportionally with a fixed multiplicative factor, it is also known as synaptic scaling (Figure 1).

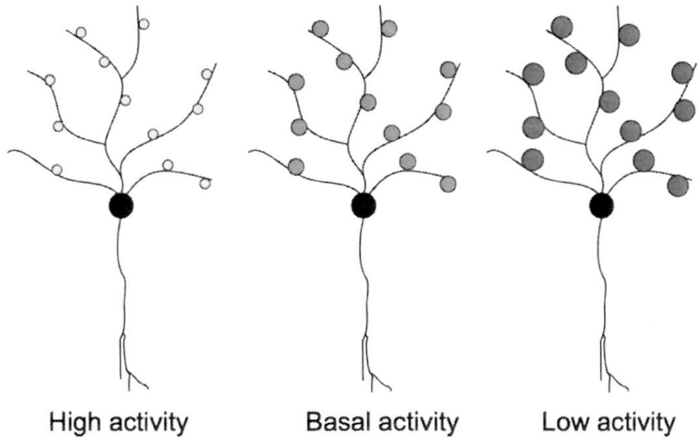

Figure 1. Homeostatic synaptic scaling. In a neuron, when activity is chronically suppressed, AMPA receptor accumulation at all synapses is increased (up scaling, right), whereas when activity is enhanced, receptor amounts are reduced at the synapses (down scaling, left). Red circles on the dendrites represent AMPAR clusters.

However, recent studies demonstrate that homeostatic regulation also occurs locally at dendrites, or even at the single synapse level (Hou et al., 2008, 2011; Turrigiano, 2008; Yu and Goda, 2009). A hallmark for Hebbian plasticity is its synapse specificity; a single synapse can be regulated specifically for a long period of time without affecting the neighboring synapses. Although relatively long-term modifications at the LTP or LTD synapses have important implications in learning and memory, the altered activity still needs to eventually be restored to basal levels in order to maintain brain stability. In this case, homeostatic regulation at single synapses is likely to play a key role in counter balancing the Hebbian plasticity. Presumably, a coupling mechanism may exist linking Hebbian and homeostatic plasticity and therefore, after a certain time, LTP activates the homeostatic mechanism to restore the synaptic activity.

Expression of Homeostatic Synaptic Plasticity

Homeostatic synaptic plasticity can be expressed either presynaptically, or postsynaptically, or both. Presynaptic release has been studied by an optical approach using an ampiphillic dye such as FM1-43. This dye has a hydrophobic tail and a positively charged polar head. When cells are treated *in vitro* with FM1-43, the tail of the molecule allows the dye to get into the plasma membrane of a cell (or more importantly a synaptic vesicle during release) and the positively charged head can't get into the lipid bilayer. As more neurotransmitter release occurs, more synaptic vesicles are retrieved back into the presynaptic terminal producing an increase in FM1-43 fluorescence. It has been shown that Ephexin, a guanine nucleotide exchange factor that interacts with ephrin ligands, is sufficient presynaptically for synaptic homeostasis. During synaptic homeostasis, Ephexin functions primarily with Cdc42, a small GTPase of the Rho-submfamily, in a signaling pathway that interacts with the presynaptic CaV2.1 calcium channel. Ephexin binds the Eph receptor (Eph) and it has been shown in Drosphila that Eph mutants disrupt synaptic homeostasis. This suggests that Ephexin/Cdc42 couples synaptic Eph signaling to the modulation of presynaptic CaV2.1 channels during the homeostatic regulation of presynaptic release (Frank et al., 2009). The primary mechanism in the expression of homeostatic plasticity is to regulate AMPAR synaptic accumulation at postsynaptic sites (Desai et al., 2002; Davis, 2006; Turrigiano, 2008; Rabinowitch and Segev, 2008; Gainey et al., 2009). During homeostatic regulation, AMPAR numbers at synapses are accordingly scaled up- or downwardly in response to activity deprivation or overexcitation, respectively, via alterations in AMPAR trafficking processes such as receptor insertion or internalization (Figure 2). When cultured cortical neurons are incubated with TTX to chronically block sodium channels and silence network activity, the synapse responds in a compensatory manner resulting in an increase in synaptic AMPAR accumulation and the strength of synaptic transmission (Perez-Otano and Ehlers, 2005; Yu and Goda, 2009; Pozo and Goda, 2010). Changes in postsynaptic surface receptor levels can result from a variety of mechanisms working at different levels.

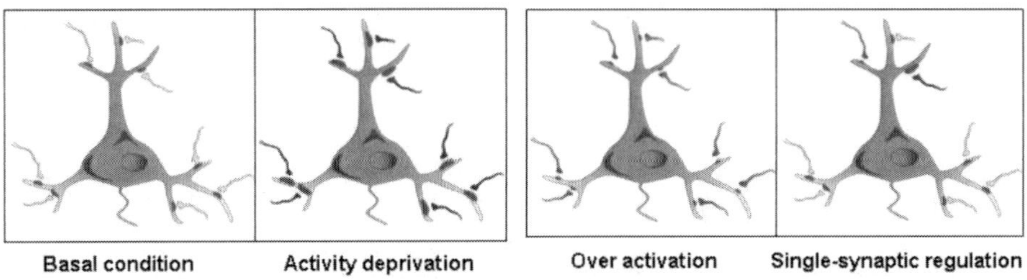

Figure 2. Global versus single homeostatic plasticity. Global activity-deprivation produces a global increase in all synaptic strengths compared to basal conditions (A and B). With global over-activation, a homeostatic response produces downscaling of synapses (C). Homeostatic plasticity can also be seen at single synapses with a selective activation or suppression of activity at a single synapse (D).

Transcriptional, post-translational, intracellular trafficking, or a combination of these have previously been shown to alter receptor surface expression levels. Although subunit synthesis determines receptor availability, and therefore can determine an initial level of control over the numbers and types of receptors present at synapses, several trafficking steps following subunit synthesis appear to operate as regulatory checkpoints for the synaptic incorporation of receptors (Barria and Manilow, 2002; Hou et al., 2008, 2011; Angonno et al., 2011; Peebles et al., 2010; Wu et al., 2011). Activation of glutamate receptors (Beattie et al., 2000; Ehlers, 2000), increasing neural network activity by membrane depolarization, or by unbalancing excitatory and inhibitory inputs to favor excitation (Lin et al., 2000) results in reductions in the synaptic receptor accumulation through receptor internalization, whereas selective activation of synaptic NMDARs can lead to facilitated AMPAR recycling and membrane insertion (Lu et al., 2001; Man et al., 2003; Park et al., 2004). We recently show that continuous activation of a single synapse causes homeostatic reduction of AMPAR surface expression via receptor internalization and degradation (Hou et al., 2011). Therefore, trafficking-dependent alterations in AMPAR synaptic localization serve as a primary mechanism not only for the expression of Hebbian type synaptic plasticity (Malenka, 2003; Malinow and Malenka, 2002; Man et al., 2000; Song and Huganir, 2002), but also for the expression of negative-feedback based homeostatic synaptic regulation (Turrigiano and Nelson, 1998; Wierenga et al., 2005; Sutton et al., 2006; Levi et al., 2008; Hou et al., 2008, 2011; Man, 2011)

A key question in synaptic homeostatic plasticity is to identify the cue molecules or signaling cascades that trigger the recruitment of AMPARs at synapses. Because calcium entry into neurons is intimately related to activity, calcium has been considered as a candidate signaling factor (Turrigiano, 2008; Desai et al., 2009). However, NMDAR-mediated calcium entry, which is critical for Hebbian synaptic plasticity, is not required for homeostatic regulation (Pozo and Goda, 2010; Vitureira et al., 2011), and calcium from voltage-gated calcium channels has been positively linked to this regulation only in a few circumstances (Vitureira et al., 2011). Work from others and our own indicate a crucial role of AMPARs that do not contain the GluA2 subunits, i.e., GluA2-lacking, calcium-permeable AMPARs (Cp-AMPARs). Suppression of neuronal activity induces the generation of Cp-AMPARs. Importantly, the homeostatic response in mEPSCs by TTX-dependent activity deprivation can be abolished when the GluR2-lacking AMPARs are specifically blocked (Sutton et al., 2006; Hou et al., 2008, Aoto et al., 2008). Interestingly, blocking of Cp-AMPARs after TTX treatment has no marked effect on the enhanced synaptic activity, indicating that GluA2-lacking AMPARs are required for the induction, but not the expression, of homeostatic synaptic plasticity. Recent studies have identified multiple molecules and pathways that are involved in homeostatic synaptic regulation (See *Molecules and signaling pathways in homeostatic synaptic plasticity*).

Local and Single Synaptic Homeostatic Plasticity

Homeostatic plasticity has been believed to be executed via global scaling of AMPARs at all synapses in a neuron or the entire network, but recent studies indicate that it can also be expressed locally on individual or a small group of synapses (Hou et al., 2011; Turrigiano, 2008; Yu and Goda, 2009; Goold and Nicholl, 2010) (Figure 2). To visualize locally controlled homeostatic scaling, surface staining and fluorescent imaging techniques are typically utilized. Utilizing microperfusion to locally apply antagonists, Sutton et al. (2006) show in dissociated hippocampal neurons that rapid synaptic scaling of AMPAR currents, induced by the suppression of network activity, was regulated by local dendritic protein synthesis in an NMDAR-dependent manner. It is also shown that TTX perfusion onto the soma can increase dendritic GluA2 levels but not when it is applied locally to the dendrites (Ibata et al., 2008). This suggests that different homeostatic mechanisms can differentially regulate AMPARs and that these mechanisms could function within distinct neuronal sub-compartments. Hou et al. (2008) first described homeostatic regulation at single synapses. In cultured hippocampal neurons, selective inhibition of the activity of an individual synapse results in a homeostatic increase in postsynaptic AMPAR expression. Recently, using light controlled glutamatergic receptors (Szobota et al., 2007) to precisely stimulate the activity of an individual presynaptic terminal, Hou et al. (2011) demonstrates that the level of AMPARs at the excited synapses is selectively down-regulated via receptor internalization and proteasomal degradation. This set of experiments indicates that homeostatic plasticity is utilized by individual synapses to maintain synaptic functional stability.

AMPA Receptor Trafficking in Synaptic Plasticity

Ionotropic glutamatergic AMPA receptors mediate the vast majority of fast excitatory synaptic transmission in the central nervous system. AMPARs are heterotetramers assembled from different combinations of four subunits, GluA1-4, the most common of which are receptors containing GluA1/GluA2 or GluA2/GluA3 (Figure 3). As a result of a unique property acquired by mRNA editing on the GluA2 subunit, AMPARs containing GluA2 subunits are impermeable to calcium. In contrast, NMDARs have high calcium permeability, with calcium influx triggering many downstream signaling pathways critical for the induction of synaptic plasticity, such as LTP and LTD. AMPARs are localized at the postsynaptic domain at a high density, and accumulation is believed to be stabilized and regulated by interactions with cytosolic scaffolding proteins such as GRIP, SAP97, PICK1, Stargazin and PSD-95. These AMPAR-associated proteins usually contain one or more PSD domains through which they interact with the intracellular C-terminus of GluAs to regulate AMPAR synaptic targeting, intracellular trafficking, as well as channel function (Tomita et al., 2001; Song and Huganir, 2002; Kim and Sheng, 2004).

A large number of studies have established that AMPARs are not passively localized at the postsynaptic domain. Rather, they are very dynamic, trafficking constantly between the plasma membrane and intracellular compartments (Bredt and Nicoll, 2003; Malenka 2003; Malinow, 2003; Sheng and Hyoung Lee, 2003; Collingridge et al., 2004). AMPARs, like other membrane proteins, are synthesized in the ER and following further modification in the Golgi, are transported to and fused with the plasma membrane, while preexisting surface AMPARs are removed through receptor endocytosis. At basal conditions, a balance between opposing processes, receptor insertion and internalization, maintains the stable cell-surface level of AMPARs. AMPARs are internalized via the clathrin-coated pit pathway, initiated by an association of the receptor subunit C-terminus with the clathrin adaptor molecule AP2 (Man et al., 2000; Lee et al., 2002). On the other hand, cell-surface insertion of AMPARs is mediated by SNARE protein interactions, a ubiquitous machinery for vesicle-based membrane fusion (Lu et al. 2001). Both receptor endocytosis and insertion seem to occur at the parasynaptic sites (Passafaro et al. 2001; Blanpied et al. 2002), likely due to the dense structure underneath the postsynaptic membrane.

AMPAR trafficking, including both exocytosis and endocytosis, is subject to activity-dependent regulation. The resulting imbalance of the two trafficking pathways causes a net change in the abundance of synaptic AMPARs, which has been demonstrated to be the cellular mechanism underlying synaptic plasticity. AMPAR internalization can be induced by application of glutamate, NMDA or AMPA to activate their respective receptors (Lin et al., 2000; Man et al., 2000), or KCl and bicuculline to enhance global neuronal network activities (Ehlers 2000). In contrast, activation of synaptic NMDARs induces AMPA surface insertion (Lu et al., 2001).

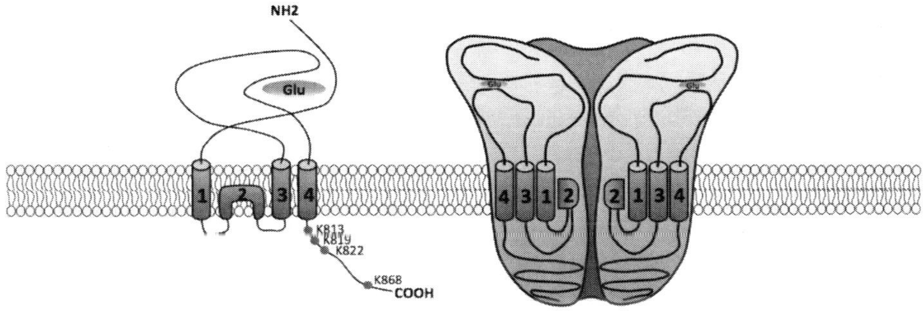

Figure 3. AMPA receptor subunits and the membrane topology. AMPARs are complexes with a variety of combinations of the GluA1-4 subunits. All subunits share the same membrane topology, with an extracellular N-terminus, an intracellular C-terminus, three transmembrane domains and an intramembrane loop. Four heterosubunits are needed to form one receptor channel, which normally contains at least one GluA2 subunit. The last several amino acids at the end of the C-terminal specifically bind to PDZ domain-containing proteins to regulate receptor trafficking. Four lysine (K) residues can be modified for ubiquitination.

A large body of data support that long-term changes in the efficacy of synaptic transmission including LTP and LTD are expressed via alterations in AMPAR trafficking and thus their synaptic expression. For instance, LTP is abolished by reagents that disrupt membrane fusion (Lledo et al., 1998); an LTP-inducing stimulation paradigm drives GluA1 into spines (Shi et al., 1999). In cultured hippocampal neurons, application of glycine, an NMDA receptor co-agonist to selectively activate synaptic NMDA receptors, results in a long-lasting enhancement in synaptic transmission via SNARE-dependent AMPARs (Lu et al.,2001; Park et al., 2004). Consistent with NMDA-induced chemical LTD (Lee et al., 1998), activation of NMDA receptors induces marked AMPA receptor endocytosis (Beattie et al., 2000; Lin et al., 2000). In hippocampal slices, disrupting the dynamin-dependent endocytotic process abolishes LTD, indicating a participation of clathrin-mediated receptor internalization (Man et al., 2000). An increasing amount of data indicates a role of glutamate transporters in the regulation of synaptic plasticity (Nicoll et al., 1998; Malinow and Malenka, 2002; Song and Huganir, 2002; Malenka, 2003; Sheng and Hyoung Lee, 2003).

GluA2-Lacking, Calcium Permeable AMPA Receptors (Cp-AMPARs)

Under physiological conditions most AMPARs contain at least one GluA2 subunit and allow only sodium influx to depolarize membrane potentials during synaptic activation. When AMPARs are composed without GluA2 subunits, the receptor channel will permit calcium in addition to sodium, causing an inward rectification in the current-voltage relationship (I-V curve). This property of GluA2 is acquired post-transcriptionally by RNA editing at the Q/R site, where a glutamine (Q) codon is replaced by an arginine (R) codon. Because the editing site is located in the channel pore, the positively charged arginine blocks calcium influx. It has been found that the expression of GluA2-lacking receptors is regulated by development, synaptic activity or pathological challenges such as ischemia or amyotrophic lateral sclerosis (ALS) (Peng et al., 2006). In early postnatal development, cortical pyramidal neurons have higher rectification in AMPAR-mediated synaptic currents, which diminishes in more mature animals, indicating a developmental switch in AMPAR composition and calcium permeability (Kumar et al., 2002). By providing an unconventional source of calcium other than NMDAR or calcium channels, GluA2-lacking AMPARs may play an important role in synaptic plasticity. Indeed, inhibitory interneuron studies demonstrate that the NMDAR-independent LTP and LTD are induced by GluA2-lacking postsynaptic AMPARs (Mahanty and Sah, 1998). Recent work shows that even in the conventional NMDAR-dependent hippocampal LTP, GluA2-lacking AMPARs are first incorporated into the synapse, which will then be replaced with GluA2-containing receptors (Plant et al., 2006). In cerebellar stellate cells which normally contain GluA2-lacking AMPARs, high frequency presynaptic activity induces a calcium-dependent increase in synaptic insertion of GluA2-containing AMPARs (Liu and Cull-Candy, 2000). These findings strongly indicate the presence of a self-regulating mechanism by which calcium, via GluA2-lacking receptors, triggers recruitment of normal calcium impermeable AMPARs to synapses. Recent studies indicate a key role of the Cp-AMPARs in the induction of inactivity-dependent homeostatic synaptic plasticity.

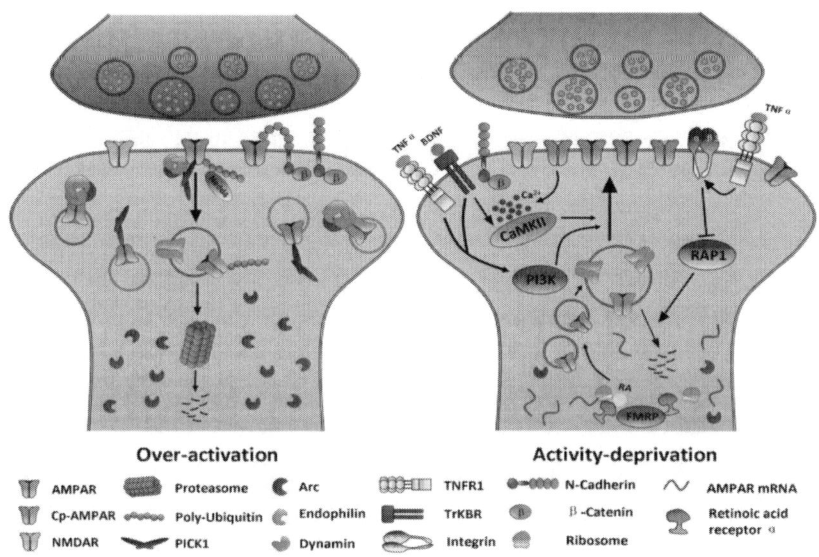

Figure 4. Regulation of homeostatic AMPAR trafficking and turnover. (Left panel) Elevated synaptic activation stimulates the expression of an immediate-early gene product, Arc, which, together with endophilin and dynamin promotes AMPAR internalization. This process also requires the association of AMPARs with PICK1, and Nedd4-mediated receptor ubiquitination and proteasomal degradation. (Right panel) Under activity deprivation, varied signaling cascades are implicated in AMPAR trafficking and synaptic localization. Activation of TNFR1 by glia-derived TNFα, or the PI3K pathway by BDNF/TrKB and TNFR1 signaling, significantly enhances the post-synaptic AMPAR levels. β3 integrin signaling inhibits the activity of RAP1, which is known to enhance AMPAR degradation. Also, calcium flux via the inactivity-induced GluA2-lacking AMPARs (Cp-AMPAR) activates the CaMKII pathway, causing AMPAR phosphorylation and insertion. In addition, retinoic acid receptor α (RARα) and FMRP lead to local synthesis of AMPARs in the postsynaptic domain.

Molecules and Signaling Pathways in Homeostatic Synaptic Plasticity

Although a change in AMPAR expression is believed to be the primary mechanism for the expression of homeostatic plasticity, molecular components and intracellular signaling pathways implicated in this type of plasticity have not been known until recently. New findings suggest the existence of complex regulatory cascades underlying the initiation, expression and regulation of a homeostatic response (Figure 4).

TNFα

Tumor necrosis factor-alpha (TNFα) is an inflammatory cytokine that is involved in inflammation, immune activation, cell death and degradation (Pery et al., 1995; Allan and Rothwell, 2001). In addition to its important functions in immune responses, TNFα is also important in maintaining the neuronal network stability (Stellwagen and Malenka, 2006; Steinmetz and Turrigiano, 2010). TNFα has been found to regulate the homeostatic up-regulation of mEPSC amplitude and post-synaptic AMPARs (Stellwagen and Malenka, 2006). When incubated with medium from TTX-treated neural cell cultures, the amplitude of mEPSCs is markedly increased. Further, the effect on mEPSCs is abolished when exogenous TNFα receptors are supplemented to the medium to scavenge TNFα, indicating a critical role of free TNFα released to the extracellular environment during activity deprivation. To further support TNFα signaling in homeostatic plasticity, TTX-dependent up-scaling in mEPSCs is completely abolished in cultured neurons or brain slices from TNFα knock-out mice. Interestingly, this abolished scaling can be rescued by co-culture of TNFα knock-out neurons with wild-type glial cells. Further, although neurons produce TNFα by themselves, co-culture of wild-type neurons with TNFα knock-out glial cells still abolishes the up-scaling in wild-type neurons (Stellwagen and Malenka, 2006). Thus, TNFα released from glia, but not neurons, plays a crucial role in the induction of homeostatic synaptic regulation. TNFα mediates homeostatic and synaptic scaling not only in cultured neural cell systems, but *in vivo* as well. In the visual cortex of TNFα deficient mice, the homeostatic regulation resulting from activity-deprivation and the ocular dominance plasticity produced by visual deprivation are both found to be impaired (Kaneko et al., 2008).

TNFα seems to mediate TTX-dependent up-scaling by regulating the trafficking and synaptic accumulation of AMPARs. TNFα induces a rapid translocation of AMPARs to the postsynaptic domain in a subunit-specific manner (Beattie et al., 2002; Stellwagen et al., 2005). After acute TNFα treatment, the delivery of AMPARs to the postsynaptic surface is enhanced significantly through downstream activation of the PI3K pathway (Leonoudakis et al., 2008). Its effects on AMPAR trafficking are mainly through TNFR1 but not TNFR2 (He et al., 2011). Of note, the enhanced delivery of the GluA1 subunit occurs faster than GluA2, leading to the generation of GluA2-lacking AMPARs. Calcium influx via these special types of AMPARs is believed to have an important role in the initiation of the homeostatic response (Man, 2011) (see *GluA2-lacking, calcium permeable AMPARs* below). In addition to enhancing AMPAR synaptic delivery, TNFα also decreases the trafficking of $GABA_A$

receptors. Thus, homeostatic regulation of neuronal activity can be achieved via re-balancing excitation and inhibition (Stellwagen et al., 2005).

CaMKs

Calcium (Ca^{2+}) is one of the most important signaling molecules in the nervous system. In neurons, Ca^{2+} transients from ligand- and voltage-gated calcium channels and intracellular calcium stores (Bloodgood and Sabatini, 2007) activate a family of Ser/Thr protein kinases known as Ca^{2+}/calmodulin-dependent protein kinases (CaMKs) (Wayman et al., 2008). Among all CaMKs, CaMKII and CaMKIV are known to be the most closely linked to synaptic transmission and different types of synaptic plasticity (Colbran and Brown, 2004; Wayman et al., 2008), due to their ability to phosphorylate AMPARs to regulate receptor trafficking (Barria et al., 1997; Hayashi et al., 2000). Although the important role of CaMKs in Hebbian type synaptic plasticity has been extensively studied (Malinow and Malenka, 2002; Wayman et al., 2008), their involvement in homeostatic regulation has not been studied until recently. By using TTX local perfusion, Ibata et al. (2008) show that somatic activity inhibition induces up-scaling of mEPSC amplitudes and increased expression of AMPARs at the post-synaptic surface (Ibata et al., 2008). This homeostatic response results from a drop in somatic Ca^{2+} levels and reduced CaMKIV activation (Ibata et al., 2008). In another study, Goold and Nicoll (2010) employed optogenetic techniques to excite the somatic area of hippocampal CA1 pyramidal neurons for 24 hours. This lasting activation leads to a depression of both AMPAR- and NMDAR-mediated currents, and a reduction in surface expression of AMPARs and NMDARs. Interestingly, in this study the researchers also identified Ca^{2+} influx and CaMKIV as the mediator of the depression effects. Since CaMKIV is a known gene transcription regulator (Soderling, 1999), both studies investigate the possible role of gene transcription in bi-directional homeostatic regulation. As expected, pharmacological abolishment of gene transcription blocks the homeostatic regulation in either direction, which demonstrates that CaMKIV regulates bi-directional homeostatic regulation via regulating gene transcription (Ibata et al., 2008; Goold and Nicholl, 2010). These studies indicate an important role for CaMKIV in cell-autonomous bi-directional homeostatic regulation. In addition to CaMKIV, other CaMK members may also participate in homeostatic regulation. Knock-down of βCaMKII in hippocampal neurons blocks NBQX-induced homeostatic increases in AMPAR expression while βCaMKII over-expression increases synaptic AMPAR levels (Groth et al., 2011).

BDNF

Brain-derived neurotrophic factor (BDNF) is broadly involved in many physiological processes in the developing or mature nervous system. BDNF is released from neurons in an activity- and calcium-dependent manner (Balkowiec and Katz, 2002). Once released, BDNF binds to the TrKB receptor to trigger a series of down-stream signaling pathways (Carvalho et al., 2008). Given its positive role in neuroprotection, BDNF is considered valuable in the management of several neurological diseases such as Hungtington's disease, epilepsy, and Alzheimer's Disease (Mattson, 2008).

The chronic presence of BDNF enhances synapotogenesis (Carvalho et al., 2008)], and controls synaptic transmission and plasticity at both glutamatergic and GABAergic synapses (Gottmann et al., 2006). In cultured neo-cortical neurons derived from rat visual cortex, BDNF was first found to mediate homeostatic down-regulation of mEPSCs (Rutherford et al., 1998). Application of exogenous BDNF blocks activity deprivation-induced homeostatic up-scaling, whereas BDNF depletion by high-affinity TrKB receptors mimics inactivity-induced homeostatic upregulaton in mEPSCs (Rutherford et al., 1998). BDNF is found to affect mEPSCs differentially in pyramidal neurons and interneurons (Rutherford et al., 1998). At synapses formed onto pyramidal neurons, exogenous BDNF attenuates the TTX-induced enhancement in mEPSC amplitudes but at synapses formed onto interneurons, BDNF enhances the quantal amplitude of mEPSCs. Intriguingly, in cultured hippocampal neurons, BDNF treatment enhances mEPSCs in excitatory synapses, likely due to enhanced AMPAR trafficking (Bolton et al., 2000; Copi et al., 2005), which is inconsistent with its role in homeostatic down-regulation. Recently, several studies show that BDNF treatment enhances AMPAR trafficking to glutamatergic synapses, both *in vitro* and *in vivo* (Caldeira et al., 2007a; Nakata and Nakamura, 2007; Li and Wolf, 2011). In cultured hippocampal neurons, Caldeira et al. (2007) show that elevated AMPAR trafficking by BDNF is receptor subunit-specific. Surface expression of GluA1 is preferentially increased in the first 30 min of BDNF incubation, leading to formation of calcium-permeable, GluA1 homomeric AMPARs. At a later stage, 30 min of BDNF incubation enhances the delivery of GluA2 and GluA3 subunits (Caldeira et al., 2007a). Enhanced AMPAR synaptic delivery requires the activation of TrKB receptors and the PI3K-Akt pathway (Nakata and Nakamura, 2007), most likely through phosphorylation of the GluA1 C-terminal at S831 (Caldeira et al., 2007a).

PI3K-Akt Pathway

The phosphoinositide 3-kinase (PI3K)-Akt pathway is known for its involvement in AMPAR trafficking, synaptic plasticity and memory consolidation both *in vivo* and *in vitro* (Lin, 2001; Sanna, 2002; Sheng, 2002; Man, 2003; Gobert, 2008). For instance, protein synthesis and AMPAR insertion in late-phase LTP require PI3K pathway activation (Man, 2003; Gobert, 2008). *In vivo*, the PI3K pathway activation is also required for the maintenance of LTP in hippocampal CA1 neurons (Gobert et al., 2008). Consistent with its important role in LTP, PI3K is also implicated in memory consolidation (Lin et al., 2001). However, in contrast to abundant evidence about its role in Hebbian synaptic plasticity, involvement of PI3K signaling in homeostatic synaptic plasticity is not clear. The PI3K cascade is potentially a good mediator in homeostatic regulation. First, the PI3K-Akt pathway is known to induce protein synthesis (Schratt et al., 2004) and AMPAR membrane insertion (Man, 2003; Qin, 2005), processes known to occur in homeostatic AMPAR up-regulation (Sutton, 2006). Second, both TNFα and BDNF, the two major factors that mediate homeostatic regulation, activate the PI3K-Akt pathway and increase AMPAR expression at the post-synaptic surface (Nakata and Nakamura, 2007; Carvalho, 2008; Leonoudakis et al., 2008). Indeed, work by Hou et al. (2008) shows a requirement of PI3K activity in homeostatic increases of synaptic AMPARs following activity suppression at single synapses (Hou et al., 2008). Consistent with this finding, a recent study also demonstrates the PI3K-Akt pathway as the mediator of homeostatic regulation. Presenilin can activate the PI3K-Akt pathway by promoting the formation of Akt-activating cadherin/PI3K complexes (Baki, 2004), and the homeostatic up-regulation is impaired in Presenilin 1 (PS1) knock-out mice (Pratt et al. 2011), suggesting a role of the PI3K cascade. Consistently, over-expression of constitutively active Akt rescues impaired synaptic scaling in PS1 knock-out mice neurons without affecting mEPSC amplitudes in wild-type neurons (Pratt et al., 2011). Together these studies clearly indicate the necessity of the PI3K-Akt pathway in homeostatic up-regulation.

N-Cadherin/ β-Catenin and Integrins

Integrins and neuronal (N)-cadherin/β-catenin are cell adhesion molecules (CAMs) that are expressed at synapses and possess important functions in synapse formation, differentiation and maturation, as well as synaptic plasticity (Dalva et al., 2007; Dityatev et al., 2010). Malfunction of CAMs can produce severe problems in the nervous system including abnormal spine morphology, decreased synapse quantity and impaired cognitive function (Chavis and Westbrook, 2001; Togashi et al., 2002; Takeichi and Abe, 2005; Dalva et al., 2007; Williams et al., 2011). N-cadherin is a Ca^{2+}-dependent homophilic adhesion protein that is present at both pre- and post-synaptic membranes (Mysore et al., 2008). It signals through Rho-family GTPases, via catenins, to control dendritic spine morphology and motility (Elia et al., 2006). N-cadherin regulates the surface expression and intracellular trafficking of AMPARs by forming a complex with β-catenin (Nuriya and Huganir, 2006; Pozo and Goda, 2010). The N-terminal domain of N-cadherin also physically interacts with the GluA2 subunit of AMPARs in the extracellular space, promoting spine growth and synaptic transmission (Sagletietti et al., 2007). By targeting AMPARs during endocytosis and exocytosis at the post-synaptic surface, N-cadherin/β-catenin complexes can regulate the bi-directional homeostatic responses of neural activity (Okuda et al., 2007). Over-expression of a dominant-negative mutant of N-cadherin results in reduced quantal AMPAR responses, whereas selective deletion of β-catenin in cultured hippocampal neurons eliminates TTX- and bicuculline-induced homeostatic regulation (Okuda et al., 2007).

Integrins are heterodimeric transmembrane molecules of which many subunits are expressed in the CNS. Previous studies have revealed multiple roles for integrins in synaptogenesis, synaptic transmission and plasticity, as well as memory formation (Chen and Grinnel, 1995; Chavis and Westbrook, 2001). β3-integrin regulates synaptic strength by modulating the surface expression of AMPARs in a subunit-specific manner. Over-expression of β3-integrin in dissociated hippocampal neurons enhances AMPAR surface expression via suppression of RAP1 (Cingolani et al., 2008), a pathway known to negatively control AMPAR trafficking during synaptic plasticity by enhancing endocytosis of GluA2 subunits (Zhu et al., 2002). Since treatment by TNFα increases surface β3-integrin, β3-integrin may be implicated in homeostatic up-regulation of neural activity by regulating AMPAR trafficking. Consistent with this, neural activity deprivation by TTX causes an increase in surface β3-integrin to a level similar to that of TNFα treatment. More directly, genetic deletion of β3-integrin completely eliminates TTX-dependent homeostatic responses in mEPSCs and AMPAR synaptic localization (Cingolani et al., 2008).

GluA2-Lacking, Calcium-Permeable AMPARs (Cp-AMPARs)

Under physiological conditions most AMPARs contain at least one GluA2 subunit and allow only sodium influx to depolarize the membrane potential during synaptic activation. When AMPARs are composed without GluA2 subunits, the receptor channel will permeate calcium in addition to sodium, causing an inward rectification in the current-voltage relationship. This property of GluA2 is acquired posttranscriptionally by RNA editing at the Q/R site at the second intramembrane loop, where a glutamine (Q) codon is replaced by an arginine (R) codon. Because the editing site is located in the channel pore, the positively charged arginine blocks calcium influx. It has been found that the expression of GluA2-lacking receptors is regulated in development, by synaptic activity or pathological challenges such as ischemia or amyotrophic lateral sclerosis (ALS) (Peng et al., 2006). In early postnatal ages, cortical pyramidal neurons have higher rectification in AMPAR-mediated synaptic currents, which diminishes in more mature animals, indicating a developmental switch in AMPAR composition and calcium permeability (Kumar et al., 2002). By providing an unconventional source of calcium other than NMDAR or calcium channels, GluA2-lacking AMPARs may play an important role in synaptic plasticity. Indeed, in the expression of hippocampal LTP, GluA2-lacking AMPARs are first incorporated into the synapse, which will then be replaced with GluA2-containing receptors (Plant et al., 2006). In cerebellar stellate cells which normally contain GluA2-lacking AMPARs, high frequency presynaptic activity induces a calcium-dependent increase in synaptic insertion of GluA2-containing AMPARs (Liu and Cull-Candy, 2000). These findings strongly indicate the presence of a self-regulating mechanism by which Cp-AMPAR-mediated calcium flux triggers recruitment of normal GluA2-containing AMPARs to synapses.

Following the induction of a homeostatic response by activity deprivation, the increase in AMPAR expression is largely dependent on GluA1, not GluA2 (Thiagarajan et al., 2005; Sutton*et al.*, 2006; Aoto et al., 2008), suggesting the formation of GluA2-lacking, calcium-permeable AMPARs. Consistent with this, AMPAR-mediated currents show inward rectification and become sensitive to Cp-AMPAR-selective antagonists philanthotoxin-433 (PhTx) or Naspm (Ju et al., 2004; Thiagarajan et al., 2005; Sutton et al., 2006; Hou et al., 2008; Aoto et al., 2008). Interestingly, multiple signaling molecules that are involved in homeostatic synaptic plasticity including TNFα, retinoic acid, Arc/Arg3.1 and β3-integrin are capable of causing imbalanced GluA1 and GluA2 regulation and Cp-AMPAR expression. The homeostatic factor TNFα is known to cause rapid membrane insertion of GluA2-lacking AMPARs (Ogoshi et al., 2005; Leonoudakis et al., 2008). In retinoic acid-mediated synaptic scaling, the increase in AMPAR surface expression is GluA1 specific, and the homeostatic response in mEPSCs is abolished by suppression of Cp-AMPAR (Aoto et al., 2008). An unbalanced regulation in AMPAR subunits is also observed in Arc/Arg3.1-mediated homeostatic regulation. Knockout of Arc/Arg3.1 results in a typical synaptic scaling of AMPAR-mediated mEPSCs. Interestingly, Arc/Arg3.1 knockout neurons reveal a significant increase in GluA1 surface expression, whereas surface GluA2 shows no change (Sheperd et al., 2006), implicating membrane addition of GluA2-lacking AMPARs. In addition, disruption of β3-integrin induces internalization of GluA2, but not GluA1 subunits, resulting in GluA2-lacking AMPARs at the cell surface. In our own study, we find that homeostatic regulation by single synaptic suppression is abolished by the application of PhTx, indicating the requirement of Cp-AMPAR signaling (Hou et al., 2008). Interestingly, the blockade of homeostatic plasticity is observed only when PhTx is applied during the early stage of activity deprivation (Hou et al., 2008), indicating that Cp-AMPARs are needed for the initiation, but not maintenance, of homeostatic synaptic regulation.

Arc/Arg3.1

Arc/Arg3.1 is an immediate-early gene product whose abundance at synapses is strictly coupled with neural activity levels. Strong synaptic activation will dramatically enhance the expression of Arc/Arg 3.1 in dendrites and spines while synaptic suppression decreases its expression (Steward and Worley, 2001; Rial Verde et al., 2006). Arc/Arg3.1 is broadly involved in different forms of synaptic plasticity including both Hebbian plasticity and homeostatic regulation (Bramham et al., 2008). Rial Verde et al. (2006) find that Arc/Arg 3.1 controls synaptic transmission strength by negatively regulating the surface expression of AMPAR expression at the post-synaptic surface. Over-expression of Arc/Arg3.1 in hippocampal neurons promotes the endocytosis of GluA2/3 containing AMPARs as well as the reduction of AMPAR-mediated synaptic current amplitude. Importantly, knock-down of Arc/Arg3.1 by siRNA abolishes this effect (Bramham et al., 2008). In Arc/Arg3.1 knockout mice, AMPARs show markedly reduced endocytosis and enhanced steady-state surface expression (Chowdhury et al., 2006; Shepherd et al., 2006). Shepherd et al. (2006) also find that Arc/Arg3.1 is involved in bi-directional homeostatic regulation of neural activity via regulating AMPAR internalization and endocytosis. In primary neuronal culture, synaptic expression of Arc/Arg3.1 is enhanced after chronic activity deprivation, while over-expression of Arc/Arg3.1 blocks the synaptic up-scaling of mEPSCs and AMPARs induced by chronic activity deprivation. Conversely, in cultured hippocampal neurons from Arc/Arg3.1 knock-out (KO) mice, either up- or down-homeostatic regulation induced by prolonged TTX- or bicuculline treatment is impaired (Shepherd et al., 2006). Activity of Arc/Arg3.1 *in vivo* has also been shown to be correlated with homeostatic scaling induced in mouse visual cortex by monocular visual deprivation (MVD) (Tagawa et al., 2005). In mouse visual cortex, Arc mRNA and protein level in the superficial layers and layer 6 is found to be dramatically manipulated by MVD (Tagawa*et al.*, 2005). Together these studies demonstrate strong evidence about the involvement of Arc/Arg3.1 in the bi-directional homeostatic regulation of neural activity through regulation of the trafficking of AMPARs. It has also been shown that Arc/Arg3.1 directly interacts with endophilin and dynamin to form post-synaptic endosomes which facilitates the endocytosis of AMPARs (Chowdhury et al., 2006). Using over-expression of Kir 2.1 (an inwardly rectifying potassium channel) in cultured pyramidal neurons, Beique et al. (2011) show that Arc is required for homeostatic up-regulation of AMPARs at single synapses (Beique et al., 2011).

Retinoic Acid and FMRP

Retinoic acid (RA), also known as Vitamin A, is best known for its role in regulating nervous system development, including neurogenesis and neuronal differentiation (Maden, 2007; Bonnet et al., 2008). Aoto et al. (2008) show that 24 hr TTX+APV treatment, a homeostatic paradigm that induces synaptic up-scaling of mEPSCs and AMPARs (Sutton et al., 2006), significantly enhances the synthesis of RA in both cultured hippocampal neurons and brain slices (Aoto et al., 2008). Application of RA rapidly increases the strength of synaptic transmission mainly through an increase in surface expression of AMPARs (Aoto et al., 2008). Effects of RA are translation-, but not transcription-dependent (Maghsoodi et al., 2008), and are occluded by TTX+APV treatment, indicating an involvement of RA signaling in homeostatic plasticity. In addition, AMPAR regulation is subunit-specific, with a preferential increase in GluA1 compared to GluA2, leading to the production of GluA2 lacking, calcium-permeable AMPARs. A recent study from the same group indicates a role of the fragile-X mental retardation protein (FMRP) in RA-induced GluA1 local translation (Soden and Chen, 2010). FMRP is a dendritic RNA-binding protein encoded by the Fmr1 gene that is involved in the down-regulation of local mRNA translation and protein synthesis (Khandajian, 1999; Mazroui et al., 2002). In Fmr1 knock-out mice, both TTX+APV-induced AMPAR homeostatic up-regulation and RA-induced local AMPAR synthesis are impaired. Over-expression of WT-FMRP, but not mutant FMRP in Fmr1 knockout neurons restores the impaired homeostatic up-scaling of mEPSCs and AMPARs. FMRP is also known for its direct regulation to the translation of homeostasis related protein Arc/Arg 3.1 (Park et al., 2008). Thus, via the effect of FMRP, homeostatic regulation may be implicated in the neurodysfunction in fragile X syndrome.

PICK1

Many proteins, most intracellularly localized, are able to associate with AMPAR subunits. PICK1, a protein kinase C-interacting protein, interacts specifically via its PDZ domain with the C-termini of GluA2 and GluA3 subunits (Xia et al., 1999). Since the same C-terminal domain of GluA2 also binds to another protein GRIP, a competition exists between PICK1 and GRIP in GluA2 binding (Dev et al., 1999; Lu and Ziff, 2005; Steinberg et al., 2006). Interestingly, phosphorylation of the Serine 880 in the GluA2 C-terminus specifically abolishes its binding with GRIP and thus favors association with PICK1 (Lu and Ziff, 2005). PICK1 is also shown to be involved in the AMPAR trafficking and surface expression. In cultured hippocampal neurons overexpression of PICK1 decreases the surface expression of GluA2 in a PDZ domain-dependent manner; whereas disruption of the interaction blocks NMDA-induced AMPAR internalization, indicating that association of PICK1 with AMPAR facilitates GluA2 endocytosis (Terashima et al., 2004; Hanley et al., 2008). Consistent with this, blockade of the binding between GluA2 and PICK1 by intracellular peptide perfusion blocks cerebellar LTD, a process known to be mediated by clathrin-dependent AMPAR internalization (Lee et al., 2002). More importantly, the overexpression of PICK1 in cultured hippocampal slices not only results in reduced GluA2 surface expression leading to formation of GluA2-lacking AMPARs, but also in the increases AMPAR-mediated mEPSCs (Terashima et al., 2004). In PICK1 knockout mice, NMDA-induced LTD is abolished in hippocampal neurons due to disrupted internalization, recycling and retention of GluA2-containing AMPARs (Steinberg et al., 2006; Volk et al., 2010). By interacting with the GluA2 subunit, PICK1 plays a key role in the plasticity involving the calcium-permeable, GluA2-lacking AMPARs (Gardner et al., 2005) In a specific type of LTP induced by cocaine exposure at the glutamatergic synapses of dopaminergic neurons in the ventral tegmental area, PICK1 directly mediates the switch of GluA2-containing to GluA2-

lacking AMPARs (Bellone and Luscher, 2006) Since calcium-permeable AMPARs serve as an important signal in homeostatic scaling (Man, 2011), these findings imply a regulatory role for PICK1 in homeostatic regulation. Indeed, a recent study shows that in cultured cortical neurons of PICK1 knockout mice, TTX-induced mEPSC up-scaling is occluded due to altered AMPAR subunit composition and aberrant trafficking of GluA2-containing AMPARs while bicuculline-induced down-scaling of mEPSC remains intact (Angonno et al., 2011).

Ubiquitin-Proteasome System

The ubiquitin-proteasome system (UPS) is a crucial proteolytic mechanism. The UPS uses a small protein of 76 amino acids, namely ubiquitin, to mark proteins destined for degradation. Following ubiquitination, a poly-ubiquitin chain is attached to the lysine residues of the target protein so that it can be recognized by the degradation machinery proteasome (Hegde, 2004). UPS components are widely distributed in a neuron from the soma to dendrites and synapses (Bingol and Schuman, 2006; Bingol et al., 2010). It regulates many important synaptic functions including synapse development, maturation and synaptic plasticity (DiAntonio et al., 2001). Given the importance of proper protein turnover in cells, UPS dysfunction is implicated in the pathogenesis of many neurodegenerative diseases (Ding and Shen, 2008).

UPS function is closely related toneural activity levels (Ehlers, 2003; Bingol and Schuman, 2006) to control the composition of post-synaptic proteins including PSD-95 (Coledge et al., 2003; Bingol and Schuman, 2004) and AMPAR-associating protein GRIP (Guo and Wang, 2007). AMPARs are directly subjected to ubiquitination, leading to their internalization and degradation (Schwarz et al., 2010; Lin et al., 2011; Lussier et al., 2011). The degradation of NMDARs is also regulated by the UPS (Kato et al., 2005). The involvement of the UPS in homeostatic regulation has been shown in recent studies (Hou et al., 2008; Schwarz et al., 2010; Jakawich et al., 2010). Jakawich et al. (2010) found that proteasome inhibitors cause a slow increase in mEPSC amplitude as well as a slow up-scaling of surface AMPAR expression, which mimics and occludes the TTX-induced homeostatic synaptic response.

Schwarz et al. (2010) found that AMPA receptor inhibition by CNQX, which has also been used as a homeostatic up-scaling paradigm, significantly attenuates AMPA receptor ubiquitination within 30min. In addition to the pharmacological study, expression of a mutant ubiquitin which blocks ubiquitin chain elongation produced similar effects (Jakawich et al., 2010). In our recent study, we utilized light-controlled glutamate receptors to selectively activate individual synapses (Hou et al., 2008). We find that single synaptic activation leads to homeostatic down-regulation of postsynaptic AMPAR abundance as a consequence of AMPAR internalization and degradation (Hou et al., 2008). This activity-dependent homeostatic AMPAR alteration is accompanied by a recruitment of polyubiquitinated protein and AMPAR E3 ligase Nedd4 (Zhang et al., 2009; Schwarx et al., 2010; Lin et al., 2011), and is blocked by proteasome inhibitors (Hou et al., 2011), strongly indicating a key role of the UPS in homeostatic plasticity. Similar enhancement of AMPAR ubiquitination has also been observed in another study in which the researchers found that bicuculline treatment, which was conventionally used as a paradigm for homeostatic down-regulation (Turrigianno et al.,

1998) induces rapid enhancement of ubiquitination on GluA2 subunits of AMPARs and results in decreased surface AMPAR expression (Lussier et al., 2011).

Homeostatic Regulation in Neurological Diseases

In the brain, the maintenance of a stable basal functional state is of fundamental importance for the execution of all the neurophysiology and higher brain functions. Therefore, disruption in homeostatic regulation could underlie altered brain activities and neurological disorders. Long-lasting changes in neuronal activity, including a misbalance of excitation and inhibition, will compromise network integrity and neuronal conditions. Furthermore, since glutamate receptors play a key role in excitotoxicity, abnormal expression of a homeostatic response may lead to calcium overload and neuron death. To date, scientists have only begun to investigate the involvement of the homeostatic response in the pathogenesis of neurological disorders, but a major role is expected to be elucidated in the near future. Homeostatic plasticity is presumably involved in the pathology of Alzheimer's and Parkinson's disease, drug addiction, autism spectrum disorders, seizure and psychiatric problems.

Alzheimer's disease (AD) is typically classified with large levels of neuronal loss as well as impaired synaptic function and loss of memory. A predominant characteristic of AD brains is the presence of amyloid beta (Aβ) peptides, which have recently been studied for their role in synaptic dysfunction. Aβ is a proteolytic cleavage product from the amyloid precursor protein (APP) and it has been suggested that the excess production of soluble small species of amyloid beta can impair synaptic function as well as the mechanisms that regulate synaptic scaling and regulation (Chapman et al., 1999; Fitzjohn et al., 2001; Hsia et al., 1999; Larson et al., 1999; Westerman et al. 2002; Li et al., 2009, 2011), and that neuronal activity can modulate the formation and secretion of Aβ in slice preparations from transgenic mice that overexpress APP (Kamenetz et al. 2003). Aβ has also been shown to selectively depress excitatory synaptic transmission onto neurons that overexpress APP, as well as neurons that do not. This depression is dependent upon NMDAR activity and can be reversed with the blockade of neuronal activity (Kamenetz et al. 2003). Synaptic depression from excessive levels of Aβ could contribute to the cognitive defects seen in AD patients and the activity dependent alterations in Aβ production could play a role in negative feedback mechanisms that normally keep neuronal hyperactivity in check. The accumulation of amyloid beta is caused by an impbalance of its production and clearance in the brain. Aβ is eliminated by a range of processes including receptor mediated transport across the blood brain barrier, microglia phagocytosis, or enzymatic degradation. Studies have shown increase levels of the cytokine TNF-α in AD brains (Gong et al., 1998; Orzylowska et al., 1999; Tarkowski et al., 1999). Increased levels of TNF-α are known to cause an upregulation of AMPARs at the postsynaptic surface thereby disregulating the brain's normal homeostatic response. Production of Cp-AMPARs will therefore become upregulated and the increased calcium could also contribute to neurotoxicity and increased neuronal loss. BDNF is found to mediate homeostatic down-regulation of mEPSCs (Rutherford et al., 1998) and recent evidence suggests that decreased levels of BDNF are associated with Alzheimer's disease pathogenesis

(Peng et al., 2009; Philips et al., 1991). This could therefore indicate that altered BDNF levels in AD disrupt normal homeostatic regulation leading to diseases progression.

Hyperphosphorylation of the microtubule associated protein tau is another hallmark of AD pathology. This hyperphoshorylated form of tau aggregates into intracellular cytoskeletal filaments called neurofibrillary tangles that can disrupt neuronal synaptic function. One common element that presents with several features of AD is disrupted neuronal calcium signaling. Increased intracellular calcium levels are functionally linked to amyloid plaques, neurofibrillary tangles and synaptic dysfunction (Chakroborty et al. 2011). A recent study shows that homeostatic plasticity is impaired in Presenilin 1 (PS1) knock-out mice (Pratt et al. 2011). Also, over-expression of constitutively active Akt can rescue the impaired synaptic scaling in neurons from PS1 KO mice, indicating an abnormality in homeostatic plasticity in AD (Pratt et al., 2011). It is possible that pathogenic factors shift the threshold of homeostatic regulation so that a normal downturn in neuronal activity triggers an upward over response, which leads to receptor over activation and elevated calcium flux, and ultimately cell death.

Parkinson's disease (PD) is the second most common neurodegenerative brain disorder after Alzheimer's disease (Forman et al. 2005). Aggregated α-synuclein (α-syn) is a characteristic pathological finding in PD, and the aggregation of α-syn into Lewy bodies is a key contributor to the dementia seen in PD. Recent evidence suggests that α-syn oligomers could be the main neurotoxic species, but the physiological mechanisms involved are still not well understood. With the application of large α-syn oligomers to normal healthy neuronal cultures, a selective enhancement of evoked AMPAR, but not NMDAR, mediated synaptic transmission is seen (Hüls et al. 2011). The spontaneous AMPAR-mediated mEPSCs show a decrease in frequency when compared to control neurons, indicating that large α-syn oligomers alter both pre- and post-synaptic mechanisms of AMPAR-mediated synaptic transmission (Hüls et al. 2011). The decreased excitatory synaptic transmission may directly contribute to nerve cell death in Parkinson's disease due to a disruption of the proper homeostatic regulatory mechanisms. For instance, low micromolar glutamate concentrations are found to be toxic in cultured neurons incubated with large α-syn oligomers. The exact function of α-syn at the synapse is still not known, however most studies support a role of α-syn in modulating vesicular presynaptic neurotransmitter release (Murphy et al. 2000; Gureviciene et al. 2007).

Mutations in the PARK2 gene, which encodes the ubiquitin ligase parkin, causes autosomal recessive juvenile parkinsonism (Kitada et al., 1998). Parkin is found throughout the brain, including the hippocampus (Horowitz et al., 1999; Stichel et al., 2000), and experiments in cultured hippocampal brain slices show disrupted excitatory synaptic function after decreasing the levels of parkin (Helton et al., 2008). Parkin associates with PDZ scaffold proteins in the postsynaptic density (Fallon et al., 2002), and parkin ubiquitinates PICK1, an AMPAR-interacting protein that is involved in AMPA receptor trafficking and homeostatic regulation (Joch et al., 2007; Angonnno et al., 2011). These interactions and the observed synaptic deficiencies suggest a disruption of homeostatic synaptic plasticity through the modulation of AMPAR trafficking to the synaptic membrane.

Huntington's disease (HD) is characterized by cognitive, motor, and psychiatric symptoms, including depression, weight loss, and movement irregularity that later develops into severe akinesia, the loss of voluntary movements (Celsi et al., 2009). HD is an autosomal dominant inherited disease and the mutant huntington gene was identified to be reponsible for the disease. The huntington protein plays a role in axonal transport, regulation of transcription, exocytosis, bioenergetic metabolism, prevention of apoptosis, and Ca^{2+} homeostasis (Cattaneo et al., 2005).

There is evidence for altered neuronal synaptic protein expression and function in HD. Reductions in neurotransmitters and proteins involved in synaptic transmission, and associated mRNAs, are prevalent in human HD brain even at early stages with little or no cell loss (Cha, 2007). Studies have demonstrated that transcription of BDNF is increased by huntington protein and reduced by mutant huntington protein (Zuccato et al., 2001). Interestlingly, reduced BDNF expression is observed in presymptomatic HD mice (Zuccato et al., 2001). Huntington protein has also been shown to enhance BDNF axonal transport via an interaction with huntington-associated protein 1 (HAP1) while the mutant form does not (Gauthier et al., 2004). BDNF is found to mediate homeostatic down-regulation of mEPSCs (Rutherford et al., 1998) and therefore reduced BDNF levels, or alterations in axonal transport, may play a pivotal role in the disruption of homeostatic plasticity in the early stages of HD.

Epilepsy and seizures are due to a change in normal neuronal activity and the result of aberrant neuronal firing. Homeostatic mechanisms are employed to regulate activity within a physiological range, but disruption of these mechanisms allows neural networks to become unstable and cytotoxic. Recently research has suggested that glial cells in the brain do not play a passive role but rather communicate constantly with neurons through gap junctions and can be activated in the brain by various injuries. Alterations in the function of activated astrocytes and microglial cells have been shown in experimental epileptic brain slice models, producing modifications in potassium channels, changes in the glutamine/glutamate cycle, alterations in glutamate receptor expression and transporters, release of neuromodulatory molecules, and the induction of molecules involved in inflammatory processes like cytokines, chemokines, or prostaglandins. (Seifert et al., 2006; Vezzani et al., 2011). Brain injury or the occurrence of seizures can activate microglia and astrocytes to release many pro-inflammatory molecules such as TNF-α, and initiate a disruption in the brain's normal homeostatic response. Pro-inflammatory molecules can alter the neuronal excitability and affect the physiological functions of glia through paracrine or autocrine actions, thereby altering the glial-neuronal communications. In experimental models, these changes contribute to decreasing the threshold to seizures and could possibly induce neuronal death (Riazi et al., 2010; Vezzani et al., 2008). The glial-neuronal interaction through these various signaling molecules can therefore play a major role in modulating homeostatic plasticity mechanisms, disrupting the brain's normal response to altered levels of activity. It has also recently been shown that transgenic mice overexpressing the mature isoform of BDNF had increased levels of anxiety and seizure indicating a disruption of a normal homeostatic response in these disease states (Papaleo et al., 2011).

Acknowledgments

This work was supported in by National Institutes of Health Grant MH 079407 (H.Y.M.).

References

Allan S. M. and N. J. Rothwell, *"Cytokines and acute neurodegeneration,"* Nat Rev Neurosci, vol. 2, no. 10, pp. 734-744, 2001.

Anggono V., R. L. Clem, and R. L. Huganir, *"PICK1 loss of function occludes homeostatic synaptic scaling,"* J Neurosci, vol. 31, no. 6, pp. 2188-2196, 2011.

Aoto J., C. I. Nam, M. M. Poon, et al., *"Synaptic signaling by all-trans retinoic acid in homeostatic synaptic plasticity,"* Neuron, vol. 60, no. 2, pp. 308-320, 2008.

Baki L., J. Shioi, P. Wen, et al., *"PS1 activates PI3K thus inhibiting GSK-3 activity and tau overphosphorylation: effects of FAD mutations,"* Embo Journal, vol. 23, no. 13, pp. 2586-2596, 2004.

Balkowiec A. and D. M. Katz, *"Cellular mechanisms regulating activity-dependent release of nativenative brain-derived neurotrophic factor from hippocampal neurons,"* J Neurosci, 22(23), 2002.

Barria A, Malinow R. *"Subunit-specific NMDA receptor trafficking to synapses."* Neuron, vol. 35, 345–353, 2002.

Barria A., D. Muller, V. Derkach, et al., *"Regulatory phosphorylation of AMPA-type glutamate receptors by CaM-KII during long-term potentiation,"* Science, vol. 276, no. 5321, pp. 2042-2045, 1997.

Beattie E. C., Carroll R. C., Yu X., Morishita W., Yasuda H., von Zastrow M. and Malenka R. C. *"Regulation of AMPA receptor endocytosis by a signaling mechanism shared with LTD,"* Nat Neurosci 3, 1291-1300, 2000.

Beattie E. C., D. Stellwagen, W. Morishita, et al., *"Control of synaptic strength by glial TNFalpha,"* Science, vol. 295, no. 5563, pp. 2282-2285, 2002.

Beique J. C., Y. Na, D. Kuhl, et al., *"Arc-dependent synapse-specific homeostatic plasticity,"* Proc Natl Acad Sci U S A, vol. 108, no. 2, pp. 816-821, 2011.

Bellone A. and C. Luscher, *"Cocaine triggered AMPA receptor redistribution is reversed in vivo by mGluR-dependent long-term depression,"* Nature Neuroscience, vol. 9, no. 5, pp. 636-641, 2006.

Bingol A. and E. M. Schuman, *"A proteasome-sensitive connection between PSD-95 and GluR1 endocytosis,"* Neuropharmacology, vol. 47, no. 5, pp. 755-763, 2004.

Bingol A. and E. M. Schuman, *"Activity-dependent dynamics and sequestration of proteasomes in dendritic spines,"* Nature, vol. 441, no., pp. 1144-1148, 2006.

Bingol A., C. F. Wang, D. Arnott, et al., *"Autophosphorylated CaMKII alpha Acts as a Scaffold to Recruit Proteasomes to Dendritic Spines,"* Cell, vol. 140, no. 4, pp. 567-578, 2010.

Blanpied T. A., Scott D. B. and Ehlers M. D. *Dynamics and regulation of clathrin coats at specialized endocytic zones of dendrites and spines*. Neuron 36, 435-449, 2002.

Bloodgood B. L. and B. L. Sabatini, *"Ca(2+) signaling in dendritic spines,"* Curr Opin Neurobiol, vol. 17, no. 3, pp. 345-351, 2007.

Bolton M. M., A. J. Pittman, and D. C. Lo, *"Brain-derived neurotrophic factor differentially regulates excitatory and inhibitory synaptic transmission in hippocampal cultures,"* J Neurosci, vol. 20, no. 9, pp. 3221-3232, 2000.

Bonnet A., K. Touyarot, S. Alfos, et al.,*"Retinoic acid restores adult hippocampal neurogenesis and reverses spatial memory deficit in vitamin A deprived rats,"* PLoS One, vol. 3, no. 10, pp. e3487, 2008.

Bramham A. R., P. F. Worley, M. J. Moore, et al.,*"The Immediate Early Gene Arc/Arg3.1: Regulation, Mechanisms, and Function,"* Journal of Neuroscience, vol. 28, no. 46, pp. 11760-11767, 2008.

Bredt D. S. and Nicoll R. A. *"AMPA receptor trafficking at excitatory synapses,"* Neuron 40, 361-379, 2003.

Burnashev, N., Monyer, H., Seeburg, P.H., and Sakmann, B. *"Divalent ion permeability of AMPA receptor channels is dominated by the edited form of a single subunit,"* Neuron.1992; 8, 189–198, 1992.

Burrone J, O'Byrne M, Murthy VN *"Multiple forms of synaptic plasticity triggered by selective suppression of activity in individual neurons,"* Nature. 2002; 420:414–418, 2002.

Caldeira M. V., C. V. Melo, D. B. Pereira et al.,*"Brain-derived neurotrophic factor regulates the expression and synaptic delivery of alpha-amino-3-hydroxy-5-methyl-4-isoxazole propionic acid receptor subunits in hippocampal neurons,"* J Biol Chem, vol. 282, no. 17, pp. 12619-12628, 2007.

Carvalho A. L., M. V. Caldeira, S. D. Santos, et al.,*"Role of the brain-derived neurotrophic factor at glutamatergic synapses,"* Cell. 2011; 147(3):615-28.

Cattaneo E., C. Zuccato, and M. Tartari. *"Normal huntingtin function: an alternative approach to Huntington's disease,"* Nat Rev Neurosci, 6: 919–930, 2005.

Celsi F., P. Pizzo, M. Brini, S. Leo, C. Fotino, P. Pinton, and R. Rizzuto. *"Mitochondria, calcium and cell death: a deadly triad in neurodegeneration,"* Biochim Biophys Acta., 1787: 335–344, 2009.

Cha, J.H. *"Transcriptional signatures in Huntington's disease,"* Prog. Neurobiol. 83, 228–248, 2007.

Chakroborty S, Stutzmann GE. *"Early calcium dysregulation in Alzheimer's disease: setting the stage for synaptic dysfunction,"* Sci China Life Sci., 54(8):752-62, 2001.

Chapman PF, White GL, Jones MW, Cooper-Blacketer D, Marshall VJ, Irizarry M, Younkin L, Good MA, Bliss TV, Hyman BT, et al. *"Impaired synaptic plasticity and learning in aged amyloid precursor protein transgenic mice,"* Nat. Neurosci. 1999, 2, 271–276, 1999.

Chavis P. and G. Westbrook,*"Integrins mediate functional pre- and postsynaptic maturation at a hippocampal synapse,"* Nature, vol. 411, no. 6835, pp. 317-321, 2001.

Chen A. M. and A. D. Grinnell,*"Integrins and modulation of transmitter release from motor nerve terminals by stretch,"* Science, vol. 269, no. 5230, pp. 1578-1580, 1995.

Chowdhury, J. D. Shepherd, H. Okuno, et al.,*"Arc/Arg3.1 interacts with the endocytic machinery to regulate AMPA receptor trafficking,"* Neuron, vol. 52, no. 3, pp. 445-459, 2006.

Cingolani L. A., A. Thalhammer, L. M. Yu, et al.,*"Activity-dependent regulation of synaptic AMPA receptor composition and abundance by beta3 integrins,"* Neuron, vol. 58, no. 5, pp. 749-762, 2008.

Colbran R. J. and A. M. Brown,*"Calcium/calmodulin-dependent protein kinase II and synaptic plasticity,"* Curr Opin Neurobiol, vol. 14, no. 3, pp. 318-327, 2004.

Colledge M., E. M. Snyder, R. A. Crozier, et al.,*"Ubiquitination regulates PSD-95 degradation and AMPA receptor surface expression,"* Neuron, vol. 40, no. 3, pp. 595-607, 2003.

Collingridge G. L., Isaac J. T. and Wang Y. T. *"Receptor trafficking and synaptic plasticity,"* Nat Rev Neurosci 5, 952-962, 2004.

Copi A., K. Jungling, and K. Gottmann,*"Activity- and BDNF-induced plasticity of miniaturesynaptic currents in ES cell-derived neurons integrated in a neocortical network,"*J Neurophysiol., 94(6):4538-43, 2005.

Dalva M. B., A. C. McClelland, and M. S. Kayser,*"Cell adhesion molecules: signalling functions at the synapse,"* Nature Reviews Neuroscience, vol. 8, no. 3, pp. 206-220, 2007.

Davis G. W.,*"Homeostatic Control of Neural Activity: From Phenomenology to Molecular Design,"* Annu Rev Neurosci, pp. 307-323, 2006.

Desai N. S., R. H. Cudmore, S. B. Nelson, et al.,*"Critical periods for experience-dependent synaptic scaling in visual cortex,"* Nat Neurosci, vol. 5, no. 8, pp. 783-789, 2002.

Dev K. K., A. Nishimune, J. M. Henley, et al.,*"The protein kinase C alpha binding protein PICK1 interacts with short but not long form alternative splice variants of AMPA receptor subunits,"* Neuropharmacology, vol. 38, no. 5, pp. 635-644, 1999.

DiAntonio A., A. P. Haghighi, S. L. Portman, et al.,*"Ubiquitination-dependent mechanisms regulate synaptic growth and function,"* Nature, vol. 412, no. 6845, pp. 449-452, 2001.

Ding M. and K. Shen,*"The role of the ubiquitin proteasome system in synapse remodeling and neurodegenerative diseases,"* Bioessays, vol. 30, no. 11-12, pp. 1075-1083, 2008.

Dityatev A., M. Schachner, and P. Sonderegger,*"The dual role of the extracellular matrix in synaptic plasticity and homeostasis,"* Nat Rev Neurosci., (11):735-46, 2010.

Ehlers M. D. "*Reinsertion or degradation of AMPA receptors determined by activity-dependent endocytic sorting,"* Neuron 28, 511-525, 2000.

Ehlers M. D.,*"Activity level controls postsynaptic composition and signaling via the ubiquitin-proteasome system,"* Nat Neurosci, vol. 6, no. 3, pp. 231-242, 2003.

Elia L. P., M. Yamamoto, K. L. Zang, et al.,p120 "*Catenin regulates dendritic spine and synapse development through Rho-family GTPases and cadherins,"* Neuron, vol. 51, no. 1, pp. 43-56, 2006.

Fitzjohn SM, Morton RA, Kuenzi F, Rosahl TW, Shearman M, Lewis H, Smith D, Reynolds DS, Davies CH and Collingridge GL, et al. *"Age-related impairment of synaptic transmission but normal long-term potentiation in transgenic mice that over-express the human APP695SWE mutant form of amyloid precursor protein,"* J. Neurosci.,21, 4691–4698, 2001.

Frank CA, Pielage J, Davis GW. *"A presynaptic homeostatic signaling system composed of the Eph receptor, ephexin, Cdc42, and CaV2.1 calcium channels,"*Neuron. 2009; 61:556–569, 2009.

Frank, C. A., Kennedy, M. J., Goold, C. P., Marek, K. W., and Davis, G. W. *"Mechanisms underlying the rapid induction and sustained expression of synaptic homeostasis,"* Neuron, 52, 663-677, 2006.

Gainey M. A., J. R. Hurvitz-Wolff, M. E. Lambo, et al.,*"Synaptic scaling requires the GluR2 subunit of the AMPA receptor,"* J Neurosci, vol. 29, no. 20, pp. 6479-6489, 2009.

Gardner S. M., K. Takamiya, J. Xia, et al.,*"Calcium-permeable AMPA receptor plasticity is mediated by subunit-specific interactions with PICK1 and NSF,"* Neuron, vol. 45, no. 6, pp. 903-915, 2005.

Gauthier, L.R. et al. *"Huntingtin controls neurotrophic support and survival of neurons by enhancing BDNF vesicular transport along microtubules,"* Cell, 118, 127–138, 2004.

Gobert A., L. Topolnik, M. Azzi, et al., *"Forskolin induction of late-LTP and up-regulation of 5' TOP mRNAs translation via mTOR, ERK, and PI3K in hippocampal pyramidal cells,"* J Neurochem, vol. 106, no. 3, pp. 1160-1174, 2008.

Gong C., Qin Z., Betz A. L., Liu X. H. and Yang G.Y. *"Cellular localization of tumor necrosis factor alpha following focal cerebral ischemia in mice,"* Brain Research, 801: 1-8, 1998.

Goold C. P. and R. A. Nicoll, *"Single-cell optogenetic excitation drives homeostatic synaptic depression,"* Neuron, vol. 68, no. 3, pp. 512-528, 2010.

Gottmann K., T. Mittmann, and V. Lessmann, *"BDNF signaling in the formation, maturation and plasticity of glutamatergic and GABAergic synapses,"* Exp Brain Res, vol. 199, no. 3-4, pp. 203-34, 2009.

Groth R. D. and R. W. Tsien, *"A role for retinoic acid in homeostatic plasticity,"* Neuron, vol. 60, no. 2, pp. 192-194, 2008.

Groth R. D., M. Lindskog, T. C. Thiagarajan,L. Li, R. W. Tsien. *"Beta Ca2+/CaM-dependent kinase type II triggers upregulation of GluA1 to coordinate adaptation to synaptic inactivity in hippocampal neurons,"* Proc Natl Acad Sci U S A, 108(2):828-33, 2011.

Guo L. and Y. Wang, *"Glutamate stimulates glutamate receptor interacting protein 1 degradation by ubiquitin-proteasome system to regulate surface expression of GluR2,"* Neuroscience, vol. 145, no. 1, pp. 100-109, 2007.

Gureviciene I, Gurevicius K, Tanila H. *"Role of alpha-synuclein in synaptic glutamate release,"* Neurobiol Dis, 28(1):83-9, 2007.

Hanley J. G., *"PICK1: a multi-talented modulator of AMPA receptor trafficking,"* Pharmacol Ther, vol. 118, no. 1, pp. 152-160, 2008.

Hayashi Y., S. H. Shi, J. A. Esteban, et al., *"Driving AMPA receptors into synapses by LTP and CaMKII: Requirement for GluR1 and PDZ domain interaction,"* Science, vol. 287, no. 5461, pp. 2262-2267, 2000.

He P., Q. Liu, J. Wu, et al., *"Genetic deletion of TNF receptor suppresses excitatory synaptic transmission via reducing AMPA receptor synaptic localization in cortical neurons,"* FASEB J, vol. 26, pp. fj.11-192716, 2011.

Hegde A. N., *"Ubiquitin-proteasome-mediated local protein degradation and synaptic plasticity,"* Prog Neurobiol, vol. 73, no. 5, pp. 311-357, 2004.

Helton T. D.,T. Otsuka, M. C. Lee, Y. Mu, M. D. Ehlers. *"Pruning and loss of excitatory synapses by the parkin ubiquitin ligase,"* Proc. Nat. Acad. Sci. USA.2008; 105, 19492–19497, 2008.

Horowitz J. M., J. Myers, M. K. Stachowiak, G. Torres. *"Identification and distribution of Parkin in rat brain,"* Neuroreport, 10, 3393–3397, 1999.

Hou Q. M., J. Gilbert, and H. Y. Man, *"Homeostatic Regulation of AMPA receptor trafficking and degradation by light-controlled single-synaptic activation,"* Neuron, vol. 72, no.5, pp. 806-818, 2011.

Hou Q., D. Zhang, L. Jarzylo, et al., *"Homeostatic regulation of AMPA receptor expression at single hippocampal synapses,"* Proc Natl Acad Sci U S A, vol. 105, no. 2, pp. 775-780, 2008.

Hsia AY, Masliah E, McConlogue L, Yu GQ, Tatsuno G, Hu K, Kholodenko D, Malenka RC, Nicoll RA and Mucke L. *"Plaque-independent disruption of neural circuits in Alzheimer's disease mouse models,"* Proc. Natl. Acad. Sci. USA, 96, 3228–3233, 1999.

Hüls S, Högen T, Vassallo N, Danzer KM, Hengerer B, Giese A, Herms J. *"AMPA-receptor-mediated excitatory synaptic transmission is enhanced by iron-induced α-synuclein oligomers,"* J Neurochem,117(5):868-78, 2011.

Ibata K., Q. Sun, and G. G. Turrigiano,*"Rapid synaptic scaling induced by changes in postsynaptic firing,"* Neuron, vol. 57, no. 6, pp. 819-826, 2008.

Jakawich S. K., R. M. Neely, S. N. Djakovic, et al.,*"An essential postsynaptic role for the ubiquitin proteasome system in slow homeostatic synaptic plasticity in cultured hippocampal neurons,"* Neuroscience, vol. 171, no. 4, pp. 1016-1031, 2010.

Jin M., N. Shepardson, T. Yang, G. Chen, D. Walsh, D. J. Selkoe. *"Soluble amyloid beta-protein dimers isolated from Alzheimer cortex directly induce Tau hyperphosphorylation and neuritic degeneration,"* Proc Natl Acad Sci U S A,108(14):5819-24, 2011.

Joch M., A. R. Ase, C. X. Chen, et al., *"Parkin-mediated monoubiquitination of the PDZ protein PICK1 regulates the activity of acid-sensing ion channels,"* Mol Biol Cell, 18, 3105–3118, 2007.

Ju W., W. Morishita, J. Tsui, et al.,*"Activity-dependent regulation of dendritic synthesis and trafficking of AMPA receptors,"* Nat Neurosci, vol. 7, no. 3, pp. 244-253, 2004.

Kamenetz F., T. Tomita, H. Hsieh, G. Seabrook, D. Borchelt, T. Iwatsubo, S. Sisodia, R. Malinow. *"APP processing and synaptic function,"* Neuron. 2003; 37(6):925-37, 2003.

Kaneko M., D. Stellwagen, R. C. Malenka, M. P. Stryker. *"Tumor necrosis factor-alpha mediates one component of competitive, experience-dependent plasticity in developing visual cortex,"* Neuron,58(5):673-80, 2008.

Kato A., N. Rouach, R. A. Nicoll, et al.,*"Activity-dependent NMDA receptor degradation mediated by retrotranslocation and ubiquitination,"* Proc Natl Acad Sci U S A, vol. 102, no. 15, pp. 5600-5605, 2005.

Khandjian A. W.,*"Biology of the fragile X mental retardation protein, an RNA-binding protein,"* Biochem Cell Biol, vol. 77, no. 4, pp. 331-342, 1999.

Kim E. and M. Sheng. *"PDZ domain proteins of synapses,"* Nat Rev Neurosci 5, 771-781, 2004.

Kitada T, Asakawa S, Hattori N, Matsumine H, Yamamura Y, Minoshima S, Yokochi M, Mizuno Y, Shimizu N. *"Mutations in the parkin gene cause autosomal recessive juvenile parkinsonism,"* Nature,392, 605–608, 1998.

Kumar S. S., A. Bacci, V. Kharazia, et al.,*"A developmental switch of AMPA receptor subunits in neocortical pyramidal neurons,"* J Neurosci, vol. 22, no. 8, pp. 3005-3015., 2002.

Larson J, Lynch G, Games D and Seubert P. *"Alterations in synaptic transmission and long-term potentiation in hippocampal slices from young and aged PDAPP mice.,"* Brain Res., 840, 23–35, 1999.

Lee H. K., Kameyama K., Huganir R. L. and Bear M. F. *"NMDA induces long-term synaptic depression and dephosphorylation of the GluR1 subunit of AMPA receptors in hippocampus,"* Neuron 21, 1151-1162, 1998.

Lee S. H., Liu L., Wang Y. T. and Sheng M. *"Clathrin adaptor AP2 and NSF interact with overlapping sites of GluR2 and play distinct roles in AMPA receptor trafficking and hippocampal LTD,"* Neuron 36, 661-674, 2002.

Leonoudakis D., P. Zhao, and E. C. Beattie,*"Rapid tumor necrosis factor alpha-induced exocytosis of glutamate receptor 2-lacking AMPA receptors to extrasynaptic plasma membrane potentiates excitotoxicity,"* J Neurosci, vol. 28, no. 9, pp. 2119-2130, 2008.

Levi S, Schweizer C, Bannai H, Pascual O, Charrier C, and Triller A. *"Homeostatic regulation of synaptic GlyR numbers driven by lateral diffusion,"* Neuron. 2008; 59, 261-273, 2008.

Li S, Hong S, Shcpardson NE, Walsh D, Shankar GM, Selkoe D. *"Soluble oligomers of amyloid beta protein facilitate hippocampal long-term depression by disrupting neuronal glutamate uptake,"*Neuron, 62:788-801, 2009.

Li X. and M. E. Wolf,*"Brain-derived neurotrophic factor rapidly increases AMPA receptor surface expression in rat nucleus accumbens,"* Eur J Neurosci, vol. 34, no. 2, pp. 190-8, 2011.

Lin A. H., S. H. Yeh, K. T. Lu, et al.,*"A role for the PI-3 kinase signaling pathway in fear conditioning and synaptic plasticity in the amygdala,"* Neuron, vol. 31, no. 5, pp. 841-851, 2001.

Lin A., Q. Hou, L. Jarzylo, et al.,*"Nedd4-mediated AMPA receptor ubiquitination regulates receptor turnover and trafficking,"* J Neurochem, vol. 119, no. 1, 27-39, 2011.

Lin J. W., Ju W., Foster K., Lee S. H., Ahmadian G., Wyszynski M., Wang Y. T. and Sheng M. *"Distinct molecular mechanisms and divergent endocytotic pathways of AMPA receptor internalization,"* Nat Neurosci 3, 1282-1290, 2000.

Liu S. Q. and S. G. Cull-Candy,*"Synaptic activity at calcium-permeable AMPA receptors induces a switch in receptor subtype,"* Nature, vol. 405, no. 6785, pp. 454-458., 2000.

Lledo P. M., Zhang X., Sudhof T. C., Malenka R. C. and Nicoll R. A. *"Postsynaptic membrane fusion and long-term potentiation,"* Science 279, 399-403, 1998.

Lu W. and E. B. Ziff,*"PICK1 interacts with ABP/GRIP to regulate AMPA receptor trafficking,"* Neuron, vol. 47, no. 3, pp. 407-421, 2005.

Lu W., Man H., Ju W., Trimble W. S., MacDonald J. F. and Wang Y. T. *"Activation of synaptic NMDA receptors induces membrane insertion of new AMPA receptors and LTP in cultured hippocampal neurons,"* Neuron 29, 243-254, 2001.

Lussier M. P., Y. Nasu-Nishimura, and K. W. Roche,*"Activity-dependent ubiquitination of the AMPA receptor subunit GluA2,"* J Neurosci, vol. 31, no. 8, pp. 3077-3081, 2011.

Maden,*"Retinoic acid in the development, regeneration and maintenance of the nervous system,"* Nat Rev Neurosci, vol. 8, no. 10, pp. 755-765, 2007.

Maghsoodi, M. M. Poon, C. I. Nam, et al.,*"Retinoic acid regulates RARalpha-mediated control of translation in dendritic RNA granules during homeostatic synaptic plasticity,"* Proc Natl Acad Sci U S A, vol. 105, no. 41, pp. 16015-16020, 2008.

Mahanty N. K. and Sah P. *"Calcium-permeable AMPA receptors mediate long-term potentiation in interneurons in the amygdala,"* Nature 394, 683-687, 1998.

Malenka R. C. *"Synaptic plasticity and AMPA receptor trafficking,"* Ann N Y Acad Sci 1003, 1-11, 2003.

Malinow R. and R. C. Malenka,*"AMPA receptor trafficking and synaptic plasticity,"* Annu Rev Neurosci, vol. 25, pp. 103-126, 2002.

Malinow R. *"AMPA receptor trafficking and long-term potentiation,"* Philos Trans R Soc Lond B Biol Sci 358, 707-714, 2003.

Man H. Y., Lin J. W., Ju W. H., Ahmadian G., Liu L., Becker L. E., Sheng M. and Wang Y. T. *"Regulation of AMPA receptor-mediated synaptic transmission by clathrin-dependent receptor internalization,"* Neuron 25, 649-662, 2000.

Man H. Y., Q. Wang, W. Y. Lu, et al.,*"Activation of PI3-kinase is required for AMPA receptor insertion during LTP of mEPSCs in cultured hippocampal neurons,"* Neuron, vol. 38, no. 4, pp. 611-624, 2003.

Man H. Y.,*"GluA2-lacking, calcium-permeable AMPA receptors - inducers of plasticity?,"* Current Opinion in Neurobiology, vol. 21, no. 2, pp. 291-298, 2011.

Marder E. and J. M. Goaillard,*"Variability, compensation and homeostasis in neuron and network function,"* Nat Rev Neurosci, vol. 7, no. 7, pp. 563-574, 2006.

Mattson M. P.,*"Glutamate and neurotrophic factors in neuronal plasticity and disease,"* Ann N Y Acad Sci, vol. 1144, no., pp. 97-112, 2008.

Mazroui R., M. E. Huot, S. Tremblay, et al.,*"Trapping of messenger RNA by Fragile X Mental Retardation protein into cytoplasmic granules induces translation repression,"* Human Molecular Genetics, vol. 11, no. 24, pp. 3007-3017, 2002.

Murphy DD, Rueter SM, Trojanowski JQ, Lee VM. *"Synucleins are developmentally expressed, and alpha-synuclein regulates the size of the presynaptic vesicular pool in primary hippocampal neurons,"* J Neurosci. 20(9):3214-20, 2000.

Mysore S. P., C. Y. Tai, and E. M. Schuman,*"N-cadherin, spine dynamics, and synaptic function,"* Front Neurosci, vol. 2, no. 2, pp. 168-175, 2008.

Nakata H. and S. Nakamura,*"Brain-derived neurotrophic factor regulates AMPA receptor trafficking to post-synaptic densities via IP3R and TRPC calcium signaling,"* FEBS Lett, vol. 581, no. 10, pp. 2047-2054, 2007.

Nicoll R. A., Oliet S. H. and Malenka R. C. *"NMDA receptor-dependent and metabotropic glutamate receptor-dependent forms of long-term depression coexist in CA1 hippocampal pyramidal cells,"* Neurobiol Learn Mem 70, 62-72, 1998.

Nuriya M. and R. L. Huganir,*"Regulation of AMPA receptor trafficking by N-cadherin,"* J Neurochem, vol. 97, no. 3, pp. 652-661, 2006.

Ogoshi F., H. Z. Yin, Y. Kuppumbatti, et al.,*"Tumor necrosis-factor-alpha (TNF-alpha) induces rapid insertion of Ca2+-permeable alpha-amino-3-hydroxyl-5-methyl-4-isoxazole-propionate (AMPA)/kainate (Ca-A/K) channels in a subset of hippocampal pyramidal neurons,"* Exp Neurol, vol. 193, no. 2, pp. 384-393, 2005.

Okuda T., L. M. Y. Yu, L. A. Cingolani, et al.,*"beta-Catenin regulates excitatory postsynaptic strength at hippocampal synapses,"* Proceedings of the National Academy of Sciences of the United States of America, vol. 104, no. 33, pp. 13479-13484, 2007.

Orzylowska O., Oderfeld-Nowak B., Zaremba M., Januszewski S. and Mossakowski M. "*Prolonged and concomitant induction of astroglial immu- noreactivity of interleukin-1β and interleukin-6 in the rat hippocampus after transient global ischemia*," Neuroscience Letters, 263: 72- 76, 1999.

Papaleo F., Silverman J. L., Aney J., Tian Q., Barkan C. L., Chadman K. K., Crawley J. N. "*Working memory deficits, increased anxiety-like traits, and seizure susceptibility in BDNF overexpressing mice,*" Learn Mem, 18(8):534-44, 2011.

Park S, Park JM, Kim S, Kim JA, Shepherd JD, Smith-Hicks CL, Chowdhury S, Kaufmann W, Kuhl D, Ryazanov AG, Huganir RL, Linden DJ, Worley PF. *"Elongation factor 2 and fragile X mental retardation protein control the dynamic translation of Arc/Arg3.1 essential for mGluR-LTD,"* Neuron., 59(1):70-83, 2008.

Park, M, Penick, EC, Edwards, JG, Kauer, JA, and Ehlers, MD. *"Recycling endosomes supply AMPA receptors for LTP,"* Science, 305, 1972-1975, 2004.

Passafaro M., Piech V. and Sheng M. *"Subunit-specific temporal and spatial patterns of AMPA receptor exocytosis in hippocampal neurons,"* Nat Neurosci 4, 917-926, 2001.

Peebles CL, Yoo J, Thwin MT, Palop JJ, Noebels JL, Finkbeiner S. *"Arc regulates spine morphology and maintains network stability in vivo"* Proc Natl Acad Sci U S A. 2010, 107(42):18173-8, 2010.

Peng P. L., X. Zhong, W. Tu, et al.,*"ADAR2-dependent RNA editing of AMPA receptor subunit GluR2 determines vulnerability of neurons in forebrain ischemia,"* Neuron, vol. 49, no. 5, pp. 719-733, 2006.

Peng S., Garzon D. J., Marchese M., Klein W., Ginsberg S. D., Francis B. M., Mount H. T., Mufison E. J., Salehi A. and Fahnestock M. "*BDNF mRNA is decreased in the hippocampus of individuals with Alzhemier's disease*," Neuron, 7:695-702, 1991.

Perez-Otano and M. D. Ehlers,*"Homeostatic plasticity and NMDA receptor trafficking,"* Trends Neurosci, vol. 28, no. 5, pp. 229-238, 2005.

Perry V. H., M. D. Bell, H. C. Brown, et al.,*"Inflammation in the nervous system,"* Curr Opin Neurobiol, vol. 5, no. 5, pp. 636-641, 1995.

Philips H.S., Hains J. M., Amaninis M., Laramee G. R., Johnson S. A., Winslow J. W., "*Decreased brain derived neurotrophic factor depends on amyloid aggregation state in transgenic mouse models of Alzheimer's disease*," J Neurosci, 29:9321-9329, 2009.

Pisani V, Madeo G, Tassone A, Sciamanna G, Maccarrone M, Stanzione P, Pisani. *"Homeostatic changes of the endocannabinoid system in Parkinson's disease,"* Mov. Disord, 26(2):216-22, 2011.

Plant K., K. A. Pelkey, Z. A. Bortolotto, et al.,*"Transient incorporation of native GluR2-lacking AMPA receptors during hippocampal long-term potentiation,"* Nat Neurosci, vol. 9, no. 5, pp. 602-604, 2006.

Pozo K. and Y. Goda,*"Unraveling mechanisms of homeostatic synaptic plasticity,"* Neuron, vol. 66, no. 3, pp. 337-351, 2010.

Pratt K. G., E. C. Zimmerman, D. G. Cook, et al.,*"Presenilin 1 regulates homeostatic synaptic scaling through Akt signaling,"* Nat Neurosci, vol. 19, no. 9, pp. 1112-1114, 2011.

Pratt KG, Watt AJ, Griffith LC, Nelson SB, Turrigiano GG. *"Activity-dependent remodeling of presynaptic inputs by postsynaptic expression of activated CaMKII,"* Neuron, 39(2):269-81, 2003.

Qin Y., Y. Zhu, J. P. Baumgart, et al.,*"State-dependent Ras signaling and AMPA receptor trafficking,"* Genes Dev, vol. 19, no. 17, pp. 2000-2015, 2005.

Rabinowitch I. and I. Segev,*"Two opposing plasticity mechanisms pulling a single synapse,"* Trends Neurosci, vol. 31, no. 8, pp. 377-383, 2008.

Rial A. M. Verde, J. Lee-Osbourne, P. F. Worley, et al.,*"Increased expression of the immediate-early gene arc/arg3.1 reduces AMPA receptor-mediated synaptic transmission,"* Neuron, vol. 52, no. 3, pp. 461-474, 2006.

Rutherford L. C., S. B. Nelson, and G. G. Turrigiano,*"BDNF has opposite effects on the quantal amplitude of pyramidal neuron and interneuron excitatory synapses,"* Neuron, vol. 21, no. 3, pp. 521-30, 1998.

Saglietti L., C. Dequidt, K. Kamieniarz, et al.,*"Extracellular interactions between GluR2 and N-cadherin in spine regulation,"* Neuron, vol. 54, no. 3, pp. 461-477, 2007.

Schratt G. M., E. A. Nigh, W. G. Chen, et al.,*"BDNF regulates the translation of a select group of mRNAs by a mammalian target of rapamycin-phosphatidylinositol 3-kinase-dependent pathway during neuronal development,"* J Neurosci, vol. 24, no. 33, pp. 7366-7377, 2004.

Schwarz L. A., B. J. Hall, and G. N. Patrick,*"Activity-dependent ubiquitination of GluA1 mediates a distinct AMPA receptor endocytosis and sorting pathway,"* J Neurosci, vol. 30, no. 49, pp. 16718-16729, 2010.

Sheng M. and Hyoung Lee S. *"AMPA receptor trafficking and synaptic plasticity: major unanswered questions,"* Neurosci Res 46, 127-134, 2003.

Shepherd J. D., G. Rumbaugh, J. Wu, et al.,*"Arc/Arg3.1 mediates homeostatic synaptic scaling of AMPA receptors,"* Neuron, vol. 52, no. 3, pp. 475-484, 2006.

Shi S. H., Hayashi Y., Petralia R. S., Zaman S. H., Wenthold R. J., Svoboda K. and Malinow R. *"Rapid spine delivery and redistribution of AMPA receptors after synaptic NMDA receptor activation,"* Science 284, 1811-1816, 1999.

Soden M. E. and L. Chen,*"Fragile X Protein FMRP Is Required for Homeostatic Plasticity and Regulation of Synaptic Strength by Retinoic Acid,"* Journal of Neuroscience, vol. 30, no. 50, pp. 16910-16921, 2010.

Song I. and Huganir R. L. *"Regulation of AMPA receptors during synaptic plasticity,"* Trends Neurosci 25, 578-588, 2002.

Steinberg J. P., K. Takamiya, Y. Shen, et al.,*"Targeted in vivo mutations of the AMPA receptor subunit GluR2 and its interacting protein PICK1 eliminate cerebellar long-term depression,"* Neuron, vol. 49, no. 6, pp. 845-860, 2006.

Steinmetz C. C. and G. G. Turrigiano,*"Tumor necrosis factor-alpha signaling maintains the ability of cortical synapses to express synaptic scaling,"* J Neurosci, vol. 30, no. 44, pp. 14685-14690, 2010.

Stellwagen D. and R. C. Malenka,*"Synaptic scaling mediated by glial TNF-alpha,"* Nature, vol. 440, no. 7087, pp. 1054-1059, 2006.

Stellwagen D., E. C. Beattie, J. Y. Seo, et al.,*"Differential regulation of AMPA receptor and GABA receptor trafficking by tumor necrosis factor-alpha,"* J Neurosci, vol. 25, no. 12, pp. 3219-3228, 2005.

Steward O. and P. F. Worley,*"Selective targeting of newly synthesized Arc mRNA to active synapses requires NMDA receptor activation,"* Neuron, vol. 30, no. 1, pp. 227-240, 2001.

Stichel CC, Augustin M, Kühn K, Zhu XR, Engels P, Ullmer C, Lübbert H. *"Parkin expression in the adult mouse brain,"* Eur J. Neurosci, 12, 4181–4194, 2000.

Sutton M. A., H. T. Ito, P. Cressy, et al.,*"Miniature neurotransmission stabilizes synaptic function via tonic suppression of local dendritic protein synthesis,"* Cell, vol. 125, no. 4, pp. 785-799, 2006.

Szobota S., P. Gorostiza, F. Del Bene, et al.,*"Remote control of neuronal activity with a light-gated glutamate receptor,"* Neuron, vol. 54, no. 4, pp. 535-545, 2007.

Tagawa, Y., Kanold, P.O., Majdan, M. & Shatz, C.J. *"Multiple periods of functional ocular dominance plasticity in mouse visual cortex,"* Nat. Neurosci, 8:380–388, 2005.

Tang T. S., E. Slow, V. Lupu, et al. *"Disturbed* Ca^{2+} *signaling and apoptosis of medium spiny neurons in Huntington's disease,"* Proc Natl Acad Sci USA, 102: 2602–2607, 2005.

Tarkowski E., Blennow K., Wallin A. and Tarkowski A. "*Intracerebral production of tumor necrosis factor-α, a local neuroprotective agent, in Alzheimer disease and vascular dementia*," J Clin Immunol, 19: 223-230, 1999.

Terashima A., L. Cotton, K. K. Dev, et al.,*"Regulation of synaptic strength and AMPA receptor subunit composition by PICK1,"* J Neurosci, vol. 24, no. 23, pp. 5381-5390, 2004.

Thiagarajan T. C., M. Lindskog, and R. W. Tsien,*"Adaptation to synaptic inactivity in hippocampal neurons,"* Neuron, vol. 47, no. 5, pp. 725-737, 2005.

Tomita S., Nicoll R. A. and Bredt D. S. *"PDZ protein interactions regulating glutamate receptor function and plasticity,"* J Cell Biol 153, F19-24, 2001.

Turrigiano G. G. and S. B. Nelson,*"Homeostatic plasticity in the developing nervous system,"* Nat Rev Neurosci, vol. 5, no. 2, pp. 97-107, 2004.

Turrigiano G. G., K. R. Leslie, N. S. Desai, et al.,*"Activity-dependent scaling of quantal amplitude in neocortical neurons,"* Nature, vol. 391, no. 6670, pp. 892-896, 1998.

Turrigiano G. G.,*"The self-tuning neuron: synaptic scaling of excitatory synapses,"* Cell, vol. 135, no. 3, pp. 422-435, 2008.

Vitureira N., M. Letellier, and Y. Goda,*"Homeostatic synaptic plasticity: from single synapses to neural circuits,"* Curr Opin Neurobiol, pp. 1189-1194, 2011.

Volk L., C. H. Kim, K. Takamiya, et al.,*"Developmental regulation of protein interacting with C kinase 1 (PICK1) function in hippocampal synaptic plasticity and learning,"* Proc Natl Acad Sci U S A, vol. 107, no. 50, pp. 21784-21789, 2010.

Wayman G. A., Y. S. Lee, H. Tokumitsu, et al.,*"Calmodulin-kinases: modulators of neuronal development and plasticity,"* Neuron, vol. 59, no. 6, pp. 914-931, 2008.

Westerman MA, Cooper-Blacketer D, Mariash A, Kotilinek L, Kawarabayashi T, Younkin LH, Carlson GA, Younkin SG and Ashe KH. *"The relationship between Abeta and memory in the Tg2576 mouse model of Alzheimer's disease,"* J. Neurosci, 22, 1858–1867, 2002.

Wierenga C. J., K. Ibata, and G. G. Turrigiano,*"Postsynaptic expression of homeostatic plasticity at neocortical synapses,"* J Neurosci, vol. 25, no. 11, pp. 2895-2905, 2005.

Williams M. E., S. A. Wilke, A. Daggett, et al.,*"Cadherin-9 regulates synapse-specific differentiation in the developing hippocampus,"* Neuron, vol. 71, no. 4, pp. 640-655, 2011.

Wu J, Petralia RS, Kurushima H, Patel H, Jung MY, Volk L, Chowdhury S, Shepherd JD, Dehoff M, Li Y, Kuhl D, Huganir RL, Price DL, Scannevin R, Troncoso JC, Wong PC, Worley PF. *"Arc/Arg3.1 regulates an endosomal pathway essential for activity-dependent β-amyloid generation,"* Cell, 147(3):615-28, 2011.

Xia J., X. Zhang, J. Staudinger, et al.,*"Clustering of AMPA receptors by the synaptic PDZ domain-containing protein PICK1,"* Neuron, vol. 22, no. 1, pp. 179-187, 1999.

Yu L. M. and Y. Goda,*"Dendritic signalling and homeostatic adaptation,"* Curr Opin Neurobiol, vol. 19, no. 3, pp. 327-335, 2009.

Zhang A. W., Q. M. Hou, M. Wang, et al.,*"Na,K-ATPase Activity Regulates AMPA Receptor Turnover through Proteasome-Mediated Proteolysis,"* Journal of Neuroscience, vol. 29, no. 14, pp. 4498-4511, 2009.

Zhao C, Dreosti E, Lagnado L. *"Homeostatic synaptic plasticity through changes in presynaptic calcium influx,"* J. Neurosci, 31(20):7492-7496, 2011.

Zhu J. J., Y. Qin, M. Zhao, et al.,*"Ras and Rap control AMPA receptor trafficking during synaptic plasticity,"* Cell, vol. 110, no. 4, pp. 443-455, 2002.

Zuccato, C. et al. *"Loss of huntingtin-mediated BDNF gene transcription in Huntington's disease,"* Science 293, 493–498, 2001.

In: Horizons in Neuroscience Research. Volume 15
Editors: Andres Costa and Eugenio Villalba
ISBN: 978-1-63321-523-8

Chapter 3

The Psychosomatic Separation of Noradrenaline at Brain and Adrenaline at Blood is a Homeostatic Lame Axis Allowing that the Fight-or-Flight Stressing of Metabolic Controls could Turn-on Related Diseases

A. Bennun
Graduate School, Rutgers University (R) and CONICET, NJ, US

Abstract

Integration between psyche and soma was analyzed at the molecular level. Signaling by sensor neurons activates the fight-or-flight response through adenylyl cyclases (AC) "In common" signaling resulting from depletion of chelating metabolites by the tendencies of ionic equilibrium to increase free Mg2+ vs the decrease of inhibitors CaATP, ATP4-, and Ca2+. Mg2+-dependent responsiveness of AC to noradrenaline in brain and adrenaline on the tissue integrates hormone-dependent circadian rhythms. The response to stress of the hypothalamic-pituitary-adrenal (HTPA) axis is mediated by hypothalamic glucose sensor neurons, which signal increasing the adrenal glands secretion of adrenaline and cortisol. These on the pancreas stimulate secretion of glucagon and inhibit the release of insulin, distorting the pulsatile rhythm of glucose metabolism. Thus, flattens the normal circadian modulation of glucose availability to brain. At the tissue network, supporting brain, the rate of ATP generation and consumption could exceed the modulatory limits of Energy Charge. Thus, opposes the ATP-requiring regeneration of glycogen and fat reserves. The lipolytic hormones: adrenaline, ACTH, glucagon, etc. releases free fatty acids from fat tissue to support liver's gluconeogenesis supporting the glucose needs of brain. The basal AC activity of fat cell shows a Mg2+ saturation curve of cooperativity value "n" of about two because saturation by Mg2+ an activator sums-up with desaturation of ATP4- an inhibitor of the enzyme. The effect of hormones increases "n" to about 4 implicate activation of additional Mg2+ saturation sites, mediating in the membrane for the bonding of AC and the associated G-protein receptor. Cortisol that crosses the blood-brain barrier (BBB) exerts a negative feedback on the HTPA axis, but adrenaline does not, allowing persistence of its stimulatory effect on the HTPA axis to favor an uncontrolled secretion of cortisol. Persistent adrenaline-dependent inhibition of insulin secretion interferes with its capability to maintain a circadian rhythm, controlling availability of glucose to muscle, and brain. Parkinson and schizophrenia share on muscle-tremor symptomatology, which relates to depletion of the ATP required for muscular sliding. The overexposure to adrenaline has an inactivatory effect on the adrenaline-AC complex, which is considerably less stable, than the complexes of AC with either glucagon or ACTH. The lack of an adrenaline feedback could have been imposed by an evolutionary tendency of mind/brain dominance over soma/body itself. Thus, fight-or-flight response became a selection path for the development of an alternatively responsiveness adapted to environmental changes formative of long-term memory. Hence, the confluence of the sense of danger and metabolic mobilization resulted in a NA in brain vs adrenaline in blood conforming and axis leading to the corpus coeruleus neurons with their long axons to reach many cerebral regions to expand harmonize the range of perception and response. The presynaptically discharged NA acts postsynaptically on the Mg2+ sympathetic receptor or on the Ca2+-dependent alpha-2-receptor for a separated AC responsiveness to an activator free Mg2+ or inhibitor Ca2+. The Mg2+ incremented cAMP became a ligand of the cellular transcription factor, cAMP response element-binding protein (CREB). This one is generally accepted to participate in the plasticity of the neuronal circuits involved in long-term memory. Thus, allows an integrative role to the brain adrenergic system, synchronizing the glutamate-mediated neurotransmission. Additionally, AC variable responsiveness to NA also fits with the effect of Ca2+ vs Mg2+ in locally adjusting the glucose permeability of each capillary-neuronal BBB. The evolutionary increment of brain capabilities were supported by the role of astrocytes to keep a homeostatic temperature. This cell drives a rapid flux of H-bond depleted water cluster, out of the system, adjusting turnover to 3.7 times of the 160 ml cerebrospinal fluid per day. This interdependent network potentiates the adrenergic receptors to endow the brain with a potent working memory, but extremely sensitive to stress. Acute-panic could be associated to a lack of a promptly reuptake of NA, allowing a symptomatology of anxiety followed by depression, due to the exposure of the NA-AC complex, to the time-dependent inactivatory effects of NA on the AC enzyme.

Introduction

Long term memory requires the construction of new synapses [1] [2] Changes in synaptic strength participate in learning and memory. The process involves neuro-transmitters, receptors and new synapses generating pathways which reinforce the strength of connectivity between neurons. Noradrenalin (NA) binding activates various subtypes of adrenergic receptors, for a differential role of signaling pathways, which functionally operate by inducing synthesis of second messengers.

The *striatum* and the hippocampus have been shown by Functional Magnetic Resonance Imaging (FMRI), to interact cooperatively to support episodic memory formation [3]. New long-lasting details of specific events could be lost to people that show amnesia, a dysfunction that can be related to hippocampus damage [4].

The *cerebral cortex* is capable of semantic memory, the ability to learn facts and relationships mediated by changes in the connections between cells, related by their conducting of specific types of information [5].

The ventromedial *hypothalamus* (VMH) has been shown to respond to catecholamine released into the glucose-sensing region, through a counter-regulatory response, to hypoglycemia by stimulation of the beta-2-adrenergic-receptor (B2AR) [6].

Emotional charged events have been related to activation of the locus coeruleus release of noradrenalin [7], activating a second messenger system [8]. Noradrenalin activates AC in membrane preparations obtained from rat's cerebral cortex, corpus striatum and hypothalamus [8]. However, over-exposure to noradrenalin of AC during pre-incubation, an assay previous to the one measuring the activatory effects of the neurotransmitter [9], results in inactivation of the enzyme.

Hence, it could be inferred that noradrenalin-dependent psychosomatic damage, lead in the course of an evolutionary selection process to a costly protection mechanism. This one, turns-off the enzyme rather than allowing a too perdurable permanence of the organism into the fight or flight state. The results of the experimentally obtained prolonged association of NA with its receptor may correspond to a stressing of the quaternary structure, during the activatory state [9]. This catecholamine-dependent inactivatory event was prevented by a pre-incubation condition, in which the presence of MgATP protects from NA, by the formation of an enzyme substrate complex. Hence, the NA-AC complex manifests thermic instability, whereas the NA-AC-MgATP allows to protect the enzyme activity, but to the cost of decreasing its responsiveness to a catecholamine activatory effect [8] [9] [10].

Stress mediated by increasing adrenalin secretion, usually shows a pattern of hyperfunction follow by hypofunction, which when the stressed organ is brain it could be associated with a symptomatology of persistent anxiety followed by depression [9].

Under prolonged stress, the synthesis of new AC, may not match the rate of noradrenalin-dependent inactivation. The viability of functional synapses may be controlled by the speed of synthesis of the proteins, which make-up the hormonal receptor sites and/or AC-itself, versus the rate of inactivation or destruction of AC. This generalized molecular modeling for the evolution of psychosomatic diseases [9] could be tested, with regard to more specific physiological conditions and permit prognosis of dysfunction, allowing preventing treatments. Already medical applications have been developed, based in the use of Beta blockers to interfere with a stressful condition, attenuating the memory of a traumatic event.

Characterization of the NA activation of brain AC indicated that could participate in the formation of molecular memory pathways [8] [11]. This proposition was extended to show its role in cAMP-dependent pathways, involved in the development of behavior patterns [12] [13] [14] [15] [16]. In absence of noradrenalin AC was shown to have a basal activity subject to up-regulation by Mg^{2+} and Mn^{2+} ions and their ATP complexes MgATP and MnATP as substrates [11]. Na^{+}-pump activity changes ionic equilibrium by the neuronal transfer out of Na^{+} coupled to K^{+} intake. Additional changes, relate to ATPase activity splitting ATP^{4-} into ADP^{3-} and P_i^{2-}, with an ionic equilibriums allowing the endogenous release of free Mg^{2+} ions. The breakdown of ATP also decreases the concentration ATP^{4-} and CaATP, which were shown to act as strong inhibitors of AC activity [11]. More recent studies obtained from the molecular, behavioral cognitive, signal transduction, physiologic and pathological research, has shown the relevance of cAMP-mediated long-term memory [17] [18] [19] [20] [21] [22] [23] [24] [25].

It could be expected that a large number of the pathways associated with neuronal activity, could be integrated into the complex process of memory formation and consolidation, by an "in common" coupling by the interdependence of signal and ATP-energy transduction mechanisms [26] [27] [28] [29].

Up-regulation of AC by ionic equilibrium in dependence of changes in ATP concentrations, leaded to investigate possible metabolic-ionic feedbacks related to the energy requirement of the NA-AC system. Stimulated neurons require O_2 to maintain the supply of ATP for the Na^+-pump. O_2 increases turnover of both electron transport chain and Krebs cycle, which occurs through the increase in dicarboxylic and tricarboxylic acids concentrations in mitochondria. Effect which increases Ca^{2+} retention in mitochondria and up-regulate the Na^+/K^+-ATPase-AC system. The relationship between the rate of aerobic glycolysis and neuronal work was therefore examined, in the context of the signal transduction of tissue networking. The latter, allows a large increment in the rate glucose and O_2 uptake by stimulated brain areas, even when homeostatically sustaining glucose levels in blood.

Methods

Preparations

Female Sprague-Dawley rats weighing 210-300g were ovariectomized to decrease interference by estrus cycles [8] and decapitated at 4°C two weeks after surgery. The brains were dissected in the cold (4°C) to isolate and separately collect the hypothalamus, cortex and striatum from groups of 10 to 15 animals [8] [30]. The total wet weight of each pool of tissues was colected in 20 ml of ice-cold 2 mM Tris-HCI containing 10 mM NaCl, 10 mM KCl and 5μM EDTA, buffer at pH=7.4. These were rinsed twice with 15 ml/g wet weight of tissue and homogenized using a motor-driven (Inframo, Wayne, New Jersey) Teflon homogenizer at 400 rpm for 2 min. Each homogenate was centrifuged at 4°C and 6,000 g_{av} in a RCA Sorvall, head SS-34 for 20 min. The resulting supernatants were discarded and the pellet washed, resuspended and centrifuged again once. Aliquots stored at -70°C. Protein concentration for each membrane preparation was determined by the method of Lowry (Lowry et al., 1951) using bovine serum albumin as the standard. Particle suspensions and reaction mixtures were tested, with a calcium ion electrode for Ca^{2+} contamination with negative results.

Assay of Adenylyl Cyclase Activity

Adenylyl cyclase (AC), systematic name: ATP diphosphate-lyase (cyclizing, 3´, 5´-cyclic-AMP-forming), EC 4.6.1.1, in the membrane preparations of cortex, corpus striatum and hypothalamus were kinetically characterized. The tissues were selected in expectation of finding differential kinetic properties, which could be related to specific areas of rat´s brain [8].

Defrosted particulate preparations (from 1.7 to 6 μg protein/assay) were re-suspended in a standard reaction mixture, containing 20 mM Tris-HCl buffer (pH 7.4), 5 mM theophylline,

$MgCl_2$, plus ATP disodium salt, from equine muscle, with GTP content of less than 2% and traces of other nucleotides from Sigma. This ATP/GTP mixture was used when neurotransmitter dependent activation was desirable as in the experiments reported in figure 1 a) and b) and table 1. The ATP disodium salt from P-L Biochemicals, Inc (Milwaukee, Wisconsin) was used for the assay of AC basal activity. Other additions indicated in the figures legends.

After incubating in a shaking bath at 37°C for 15 min the reaction was terminated by heating in a boiling water bath for 5 min. Time-dependence studies using the membrane preparations, showed linear responses for no less than 30 min.

Cyclic AMP formed by adenylyl cyclase activity was assay by the Gilman method [31] modified by Brown et al. [32] according to publish specifications [33] but with cyclic $[^3H]$AMP increased to 20,000 c.p.m./assay. These membrane preparations show an ATPase activity [34] of only 0.1 nmol P_i/h/μg-protein [8].

Results were calculated by formula, (C- b)/(T- b), C: c.p.m. from the assay of the enzyme activity, T: c.p.m. from total binding (zero-time blanks) and b: c.p.m. in the millipore filters (moisture blanks). Calibration curves are linear for % of C- b/T- b versus log cAMP without interfering ATP or other reagents. Results mean values of triplicates or more: nanomoles of cAMP formed per hour per milligram of protein. The standard error of the mean (S.E.M.) did not exceed 6% [35].

Results

Previous kinetic studies of the noradrenalin-activated adenylyl cyclase (AC), evidenced that the neurotransmitter response could be differentiated from a basal response, according to their different Vmax, which could explain memory at two differentiable levels. The noradrenalin dependent activity of AC, could be induced by events with strong emotional connections, related to the fight or flight reaction. The activity of the neurotransmitter-enzyme complex correlates to a reinforcement over the basal activity one elicited by free Mg^{2+}- or Mn^{2+}-dependent activation of the basal state of the enzyme [9] [10]. However, prolonged effects may elicit a CNS excited or anxious state which prolonged results in a depressive state [9] [36]. The latter, could be related to memory deficiencies so emotional potentiation of memory capability became limited because under stress condition collapses.

The present studies were directed: I) To elucidate the kinetic properties of AC relevant to modeling mechanism: a) for NA-dependent emotional eliciting and reinforcement of AC-dependent long-term memory, b) for continuous memory processing by activation of the basal AC activity by free divalent metals; II) Characterization of hemoglobin red cell-system as a sensor of glucose levels in blood, and responsiveness for O_2-release matching glucose uptake III) To detect a metabolic feedbacks of a tissue network, capable to shift the rates of O_2 and glucose consumption, between hormonal activated and the basal memory processing stage. The matching of the rates of brain glucose consumption and tissue oxygenation requires “in common” molecular signaling. This one should operate within a network, by a healthy physiology resolving through the sensor role of the red cell-Hb system, the matching of brain metabolism with the homeostatic need of balancing the metabolic flows, within the liver fat-cell subsystem.

Relevant Brain AC Kinetic Properties for Emotional Reinforcement and Basal Memory Processes

Purified enzymes preparations from rat's tissues were obtained for adenylyl cyclase of fat cells, brain and insulin receptor tyrosine kinase (IRTK) from liver. This preserved their allotopically associated receptors by not extend purification to an extraction procedure to separate the enzyme from the membrane [34]. These enzyme-membrane systems were shown to preserve modulatory properties required for their physiological role.

The preparations of AC [ATP pyrophosphate-lyase (cyclizing) (EC 4.6.1.1)] obtained from corpus striatum, hypothalamus and cortex of the brains of rats show that noradrenalin (nor-epinephrine) activation is dependent of the concentration of magnesium ion (Mg^{2+}).

Impulse transmission and/or noradrenaline binding the receptor induce the depolarization of the plasma membrane. Neuronal transmission is terminated by the neurotransmitter reuptake linked to Na^{+} entry. The β-receptor is present in the hormone-binding subunits of the enzyme AC. When noradrenaline binds to its receptor, this one becomes activated to bind G-protein, which is activated by GTP. As has been reported elsewhere, the neurotransmitter activated enzyme, increases Vmax. The activated enzyme generates endogenous cAMP, which rather than acting as a messenger between one neuron and another, has the capability to induce changes in the function of ion channels within the membrane. The cAMP has effect on gene expression in the nucleus of the cell and in the processes of growth and development. Changes on the genetic material of cells may lead to long-term alterations of behavior. Second messengers play a role in the manufacture and release of neurotransmitters, intracellular movements, and a carbohydrate metabolism in the brain.

Table 1. Kinetic characterization of adenylyl cyclase obtained from specifics brain areas

Particulate Membrane preparations	Km(MgATP) [mM]	Cooperativity (n)
Cortex	2.2	1.4
Striatum	2.0	1.1
Hypothalamus	2.0	1.1

The values obtained do not allow inferring differences between the responsiveness to the substrate of the basal adenylyl cyclase activity.

Hence, Table 1 shows that allotopic enzyme-membrane properties are similar for the brain regions tested.

Fig.1.a) shows the activatory effect of noradrenaline versus basal on the AC of cortex increasing by 300% basal Vmax. The cortex enzyme shows cooperativity for total-magnesium in the presence of 0.1mM-noradrenaline with a value $n(Mg_T)=3.9$.

Fig.1.b) shows that the AC basal cooperativity decreases from $n(Mg_T)=2.9$ and by addition of noradrenaline plus $CaCl_2$ decreasing to $n(Mg_T)=2.5$ and decreased Vmax to 1/3.

Comparative results on the effect of noradrenaline in adenylyl cyclase of corpus striatum are reported in table 2.

In table 2, the AC of corpus striatum at basal conditions and for 0.1mM-noradrenaline and +0.1mM-noradrenaline+0.3mM $CaCl_2$, show similar affinity $S_{0.5}(Mg_T)$ and $S_{0.5}(Mg^{2+})$ of about 9mM. The affinity for the inhibitory specie $S_{0.5}$ (ATP^{4-})=0,1mM. Hence, noradrenalin mainly increases Vmax but does not change by much the enzyme affinity for either Mg^{2+} or ATP^{4-}. The AC of corpus striatum cooperativity for 0.1mM-noradrenaline n=4.4. Cooperativity for basal n=2.7, since for basal $n(Mg^{2+})$, $n(ATP^{4-})$ and n(MgATP) show values close to 1.

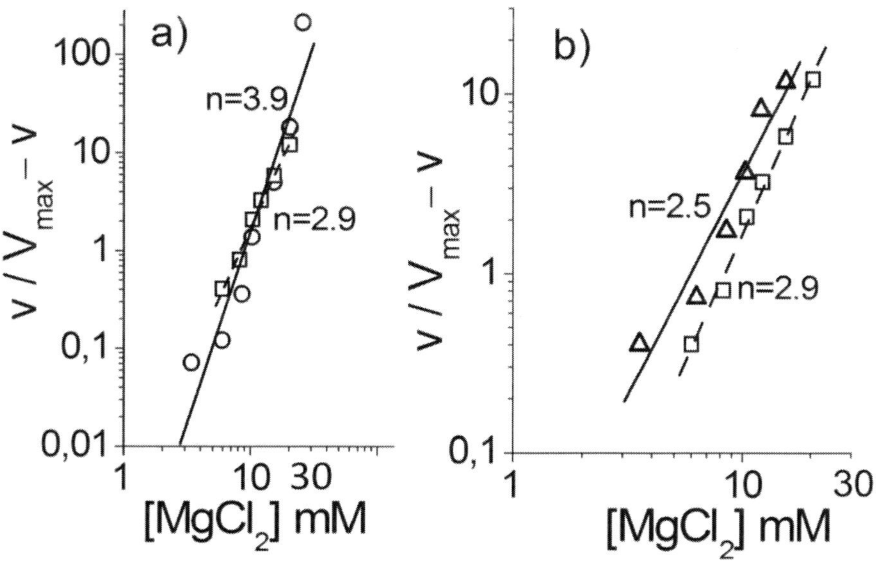

Figure 1. a) and b). Control by Mg_T of the responsiveness of noradrenaline activated adenylyl cyclase. Samples of an EDTA-particles preparation from cortex containing 100μg of protein were incubated at 37°C for 1hour, in 40mM-Tricine/Tris buffer, pH=7.4, 6.67mM-caffeine, constant 1mM-ATP(Sigma purified from horse muscle), in a final 0.5ml with the indicated concentration of $MgCl_2$ (Mg_T). Vmax for each curve was determined at the indicated $MgCl_2$ concentration. Activity reported: nano-moles of cAMP formed/h per mg of membrane protein. Fig.1.a). $MgCl_2$ saturation curves: o) +1mM-adrenaline, at 30mM [$MgCl_2$] were determined Vmax values, 212 ±14(4);Δ) basal, 78±7(4). Fig.1.b) $MgCl_2$ saturation curves: basal, at 15 mM [$MgCl_2$], 78±7(4); Δ) +1mM-adrenaline +0.3 mM $CaCl_2$ at 25mM [$MgCl_2$], 27±2(4).

Table 2. Effects of Mg^{2+} and ATP^{4-} on the response to noradrenaline and calcium of adenylyl cyclase of corpus striatum

Kinetic Constants	Basal	0.1mM-Noradrenaline	0.1mM-noradrenaline +0.3mM $CaCl_2$
Mg_T	n=2.7	n=4.4	n=2.3
Mg_T	$S_{0.5}$=8mM	$S_{0.5}$=9mM	$S_{0.5}$=9.4mM
Mg^{2+}	n= 1.1	n=1.9	n=1.3
Mg^{2+}	$S_{0.5}$=4.7mM	$S_{0.5}$=4.6mM	$S_{0.5}$=4.6mM
ATP^{4-}	n=1.1	n=1.9	n=1.3
ATP^{4-}	$S_{0.5}$=0.1mM	$S_{0.5}$=0.1mM	$S_{0.5}$=0.1mM

To investigate the individual and separate effects of free Mg^{2+} and of ATP^{4-}, their concentration were determined by multiple-equilibrium equations.

Hence, the values obtained for basal activity n(Mg^{2+}), are influenced by the summation to the activatory effect of Mg^{2+}, and the disappearance of the inhibitory effect of ATP^{4-}.

The enzyme saturation system, shows a n(adrenaline) values of about 4, because the neurotransmitter potentiates Mg^{2+} mediated binding to the membrane. The hormonal receptor for the expression of properties dependent of its position in the membrane, allotopic, may require Mg^{2+} [34] for the activatory binding of G-protein. The enzyme shows a single active site, therefore, requires Mg^{2+} and MgATP for basal activity, and a separate for the binding of an adrenaline receptor.

The applicability of the random of the RARE BiBi model is confirmed by the ability of the model to predict a change of Hill cooperativity from a basal value of n=2.6 to 4.0 in the presence of noradrenaline (Fig. 1 and Table 2). This indicates that excess ions Mg^{2+} over that required for the saturation of sites, for substrate formation and for the Mg^{2+} regulatory activation. Hence, an additional saturation by ionic Mg^{2+} is an obligatory requirement for the manifestation of third site linked to the responsiveness to hormonal dependent activation [8] [33] [35]

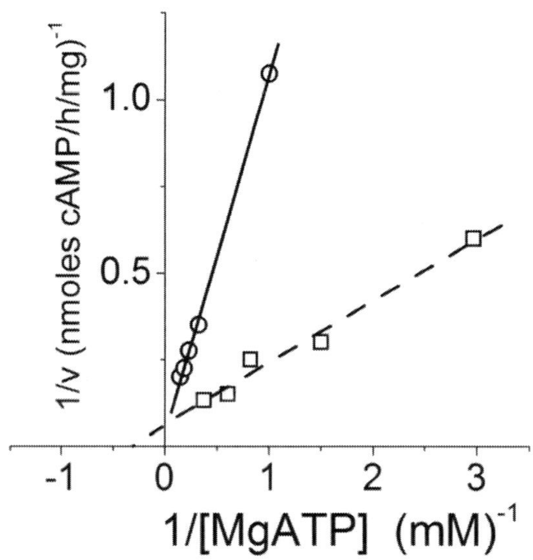

Figure 2. Lineweaver-Burke plot. Effect of ATP^{4-} on the substrate saturation curve of adenylyl cyclase particulate preparation from corpus striatum. Portions of an AC particulate preparation containing 2.5 μg of protein were incubated. Basal, MgATP saturation curve, at 5mM MgATP, Vmax: 9nmol cAMP/h/per mg of protein; o) basal+2.5 mM ATP^{4-} at 9mM MgATP, $Vmax_{app}$: 7nmol cAMP/h/per mg of protein constant equilibrium concentration adjusted according to table 3.

Lineweaver-Burk plots of adenylyl cyclase activity as a function of substrate concentration at different concentrations of Mg_{2TOTAL} show a noncompetitive behavior, indicating that Mg^{2+} interacts with a distinct regulatory site differentiable from the active site [11]. If the data of Fig. 5 is plotted according to Lineweaver-Burk, the addition of CaATP shows Ki=0.015mM. However, Mg^{2+} (Mg_T) can easily displace Ca^{2+} from ATP. The equilibrium favors that Mg^{2+} can easily displace Ca^{2+} from ATP, and reduce the concentration of CaATP. Since Ka (MgATP)=$-20\times10^3M^{-1}$ versus Ka(CaATP)=$-9\times10^3M^{-1}$.

Table 3. Equilibrium concentrations of MgATP, ATP^{4-} and Mg^{2+} under experimental conditions where ATP^{4-} is kept constant and MgATP is increased. Equilibrium concentrations, adjusted from the calculated ionic equilibrium [37] [38]

Mg^{2+} (mM)	ATP^{4-} (mM)	MgATP (mM)	[MgATP]/[Mg^{2+}]
0.0058	2.20	0,244	42.1
0.0229	2.23	0.976	42.6
0.0455	2.25	1.950	43.4
0.0896	2.29	3.910	43.4
0.1110	2.31	4.880	43.9
0.1320	2.33	5.860	44.0
0.2000	2.33	8.900	44.0

Evaluation of the responsiveness of the enzyme indicates that there is a requirement that the concentration of free Mg^{2+} exceeds that of MgATP to activate the enzyme. The AC response appear coupled to the operation of the Na^{+} gates augmenting free ionic Mg^{2+}. The gates associated ATPase reaction releases free Mg^{2+} because since ADP^{3-} has a much weaker association constant for the divalent metal than ATP^{4-}.

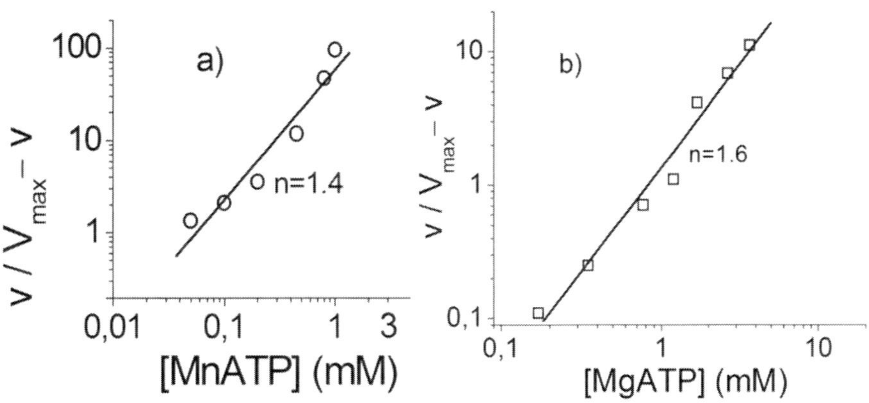

Figure 3. a) MnATP substrate saturation curve and 3.b) MgATP substrate saturation curve. Portions of adenylyl cyclase particulate preparation from corpus striatum containing 6μg proteins were incubated in the conditions for assay of basal activity described for fig.1. Activity reported: nmol cAMP formed/h per mg of membrane protein. Fig.3.a), MnATP saturation curve: o) At 1mM MnATP reached Vmax value: 19.2nmol cAMP, cooperativity value n(MnATP)=1.4, Km(MnATP)=0.05mM. Fig.3.b), MgATP saturation curve: At 3.6mM MgATP reached Vmax: 8.2nmol cAMP, cooperativity value n(MgATP)=1.6, Km(MgATP)=0.84mM.

The Hill Plot in fig. 3 was obtained from saturation curves with a Vmax(MnATP): 19.2nmol cAMP/h/mg, a considerable higher activity than the corresponding Vmax(MgATP): 8.2nmol cAMP/h/mg.

Comparing both Km, it could be shown a marked increase for the basal activity of adenylyl cyclase in the presence of MnATP, with an affinity increment of over 17 times than the Km MgATP. Moreover, Vmax was increased for the substrate MnATP 2.3 times over that of Vmax MgATP. Equilibrium conditions for MnATP (log Ka=4.6) allowed to increase the concentration of MnATP at negligible concentration of free ionic Mn^{2+}. Hence, it could be infer that if both substrates are separately compartmentalized the adenylyl cyclase consuming MnATP would be able to produce cAMP at much higher rates than in the presence of MgATP alone. Additionally, the adenylyl cyclase in the presence of substrate MnATP manifests generation of cAMP independently of the already discussed free Mg^{2+} activatory requirement (fig.1), controlling responsiveness at the neurotransmitter and basal levels. Hence, it could be infer the existence of neurons or synaptic structures able to orchestrate the response of adenylyl cyclase utilizing MgATP, with a control role in the generation of neuronal memory nets.

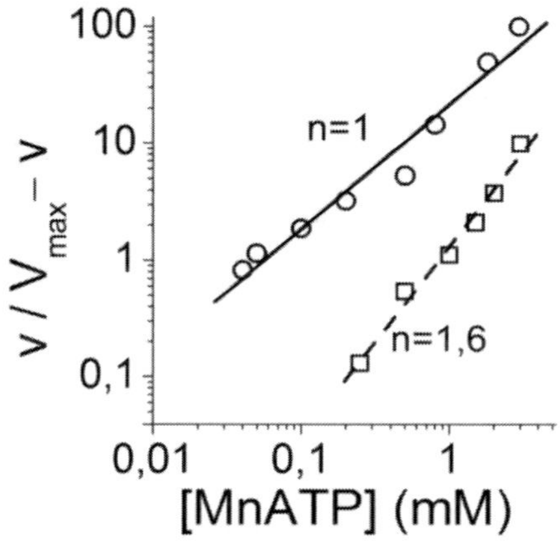

Figure 4. Effect of inhibitor ATP^{4-} on the substrate MnATP saturation curve of corpus striatum adenylyl cyclase. 6μg protein portions of an adenylyl cyclase-membrane preparation were incubated at 37°C as indicated in the legend to fig.1 o) Basal; with the indicated increasing equilibrium concentration of MnATP at 1mM MnATP reached Vmax: 19.2nmol cAMP/h/per mg; basal+inhibitor 5.5mM ATP^{4-} at 4mM reached Vmax: 15.2nmol cAMP/h/per mg.

The increasing concentration of $[MnATP]/[Mn^{2+}][ATP^{4-}]=4\times10^4$, implies that equilibrium maintains negligible concentration of free Mn^{2+}. Hence, the addition to the basal curve of ATP^{4-} maintains its equilibrium concentration well above that in the basal curve. However, the ATP^{4-} inhibitory-curve approaches the basal at the higher maximal concentration, showing unchanged Vmax.

Kmapp(MnATP) changes from 0.075 to 1.5 mM (Table I). The calculated $Ki(ATP^{4-})$=0.25mM similar to the value obtained for substrate MgATP, $Ki(ATP^{4-})$=0.27mM. Hence, the presence of MnATP does not alter the affinity for adenylyl cyclase for ATP^{4-}, ruling out, indirect modulation by Mn^{2+} on the affinity of the enzyme for ATP^{4-}.

Fig.5 shows that addition of CaATP produce competitive inhibition decreasing affinity from 2mM to 20mM Kmapp(MgATP) [11].

Energy charge [39] is described as the intracellular relationship of [ATP]+1/2[ADP]/([ATP]+[ADP]+[AMP]), can have a value ranging from 0 (all AMP) to 1(all ATP). This index could be given a role in ionic equilibrium by its relationship to the control of the availability of free divalent metals. The latter, related to the stronger binding constants of ATP^{4-} for Mg^{2+} and Ca^{2+} than for ADP^{3-} and AMP^{2-}.

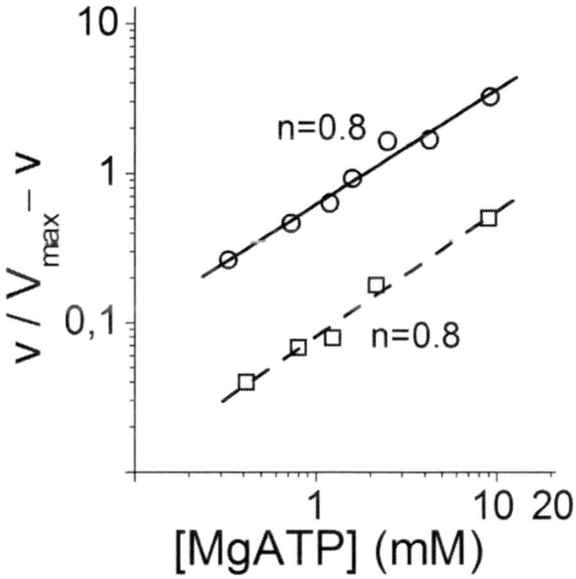

Figure 5. Effect of inhibitor CaATP on the substrate MgATP saturation curve of hypothalamic adenylyl cyclase. 3.3μg protein portions of a hypothalamic-membrane preparation were incubated at 37°C as described in the legend to fig.1 with increasing equilibrium concentrations of 0.41 to 9.3 mM MgATP: MgATP saturation curve: o) Basal at 9.3mM MgATP Vmax: 13.2nmol cAMP/h/mg; □) basal+inhibitor 0.37±0.05mM CaATP at 9.3mM MgATP Vmax:5.8nmol cAMP/h/mg. Cooperativity value: n, computational obtained were included in the figure.

Table 4. Kinetic values from Hill and Lineweaver and thermodynamics constant from Fig. 2, 4 and 5 for the indicated E-Complexes

Substrate Saturation	Additions	Enzyme Complex	Km (mM)	Kd (mM)	Ki (mM)	Kf×10^4 (M^{-1})	Cooperativity (n)	ΔG° (Kcal)
MnATP	---	E-MnATP	0.065	---	---	1.43	0.8	5.6
MgATP	---	E-MgATP	1.05	---	---	0.10	1.1	4.1
MgATP	+0.5mM CaATP	E-CaATP	---	---	0.015	6.67	0.9	6.5
---	---	E-Ca	---	0.02	---	5.0	---	6.4
MgATP	+2.26mM ATP^{4-}	E-ATP^{4-}	---	---	0.27	0.37	1.49	4.8
MgATP	+2.2mM Mg^{2+}	E-Mg	---	4.7	---	0.021	1.1	3.2
MnATP	+5.5mM ATP^{4-}	E-Mn	---	3.8	---	0.026	---	3.3

Kd dissociation constant; Kf formation constant for enzyme complexes; ΔG°, Gibbs free energy. Kd values [37] [38] change by only 9% when ionic strength changes 100%. Hence, Kd values where affected by considerably lower deviations

In table 4 is reported that Km(MnATP)=0.065mM versus Km(MgATP)=1.05mM, correspond to 16 times lower affinity of the basal enzyme for the substrate MgATP than for MnATP. Table 4 dissociation constants, show an ionic equilibrium order Kd(Mg^{2+})= 4.7mM>Kd(Mn^{2+})=3.8, indicating that Mg^{2+} can only at higher concentrations than Mn^{2+} displace the latter from the divalent metal activatory site.

The activity order MnATP>MgATP and both ATP^{4-} and CaATP act as inhibitors. Since Ki(ATP^{4-})=0.27 and Ki(CaATP)=0.015mM the competitive affinity order CaATP>MnATP>ATP^{4-}>MgATP.

Table 4, indicates that at lower concentration of excess ATP^{4-} the greater Ka of the nucleotide for either Mn^{2+} or Mg^{2+} than for Ca^{2+} could maintain the enzyme active. However, at higher concentration of excess ATP^{4-} the greater affinity (Ki=0.01) of the enzyme for CaATP, could predominate reducing the availability of cAMP for sustainable memory persistence. Hence, eradication of particular memory, confer to the system plasticity to re-conform synaptic connectivity for acquisition of a new memory or reinterpretation of the preceding one.

However, Mg^{2+} can easily displace Ca^{2+} from ATP and reduce the concentration of CaATP, because the association constant: Ka(MgATP)=20×10^3 M^{-1} is greater than Ka (CaATP)=9×10^3 M^{-1} [37] [38].

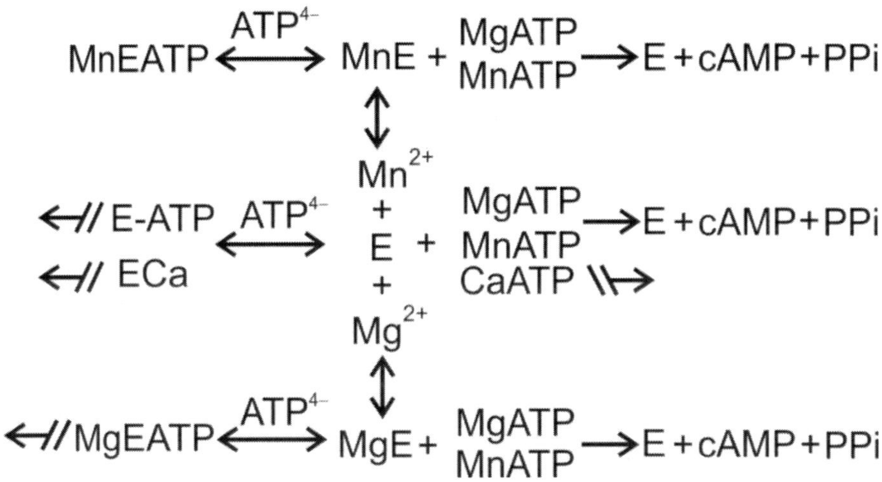

Scheme 1. Equilibriums for the RARE BiBi kinetic mechanism of the basal adenylyl cyclase. Study of initial velocity patterns gave evidence, that at the active site the substrates MgATP and MnATP by group transfer exchange PPi between ATP^{4-} and a pyrophosphate (PPi) accepting group, releasing cAMP and PPi. Mn^{2+} substitutes for Mg^{2+} for substrate formation and as an enzyme modulator. The catalytic rates and dissociation constants for Mn^{2+} and MnATP differ from Mg^{2+} and MgATP (Table 4), with active site binding by inhibitor CaATP and ATP^{4-} resulting in dead-end inhibition (\\↦).

The mutual competitive inhibition of ATP^{4-} and CaATP on the substrates, MgATP and MnATP, and a constant cooperative value of about one indicate that the ionic and divalent metal complexes of ATP, bind to the same site on basal adenylyl cyclase.

A requirement of cAMP leading to increment neuronal connectivity implies that the level of AC activity may play a partial role of neuronal plasticity [40] [41]. This process results from a perdurable lack of cAMP to drive the creation and/or reinforcement required to maintain synaptic connectivity. The requirement to control cAMP levels indicates the applicability of AC kinetics, for modeling the versatility of AC responsiveness. Enzyme plasticity is implied in the scheme 1, showing a random BiBi sequential group transfer mechanism (RARE BiBi) with dead-end inhibition, since the binding at the active site of CaATP or ATP^{4-} results in dead-end inhibitions.

In the intracellular fluid at 1mM $[Mg^{2+}]$ is considerable higher than 1μM Ca^{2+} [42]. The ratio is practically constant, except for certain pathological states [43]. However, the equilibrium dynamics of divalent metal existing as Mg^{2+} and MgATP, as well as Mn^{2+} and MnATP fluctuate. The latter, due to accumulation of endogenous metabolite intermediates like ATP^{4-}, amino acids, dicarboxylic acids, and organic phosphates, which can complex Mg^{2+} or Mn^{2+}. The cytoplasmic concentration of Ca^{2+} is in a dynamic state by the same metabolic variables. Additionally, 1mM extracellular and 10mM mitochondrial concentration of Ca^{2+} allow import and export of Ca^{2+} across structural compartmentalization [44] [45].

A strong restriction to the adenylyl cyclase activation, became evident from $S_{0.5}(ATP^{4-})$ of about 0.1mM (table 4). Hence, any decrease in the rate of ATP^{4-} consumption by the Na^+ gate activity and others functions, should turn off the adenylyl cyclase at both the neurotransmitter and basal levels. It is suggested that a matching the rate of ATP production and consumption should be optimal because the brain lacks storage capability. If the rate of ATP production overcomes consumption and endogenous accumulation occurs, ATP^{4-} may compete for the divalent metal structurally present for binding receptors and enzyme to the membrane [34]. This requirement may have important implications, since neurons may need to adjust endogenous aerobic glycolysis to prevent ATP^{4-} accumulation, overcoming the stability of receptors in the membrane.

Characterization of the Dynamics of the Quaternary Structure of Hemoglobin as a Function of a Sensor Role of Glucose Levels in Blood by the Erythrocyte

The insulin-regulated transporter GLUT4 is found in adipose tissue and striated muscle. GLUT1 is present in erythrocytes and endothelial cells of barrier tissues like the Blood brain barrier (BBB), is also expressed in liver cells which require bidirectionality to uptake glucose and by having glucose 6-phosphatase to release it from glucose-6-phosphate (G6P). GLUT3 is expressed mostly in neurons. Hence, brain and liver use transporters which do not require insulin for efficient uptake of glucose.

Cerebrum cells, lack of storage pathways are nutritionally dependent of a continuous uptake of glucose and oxygenation, to support aerobic glycolysis. Under starvation ketone bodies can only partially substitute for glucose. Hence, the need for signal transduction between tissues involved, required for matching the rates of O_2 and glucose consumption. This capability allows cerebral areas to increase by 10 time's the rate of aerobic glycolysis, without overcoming the homeostatic controls of glucose level in blood.

Heterotropic allosteric modulation was shown for: 2,3-diphosphoglycerate (2,3-DPG) [46]. 2,3-DPG cross-links subunits β_1 with β_2 of deoxyHb, at the tissues low pH a central cavity of the Hb tetramer became structured as an binding site for 2,3-DPG [47]. The latter, became surrounded by the terminal amino acid, Lys 82, and the protonated imidazole rings of histidines His 143 and His 2, in the β_1 and β_2 chains, which bear two NH bonds and show overall positive charge.

The 2,3-DPG-Hb complex of much lower affinity for O_2 than isolated Hb could be regarded as the physiological T structure [48] [49]. Upon deoxygenation increase the concentration of free Mg^{2+} ion [50] showing that deoxyHb and Mg^{2+} were competitors binding reciprocally with organic phosphates [50]. At the pH of lungs, the protons dissociate from R-histidines and the imidazole ring changes from overall positive to manifest the reactivity of negatively charged nitrogen atoms. Thus, now negatively charged groups like His 143 [51] repels the similarly charged 2,3-DPG.

The R-groups participants in H^+ release or Bohr effect have been widly reported and allows to molecular characterization of proton disociated Hb as the R configuration within oxyHb. Interconvertion of deoxyHb pKa=8±0.2 to oxyHb pKa=6.4±0.2 releases Bohr protons from R-groups, within the two β chains: from Cysβ_1 and $\beta_2$93, plus that from Aspβ_1 and $\beta_2$94 [52] [53] [54].

Treatment to obtain human Hb stripped-out of 2,3-DPG and other ligands increases Hb affinity for O_2 decreasing $P_{50}(O_2)$ 2.8mmHg [53]. The fig.6.a) illustrates a Hill plot showing that addition of 2,3-DPG [55] decreases Hb affinity for O_2 by increasing P_{50}, but without affecting Hb cooperativity for O_2.

A modified Woolf–Augustinsson–Hofstee (WAH) plot of %saturation on the ordinate versus %saturation/$(pO_2)^{2.3}$ on abscissa, fig.6.b) allowed to obtain an straight lines intersection at the theoretical 100%saturation on the ordinate.

The plot indicates that under the ideal condition of increasing pO_2 until reaching a theoretical maximum O_2-saturation, the presence of a constant concentration of 2,3-DPG may not prevent the homotropic Hb response of forming oxyHb.

Hence, a mutual exclusion between pO_2 and 2,3-DPG could lead to predict, that a theoretical increase in pO_2, could reduce an inhibitory effect of a constant 2,3-DPG concentration [55].

However, under physiological conditions Hb binds into a specific bind 2,3-DPG and forms a complex in which the Hb quaternary structure, became stabilized as the deoxyHb form by the release of O_2.

Lineweaver-Burk a double reciprocal plot adapted to subtract cooperativity, by introducing the cooperativity n-value for each curve in the abscissa $1/(pO_2)^n$. The plot becomes 1/%Oxygenation versus $1/(pO_2)^n$ and allowed to obtain intersection of straight lines for a theoretical maximal $1/(pO_2)^{1.5}$ and $1/(pO_2)^{2.3}$. Hence, at a theoretical maximal pO_2 ($1/pO_2=1/\infty\rightarrow 0$) saturation of Hb by O_2 would be reached even in the absence of Zn^{2+}. Hence, the activatory effect was to decrease the control value $S_{0.5}$=1.64mmHg by addition of Zn^{2+} to $S_{0.5}$=0.5mmHg [53] [55]. The strip method to free of ligands the Hb molecule appears to reduce steric hindrance. Even so Zn^{2+} becomes a strong enhancer of striped Hb affinity for O_2. The effects of 2,3-DPG decreasing affinity (fig.6) and Zn^{2+} increasing affinity (fig.7) of Hb for O_2, suggest by allosteric considerations alone, that a single site for both actions, should be discarded. However, even if solving by classical competitive interaction could not be applied, other treatments were tested. The nullification of 2,3-DPG inhibitory (fig.6.b) and of the activatory Zn^{2+} actions (fig.7.b), became evident when the plottings allow to assume a tendency for pO_2 to reach the equivalent of infinite concentration. The figures 6 and 7 illustrate that because these ligands were at an invariant concentration, could be displaced from Hb by a competitive mechanism involving pO_2 increment.

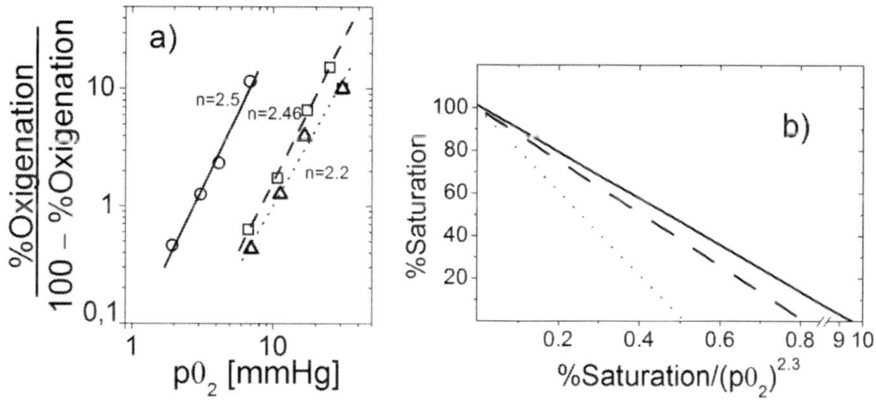

Figure 6. a) b). Graphic of the 2,3-DPG effect on the %-Oxygenation of stripped Hb. Adapted data [53] for the control: absence of 2,3-DPG (—); addition: $+4\times10^{-1}$mM of 2,3-DPG (----) and $+6\times10^{-1}$mM of 2,3-DPG (˙˙˙) were plotted according to: a) to Hill equation and b) a modified Woolf-Augustinsson-Hofstee plot in where the abscissa shows pO_2 at the exponential 2.3, value obtained from the Hill Plot. Introduction of the exponential is a solution to include cooperativity in order to transform curves into straight lines. Control values: (—) P50=2.8mmHg, cooperativity n=2.5; in presence of $+4\times10^{-1}$mM of 2,3-DPG; (----) values: P50=8mmHg, cooperativity n=2.46 and $+6\times10^{-1}$mM of 2,3-DPG (˙˙˙) P50=10mmHg cooperativity n=2.2.

Therefore, competition between O_2 with inhibitor 2,3-DPG (fig.6.b) do not ocurr at the same site but may result from this heterotropic ligand tendency to bind and stabilize one or more R-groups restricting O_2-saturation of the Heme site of Hb. Accordingly, increasing pO_2 would eventurally release the hindrance preventing O_2-binding to the Heme. The activatory effect by Zn^{2+} (fig.7.b) implicates that the effect of divalent metal is to decrease the concentration of O_2 required to remove the hindrance effect. Hence, a model (fig.8 and 9) was develop that shows that O_2 *per se* has full homotropic capability for the interconversion of deoxy- to oxyHb. Hence, competition for a modulatory site, which control hindrance, could explain Hb heterotropic reactivity with ligands decreasing (fig.6.b) or increasing (fig.7.b) affinity for O_2. These dynamics imply that the modulatory sites could not be fully covalently structured, but are rather structures supported by coordinative bonding. The affinity of Hb for the ligands Mg^{2+} and/or Zn^{2+} supports the tendency to conform two coordinative structures of Hb, which fit the theoretical and physiological findings.

The plasticity of this divalent metal or sugar phosphates modulated sites becames evident in term of a glucose sensor role for the Hb-erythrocyte system. This one, adding the physiological regulation by the endogenous metabolic state of the red cell, expressed by endogenous 2,3-DPG concentration, to the physiological ΔpO_2 between lung and CSF. Hence, when Hb becomes characterized in the sole presence of O_2 shows the quaternary changes in the structural links between T- to R- forms, which characterize homotropic interconversion. However, the Hb potential to acquire metal chelating sites allows the molecular quaternary structures T- and R-, to participate in 2,3-DPG-Hb and divalent metal-Hb complexes stabilizing changes in the O_2 affinity. These ones induced on Hb by endogenous heterotropic ligands within the red cells, allowing the addaptative response to the O_2 needs of tissues and fullfiling in brain the matching of the rates of glucose and O_2 uptake.

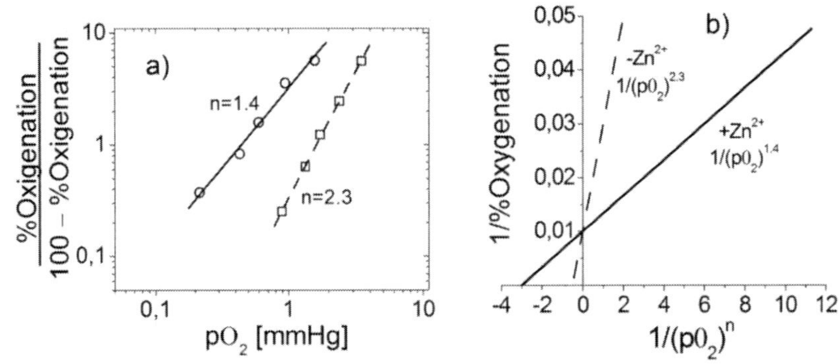

Figure 7. Graphic of the effect of Zn^{2+} on the %-Oxygenation of stripped Hb. Adapted data [53] for the control: absence of Zn^{2+} (----), values: P50=1.6mmHg; and addition of 4.47×10^{-2}mM Zn^{2+} (——), values: P50=0.5mmHg was plotted according to a) Hill equation and b) a modified doble reciprocal Lineweaver-Burk plot in were the abscissa shows pO_2 at the exponentials values obtained for the individual curves plotted according to Hill. Thus, the $-Zn^{2+}$, $1/(pO_2)^{2.3}$ and $+Zn^{2+}1/(pO_2)^{1.4}$.

The homotropic tendency of isolated Hb shows that during oxygenation a link between Hisβ146 and Aspβ94 breaks up and the imidazole ring at above pH 6 takes up a proton (alkaline Bohr Effect), which may be able to trigger the changes from T to R as a sole function of O_2 through a ΔG=3Kcal [56] [57].

The erythrocyte concentrations: 65µg/g for Mg^{2+} and 11µg/g for Zn^{2+}, lead to physiological significant forms of the coordinative complexes: $(Mg^{2+})_2$Hb, $(Zn^{2+})_2$Hb and (Mg^{2+}or Zn^{2+}) Hb stabilizing complexes of higher affinity for O_2 than the R-form of Hb [53] [55] [58] [59].

In fig.6 b) and fig.7 b), were reported plots [55] [59], both directed to use the cooperativity value n as exponential of the pressure of oxygen: $(pO_2)^n$. This treatment allows obtaining straight curves. These predicts that the extrapolation to a maximal $(pO_2)^n$, renders an homotropic dependence on the transition between T and R forms of Hb. However, it could be predicted that the physiological ligand-Hb equilibriums, within the red cell, would lead to an heterotropic mutual inclusion of O_2 and divalent metals Zn^{2+} or Mg^{2+}, with a mutual exclusion of 2,3-DPG and H^+.

The exergonic association per atom of Zn^{2+} to Hb: ΔG=–9.7Kcal [53], the tetramer has two binding sites, for either Zn^{2+} or Mg^{2+}. In the direction of mutual inclusion of divalent metal and oxygen the exergonic binding of Zn^{2+} of 19.4Kcal plus pH dependent neutralization of Bohr protons, exceeds the endergonic dissociation of 2,3-DPG from Hb of only: ΔG=5.7Kcal. Hence, in the lungs this thermodynamic coupling integrates pO_2 with other contributions for a decrease in free energy (ΔG<0).

At lower pH than 7, the increasing stability order: $(Mg^{2+})_2$-oxyHb < (Zn^{2+},Mg^{2+})-oxyHb < $(Zn^{2+})_2$-oxyHb, indicates pH-differentiable complexes with O_2 dissociation all along the body pH-axis. Hence, the multiplicity of complexes allows that deoxygenation could occur over an extended pH-range, than could be possible for any particular divalent metal complex of Hb. The pH of blood remains 7.4 but surroundings tissues allow increasingly lower pH from the trunk to the limbs. Hence, deoxygenation could follow a pattern of gradual release of O_2, first from the Mg^{2+}-Hb and thereafter from the Zn^{2+}-Hb species. However, the augment of glucose level in blood, in untreated diabetes, induces in the erythrocyte an increase of sugar phosphates. These ones by releasing the divalent metal from liganded state of Hb decrease the molecule affinity for O_2. Hence, a premature release of O_2, allows O_2-depleted Hb, to reach the body limbs and subsequence tissue anoxia results in the ulcerations characteristics of untreated diabetes.

In the red cell, at the dimerization interphases $\beta_2\alpha_1$ and $\beta_1\alpha_2$ of Hb, pO_2 is mutually inclusive with the heterotropic ligands Mg^{2+} and Zn^{2+}, and could changes the ions coordinative state from "bi-" to "tetra", and "hexa"-dentate. The more stable geometric configuration could be an octahedron, with the β_2 R-groups: His 143, Cys 93, Asp 94, His146 and the α R-group His 87, at the α_1-Heme. A symmetric structuring of the R-groups at the interphase $\beta_1\alpha_2$ will result in the formation of an oxyHb$(Me^{2+})_2$ complex, in which the quaternary form of Hb would show a lower tendency for O_2 dissociation.

Fig.8 illustrates the dynamics of divalent metals association with exclusion of 2,3-DPG in Hb and viceversa. As scheme 2 illustrates low pH inhibits 2,3-DPG mutase, allowing anaerobic glycolysis without generation of 2,3-DPG. The 2,3-DPG-depleated erythrocyte, arrieves at the lung level. These events facilitate up to an 180° turn-around of the R-groups of Hb over their α-carbon. This effect adds to the directional position of histidines another 0.5nm length, to reach the 1nm coordinative distance to the metal binding site. The chelating structure becames conformationaly stabilized by the forming of the divalent metal-Hb complex.

Simultaneously, at the lung pH=7.4 neutralization of H^+ released from the the histidines of Hb, induce a same charge rejection, favoring the release of 2,3-DPG from its state as ligand of Hb and increasing O_2-affinity allowing that the higher pO_2 by decreasing free energy favors the direction of oxygenation versus that of deoxygenation. At tissue level the change in free energy became associate to deoxygenation by the increase in the erythrocyte 2,3-DPG concentration (scheme 2) and the much lower pO_2 (fig.8).

At tissue low pH, the tendency of divalent metals to became surrounded by hydration spheres, became an exergonic contribution to their release as ligand from Hb. Effect, facilitating that β His 143 and other histidines H^+-release, to became positivelly charged and by rotating along their α-carbon, to participate in the conformation of a binding site for 2,3-DPG (fig.8). Along the blood path to the tissues, progressive accumulation of 2,3-DPG (scheme 2), would provide a constant decrease in the ratio of oxy- to deoxyHb.

Fig.8 shows that α-carbon rotation allows to couple the endergonic dissociation of 2,3-DPG with the exergonic association of Mg^{2+} or Zn^{2+}. Oxygenation in the red cell promote Mg^{2+} coordination to Cys-$\beta_2$93 inducing fit of His-$\beta_2$92, and facilitate the movement of the F-helix toward the H-helix [60] [58] [59] [61].

Thermodynamics considerations indicate that modeling of Hb has to balance oxygenation vs. deoxygenation for an energy efficiency coupling between the endergonic dissociation of 2,3-DPG from deoxyHb: ΔG=5.7KCal [53] and the binding of Zn^{2+} with a $\Delta G^{o}=-80$KJ per 2 Zn^{2+} atoms [62]. The solvation or hydration interaction between a solute like Hb and the solvent media of the cytoplasm, allows an Hb hydration sphere. The tendency of a divalent metal to acquire a hydration sphere could facilitate Me^{2+} and O_2 dissociation from oxy to deoxy.

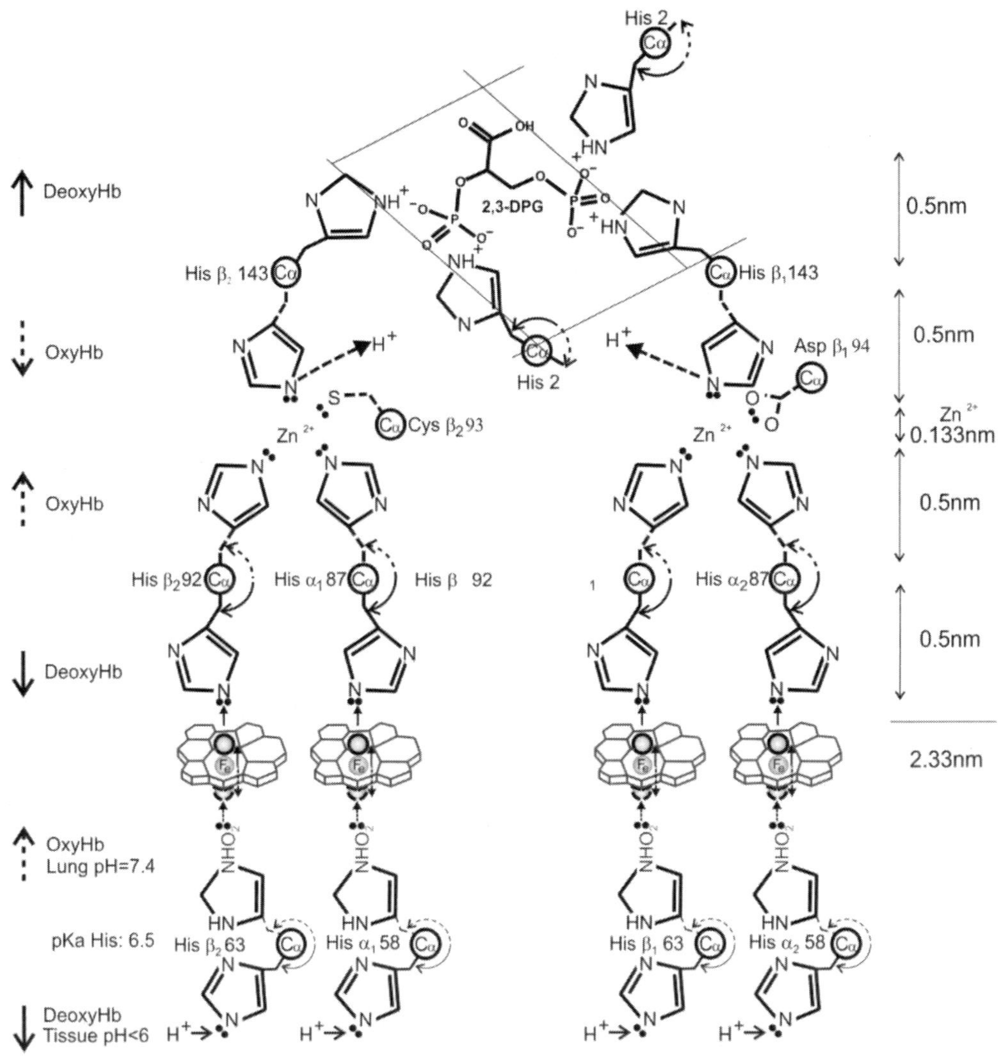

Figure 8. Model of the R-groups dynamics of Hb for mutually inclusive interactions of divalent metals and pO_2 allowing simultaneous exclusion of 2,3-DPG and H^+. According to saturation ranges of free (dissociated) divalent metals (Zn^{2+} or Mg^{2+}) could induce a coordinative-fit of ligand R-groups of Hb conforming two chelating sites. Hence, at the subunit interface β_2, α_1 and β_1, α_2 by displacement of His $\beta_2$92, His $\alpha_1$87, His $\beta_1$92 and His $\alpha_2$87 the Fe^{2+} atoms move from their position at the heme planes in the deoxyHb state (○) to allow their reactive position for oxygenation (○) and form with the 4 distal histidyl residues, $\alpha_1$58His, $\alpha_2$58His, $\beta_1$63His and $\beta_2$63His, to form H-bonds with the O_2 ligands. The oxygenation pathway (shown in broken lines: - - - -) predicts a physiological cooperative number n=2.8 for an oxygen saturation curve of Hb. This cooperativity number sums up the increase in affinity for O_2 and divalent metals Me^{2+} to a decrease in affinity for 2,3 DPG and H^+. Deoxygenation pathway (shown in solid lines: ——) the R-groups shift-out of the chelation complex and His $\beta_2$143 and His $\beta_1$143 to capture protons and the resultant imino groups attract 2,3-DPG.

The model graphics patterns a solution involving the symmetric turnarounds of both His β 92 and both His α 87, for dynamic transfer of the structural changes expanding the 2nm required for Heme to Heme interaction. Hence, fig.8 illustrates a model of mutual inclusion of divalent metals and O_2, which could lead from mono to the tetra oxygenated molecule: HbO_2

→ $Hb(O_2)_2$→ $Hb(O_2)_3$→$Hb(O_2)_4$. Each state of oxygenation, coupled by the concerted interaction between the two divalent metals ligand sites, in which the coordinatively sharing of His β_2 143 and His β_1 143 are being substracted step by step. This effect jointly with the displacement of the other ligands, that were conforming the binding site of single 2,3-DPG, allow for the gradual disappearance of an structure capable to bind 2,3-DPG. Hence, the binding of each O_2, decreasing the free energy requirements for the binding of the subsequent oxygens.

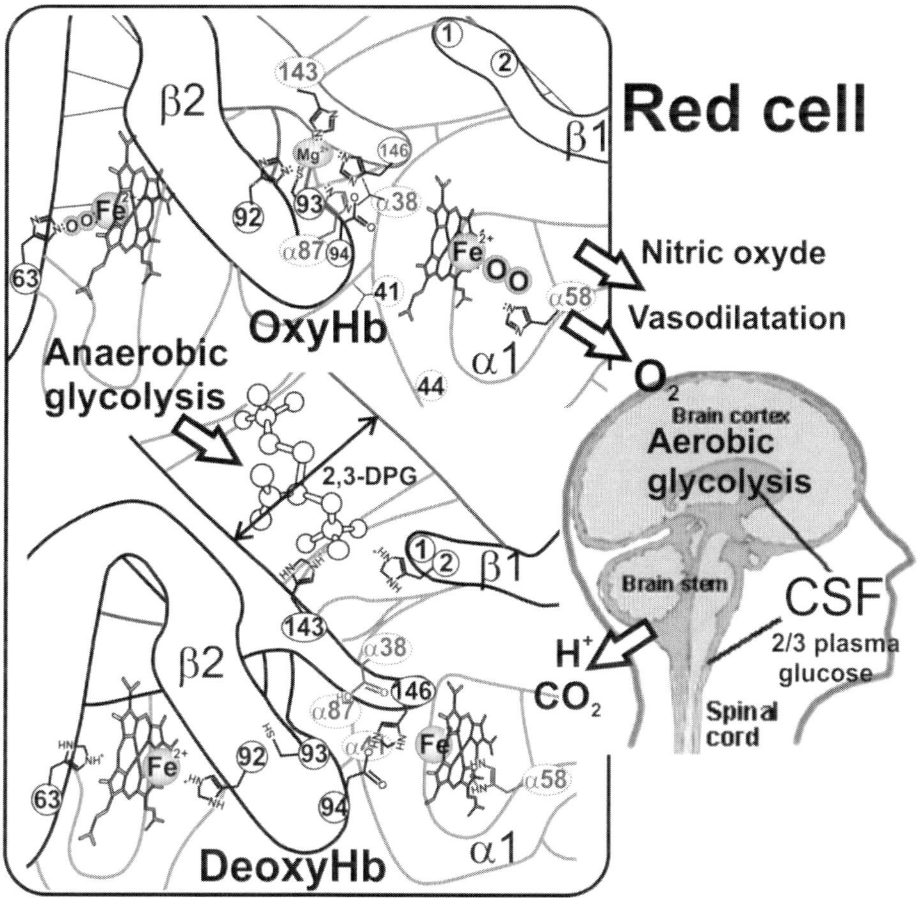

Figure 9. In the transition from oxyHb to deoxyHb, inclusion of 2,3-DPG and H^+ couples with mutual deoxygenation and dissociation of Mg^{2+} or Zn^{2+}. The illustration has been scaled for Hb projected with a diameter of 5.5nm [61]. Freedom for α-carbon rotation of the side chains implicated for their participation on the coupling model described chelated structures, were tested with the generous assistance of Fermi, G. and Milstein, C. in the large tri-dimensional model of Hb, located in the laboratory of M.J. Perutz, Cambridge, England. The figures show only the $\beta_2\alpha_1$ interface, since structural symmetry allows an equivalent set of changes at the $\beta_1\alpha_2$ interface. Peptide bonds are rigid and fixed in a plane in where two α-carbons, 3.6Å apart, rotate 180o over angles ϕ (fi) and ψ (psi). The longitudinal lengths from α-carbon to their coordinating negative atoms: His=0.5nm, Cys=0.23nm and Asp=0.39nm; Magnesium Radius: atomic=0.150 nm and ionic (2+ ion)=0.086 nm; Zinc radius: atomic=0.135 nm and ionic (2+ ion)=0.088 nm; Iron radius: atomic=0.140 nm and ionic (2+ ion)=0.077 nm.

The Role of BBB in Modulation of Red Cell Control of Hb Deoxygenation

Figure 9 illustrates brain as "bathing" in the cerebrospinal fluid (CSF) within the subarachnoid space and the ventricular system, around and inside the brain and spinal cord. The blood-brain barrier (BBB), show capillaries without fenestrations or pinocytotic (transportation) vesicles, and have tight junctions that almost fuse adjacent cells.

Efficient mechanisms guard the constancy of the CSF pH 7.30–7.36 units, even in non-respiratory and, to some extent, in respiratory acid-base disturbances. The pH of blood is maintained around 7.4, and buffering is required to maintain homeostasis. The blood contains Na^+, K^+, Ca^{2+} and Mg^{2+} cations, and Cl^-, PO_4^{3-} and SO_4^{2-} anions, which play a role in buffering. In surrounding peripheral tissues pCO_2 is about 50mm Hg, which allows diffusion into plasma and erythrocytes at about 40 mm Hg. Into the erythrocyte, the H^+ dissociated from HCO_3^- ion is buffered by hemoglobin. The production of H^+ within erythrocytes and their subsequent buffering by hemoglobin, results in a reduced affinity of hemoglobin for oxygen. This leads to a release of O_2 to the peripheral tissues, a phenomenon is termed the Bohr Effect, which is potentiated by the change in the endogenous 2,3-DPG concentration in the erythrocyte (scheme 2).

The presence of chlorides ions: 120-130 mEq/l may not only play an isotonic role, because across the CSF barrier may help dissociation of the divalent metals from oxyHb and facilitate deoxygenation. Cerebral blood flow may be constricted at high CSF pressure. Disorders of CSF flow may also affect the intracranial blood flow, with subsequent neuronal and glial vulnerabilities.

CSF contains < 10-12 U/ml of adenosine deaminase that breaks down adenosine into Inosine and ammonia. Adenosine, has inhibitory effects over neuronal activity, sedates and increments the non-rapid eye movement sleep (NREM) in stage IV. Caffeine blocking the adenosine binding to the receptor decreases sleeping tendency.

A Mechanism for Hb-Deoxygenation at Neutral pH

By incubating cells in media containing acetate, pyruvate, etc., the total concentrations of acid-soluble organic phosphate accumulation fractions depends on the extracellular phosphate level [63] [64] [65] [66]. In the human erythrocyte incubation with pyruvate and inosine, the total concentration of acid-soluble organic phosphates in the red cells rises from 14 to 55 μmol/g red cells. The organic phosphate other than 2,3-DPG can be elevated in vitro to 24 μmol/g [66].

The CSF pH=7.4 prevents pH-changes at the brain level, but the red cell heightens the concentration of the sugar phosphate intermediates of anaerobic glycolysis. The concentration of 2,3-DPG became maximal at pH=7.4 because 2,3-DPG mutase becames inhibited by low pH (scheme 2) allowing the shift the from oxy- to deoxy-Hb.

The energy necessities of the red cell, devoid of mitochondria, indicate the important role of anaerobic glycolysis. The regulation of oxygen-hemoglobin erythrocyte-system have requirement for both NADPH produced by the hexose monophosphate pathway (HMP) and 2,3-DPG the most concentrated organophosphate which varies inversely with pH [67].

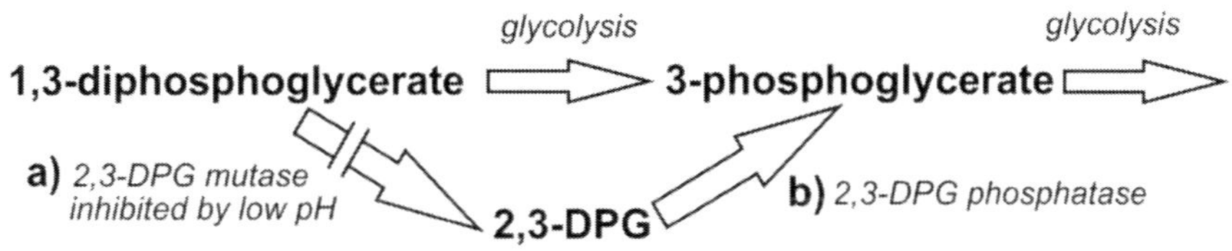

Scheme 2. Rapapport-Luebering scheme. a) diphosphoglycerate mutase transfers from a carbon one (C1) phosphoryl group of 1,3-DPG to C2 of 2,3-DPG that by action of b) diphosphoglycerate phosphatase forms 3-PG. In the erythrocyte pH is an inhibitor of 2,3-diphosphoglyceromutase. The insulin-independent glucose uptake of erythrocyte allows its endogenous sugar concentrations to be directly related to that in blood. Anaerobic glycolysis within the erythrocyte could generate 2,3-DPG, to reach is highest endogenous concentration, because the pH inside the blood capillaries surrounding the CSF, would be close to the latter homeostatic-pH 7.4. The lack of a pH differential across the BBB indicates that the displacement of the equilibrium from the oxy to deoxyHb for the continuous release of O_2 into CSF, becomes only dependent in the increment in 2,3-DPG. The exergonic binding of 2,3-DPG of $\Delta G = -5.7$Kcal favoring the release of O_2 to match oxygenation for a sustentable aerobic glycolysis of the activated brain areas.

cAMP [68] and cGMP [69] are transported into human erythrocyte (Human-RBC) against significantly large differentials of the intracellular-extracellular concentrations [70] [71]. 0.1mM Dibutyryl-cAMP and dibutyryl-cGMP, compounds were found to diffuse freely into the erythrocyte. However, even if cAMP and cGMP catabolism did occur, were not able to elevate endogenous lactate and 2,3-DPG. Absence a guanylyl or adenylyl cyclase activity in the erythrocyte or the presence of cyclic nucleotides signaling was shown [72]. However, at 0.1μM dibutyryl-cGMP inhibits lactate production, which could be a mechanism contributing to pH homeostasis at the BBB.

Control of metabolic shifts on the erythrocyte appears to rest mainly in the interconnection of glycolytic pathways with the catabolism of cyclic nucleotides. The HMP-pathway production of CO_2 is decreased by cAMP and cGMP. Effect prevented by theophylline-dependent inhibition of phosphodiesterase, the enzyme producing AMP and GMP. Inhibition of glycolysis occurs as a feedback by the mass action of ribose or ribose phosphate, as a product buildup from the oxidative branch of the HMP pathway. Hence, increase in the non-oxidative branch might increase 2,3-DPG levels (scheme 2). Inosine, could also provide relatively small increases in the pathway of 2,3-DPG [73] [74] [75] [76].

Human-RBC, hemolysates show an increment of $^{14}CO_2$ released from the consumption of [1-^{14}C]-glucose, within the HMP pathway by the Mn^{2+} stimulation of the redox recycling of NADPH, which contribute to pH of homeostasis because otherwise could favor lactic acid formation [77].

Structure and Function of Neuronal Regions

The structure of neuronal regions [78] allows functional differentiation of the hippocampus [79] and the striatum, which have distinct roles, but functional magnetic resonance imaging (FMRI) have revealed that successful memory require greater activity of both, for their jointly support of episodic memory formation [80] [81] [82] [83].

Long-term potentiation (LTP) and long-term depression (LTD) of synaptic transmission, are mechanisms which vary according to the synapses and circuits involved [80] [84]. These mechanisms can contribute to experience-dependent modifications of brain function [85] [34] [86] [87] [88].

Fig.10 illustrates the relationship of noradrenaline activated AC catalyzing cAMP formation and its relationship to other second messenger-dependent systems [11] [8] [9] [86] [87] [88]. The level of cAMP is controlled by phosphodiesterase which breaks down cAMP into AMP. The entry of calcium into the neuronal cell inhibits AC and allows completing the cycle of a neuronal transmission pulse.

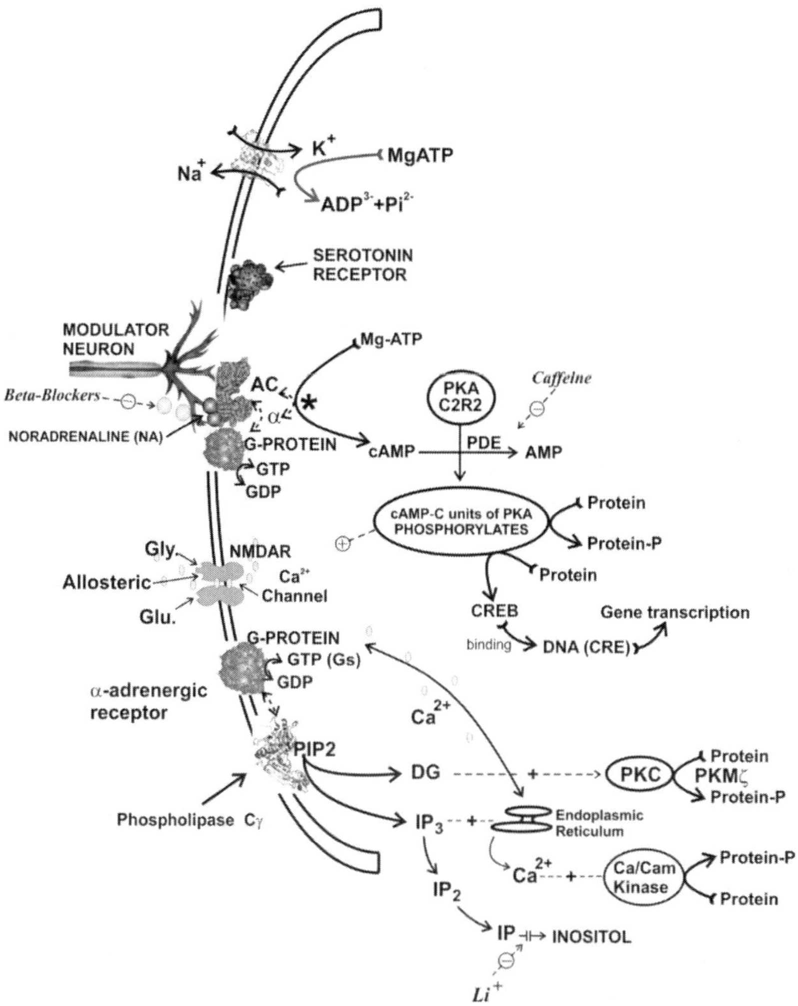

Figure 10. Integrative ionic-metabolic signaling to second messenger-dependent systems. G protein-coupled receptor (GPCRs) exchange its binding to GDP inactive for GTP active (GDP/GTP-exchange) for noradrenaline (NA) receptor coupling to AC. The GPCRs α, β, γ subunits dissociate into separate α and βγ activated complexes [*]. Once the α-subunit hydrolyzes bound GTP to GDP, it re-associates with the βγ complex for ending receptor activation. The system has a requirement for free Mg^{2+}, for AC conversion of MgATP or MnATP into endogenous cAMP. The latter, could be converted by phosphodiesterase (PDE) in AMP or could activate protein kinase-A (PKA, with subunits C2R2), by dissociating regulatory (R) from catalytic (C) units. cAMP-C acts by phosphorylating proteins and enzymes, changing these from inactive to active forms or vice versa. The WM relates to phospholipase C [17] and endogenous intracellular IP_3-mediated Ca^{2+} level [23] [24] [99].

The beta-1 adrenergic receptors interact with stimulatory Gs [86] [88] by coupling mediated through a cAMP-induced phosphorylation of protein kinase A (PKA) of the receptor [89] [90] [91] [92] [93] [94] [95]. The B2AR-channel complex by activating adenylyl cyclase forms cAMP, activating PKA to counter balancing the protein phosphatase 2 (PP2A). Sensitivity to subsequent stimulus could condition its answer, which may be functional for long-term memory formation [96] [97].

In the nervous system, based on their structural and functional characteristics, neurotransmitter receptors can be classified into two broad categories: i) Ionotropic receptors activated by opening a pore channel that allows ions flows such as Na^+, K^+, or Cl^- [98] [94] [99] ii) Metabotropic, subtype of membrane receptors at the surface or in cell vesicles, which are indirectly linked to activated ion-channels. These are associated to a second messenger signal transduction mechanisms. G protein-coupled receptors are encoded in the human genome in large numbers [96] has been detected about 350 of which 150 do not have known functions but the others has been related to hormones, growth factors and other endogenous ligands.

G-protein-coupled-receptor (GPCR) ligand binding induces a conformational modification into guanine (GEF). This activates an associated G-protein by exchanging its bound GDP for a GTP. The α-subunit of G-protein and bound GTP dissociate from the β and γ subunits. This allows interaction with endogenous signaling or target proteins directly depending on the α subunit type ($G_{\alpha s}$, $G_{\alpha i/o}$, $G_{\alpha q/11}$, $G_{\alpha 12/13}$) [88] [96].

The glutamate agonist N-methyl D-aspartate conforms a receptor system (NMDAR) multiple isoforms, to distinct distribution and functional properties. The ion channel in NMDAR is blocked by extracellular Mg^{2+}, the flow of Na^+- and Ca^{2+}-in and K^+-out of the neuron, is activated by voltage and gating by ligands glutamate (Glu.) and glycine (Gly.) [25] [100] [101] [102].

The Na^+/K^+ pump binds 3 intracellular Na^+ ions. ATP becomes hydrolyzed by phosphorylation of an aspartate residue and subsequent release of ADP. cAMP up-regulates the Na^+/K^+-ATPase. The process includes the ligands of the Gs- protein-coupled receptor (GPCRs). A decrease in cAMP and the effect of Gi-GPCRs down-regulates the enzyme [100] [101] [102] [103].

The phosphorylated form of the pump has a low affinity for Na ions a conformational change in the pump transfers the three Na^+ ions, out of the cell for every two extracellular K^+ ions. The dephosphorylated likeness of the pump, has a higher affinity for Na^+ ions than K^+ ions, so the two bound K^+ ions, are released endogenously. The electrogenic transmembrane Na^+/K^+-ATPase-pump-system consumes ATP^{4-} generating ADP^{3-} plus inorganic phosphate (Pi^{2-}). The reaction modifies ionic equilibrium, by releasing free Mg^{2+} which activates basal AC and responsiveness to l-noradreanline [9] [11]. Up-regulation of the ATPase by the cAMP formed are therefore synchronized with a nerve impulse. The depolarization phase by increasing Ca^{2+}, within the cytoplasmic side of the membrane, could form CaATP which could turn-off AC activity [11] [33].

The carrier, Na^+-Ca^{2+} exchanger (NCX) uses the Na^+ gradient generated by the Na^+-K^+ pump, to transfer out the endogenous Ca^{2+}. A slowing down of the Na^+-K^+ pump increases intracellular Ca^{2+}. The site level at which $[Ca^{2+}]$I regulates NCX activity is different from the one required for Ca^{2+} transport [103] [104]. Sub-micromolar endogenous concentrations 0.1–0.3 μM Ca^{2+} are needed to activate the anti-porter and removal of Ca^{2+} ions completely blocks NCX activity [103] [104] [105]. ATP participates in the NCX molecule phosphorylation process by protein kinase A (PKA) and protein kinase C (PKC) [106]. The phosphorylation of NCX increases affinity for both endogenous Ca^{2+} and exogenous Na^{+}, and decreases its inhibition by endogenous Na^{+}. Nitrogen oxide (NO) is able to stimulate NCX in the reverse mode of operation in neuronal preparations and astrocytes through a cGMP-dependent mechanism [107].

A decrease in the intracellular pH=0.4 can induce a 90% inhibition of NCX activity [107] [108], increasing intracellular Ca^{2+} and inhibiting AC. Acting as a phosphoryl donor molecule. ATP may increase the activity of the exchanger in a number of ways [109]. ATP cellular depletion inhibits NCX1 and NCX2 isoforms but does not affect NCX3 activity [110].

A dual-store memory model predicts that a short-term memory allows capturing events that can be recalled easily for about 30s. Thus, should differ structurally and functionally from long term memory. A single-store retrieved model predicts biophysics contexts, which do not differentiate between the initial events of short versus long-term memory.

Working memory (WM) behavior is a process involving retention of information for a short period of time. GPCRs also couple to G-protein activation of phospholipase C-γ (PLCγ). The activated form hydrolyzes membrane phospholipids into fatty acids and other lipophilic substances, there are four major classes A, B, C and D. PLCγ attack of substrate phosphatidylinositol 4, 5-bisphosphate (PIP2) increases the levels of inositol 1,4,5-triphosphate (IP3), the primary mechanism for Ca^{2+} release, and diacylglycerol (DAG) (fig.10).

Downstream signaling proteins are phosphorylated on serine and threonine by PKA and DAG-activated protein kinase C (PKC) leading to alterations in their activities. Membrane-associated and intracellular tyrosine kinases phosphorylate specific tyrosine residues on target enzymes and other regulatory proteins. Synaptic plasticity and neuroprotection involves DNA sequences with a CRE motif, which interact with a cAMP response element binding protein (CREB) to increase or decrease transcription. The binding activity of cAMP on PKA is rapidly suppressed and the activation of CREB-mediated transcription of genes rapidly diminishes [111].

The system constitutes a convergence point for many signaling cascades. These are related in structure and function to CREM and activating transcription (ATF-1), involved in long-term memory formation and neuronal plasticity [111] [112] [113] [114] [115]. Expressed CREB proteins include Brain-derived neurotrophic factor (BDNF), tyrosine, and many neuropeptides, like somatostatin, enkephalin, VGF and corticotropin [114] [115].

Dopamine regulates hormonal responses, directing the hypothalamus to manufacture hormones and store then in the pituitary gland [106] [107]. Dopamine activates in brain a per-durance phase, with a control role of the persistence process, remembrance required for long-term memory storage [70]. It also controls muscle movements, patients with Parkinson's disease have practically no dopamine. Several psychiatric disorders are related, example an excess of dopamine in the frontal lobules of the brain has been implicated in schizophrenia, and abnormalities on the cerebral dopaminergic centers of psychopaths. The noradrenaline-dependent acceleration of the destruction-inactivation rate of AC is tempered by glucose metabolite [116]. Induction of deficiency of adrenalin in blood by Bilateral Electro-coagulation of the Adrenal Medulla (BEAM), has been reported as a restorative treatment, effective for Disappearance of Parkinson Symptoms [116].

The relevance of the energy charge as the intracellular mechanism supporting memory formation has to be evaluated, within the more inclusive control by extracellular hormones and synaptic neurotransmitters.

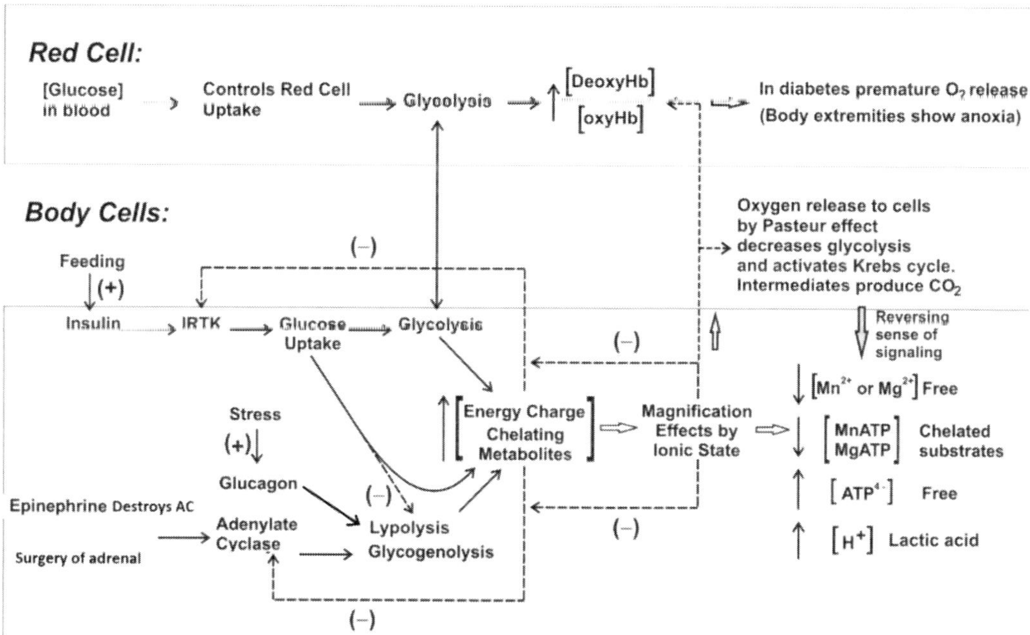

Figure 11. Integration of hormonal responsiveness of AC and IRTK to metabolic-ionic feedback. The metabolic relationships of fat cells releasing free fatty acids and liver, glucose and ketone bodies respectively, within a network idealized by an "in common" feedback controlling of hormone activity by the metabolite modulated availability of free Mg^{2+} or Mn^{2+}.

Free Mg^{2+} and Mn^{2+} metabolic feedbacks control an "in common" tissue network, supporting the bioenergetics of brain processes in fat cells [33] [36], cAMP protein kinase activation of hormone-sensitive triglyceride lipase releasing free fatty acid and in liver by controlling the kinetic response IRTK [58] [76]. MnATP stimulates the insulin response of IRTK phosphorylating [Val^5] angiotensin II, which potentiates the kinetic response of insulin-activated IRTK to MnATP [73] [117] [118] [119].

Conclusion

- *Long-term memory:* is involved into the affirmative process associated with emotional experiences. Is sensitive a beta-blocking of the noradrenalin of AC receptor of AC. Persistence of an activate state of AC may allow significant activation by cAMP of kinase-A, which by phosphorylation modifies the responses of enzymes and proteins within neuronal pathways.
- Additionally, free Mg^{2+} is required for binding of oxytocin to if receptor. AC dependent emotional intelligence, could also involves, oxytocin favoring attachments which involves a response to reward and the desire to obtain it.

- *Short-term memory:* the release of Ca^{2+} inhibits AC, allowing completion of a neuronal transmission impulse
- The pulse of endogenous generated cAMP controls Ca^{2+} fluxes.
- The nerve impulse releases Ca^{2+} for Ca^{2+}-actin formation regulating *Microtubule* Self-Assembly. ATP^{4-} binds Ca^{2+} releasing actin.
- cAMP activate calmodulin binding free Ca^{2+} and allows its return to the endoplasmic reticulum.
- cAMP is consumed by phosphodiesterase enzymes, sensitive to inhibition by caffeine, prolonging cAMP action.
- In short term memory a high uptake of glucose versus a low oxygenation may affect the turnover of the Na^+-pump dependent cyclic fluctuations of ATP^{4-} level.
- Noradrenaline dependent activation of Adenylate cyclase (AC) of brain is fully dependent on free Mg^{2+}, in excess of MgATP.
- In brain, the Na^+ pump consumes ATP^{4-} for the propagation of nerve impulses, releasing free Mg^{2+}. The event increases AC responsiveness because ATP^{4-} has a much higher binding constant for Mg^{2+}, than ADP^{3-} and AMP^{2-}.
- AC of brain, liver, and fat cells as well than insulin receptor tyrosine-kinase (IRTK) show 3 differentiable sites as a function of free $[Mg^{2+}]$.
- An active site for the formation of the enzymes substrate complex: MgATP.
- A basal site(s) activatory: for free $[Mg^{2+}]$ saturation and by decreasing saturation with $[ATP^{4-}]$.
- A free $[Mg^{2+}]$ requirement for the hormonal-receptor sites.
- The computation separable saturation curves, of the 3 sites, yields a joint saturation response with a cooperative value of about 2.6 resulting of the sum of activatory sites for MgATP and for Mg^{2+}, plus desaturation of one inhibitory site for $[ATP^{4-}]$.
- The affinity for O_2 of the Heme Prosthetic Group of Hb, is mutually inclusive for the binding of free Mg^{2+} or Zn^{2+}, at the R-groups of interphase $\beta_2\alpha_1$: His$\beta_2$143, His$\beta_2$92, Cys$\beta_2$93, His$\alpha_1$87 and for the respective R-groups in $\beta_1\alpha_2$. Moreover, these bindings are mutually exclusive of the H^+ and 2,3-DPG binding. Hence, the pO_2 saturation curve of Hb by inclusion of the activators Mg^{2+} plus O_2 binding and exclusion of inhibitors shows a cooperative value of about to 2.6.
- At the organismal level, the fight or flight reaction releases adrenalin in blood activating AC. Adrenalin in liver increases glucose level in blood and adrenalin plus other hormones in fat cells releases free fatty to produce in liver ketone bodies.
- An action or hormone dependent increase of glucose in blood results in an increase in sugar phosphates within the red cell decreasing free Mg^{2+} and Hb releasing O_2 that diffuses into the cerebrospinal fluid (CSF).
- The glia cells support minimal aerobic glycolysis. The activation effect on the membrane receptors of brain, by increasing free Mg^{2+} by the conception of MgATP by AC and Na^+ pump, requires a sudden increase of aerobic glycolysis to regenerate MgATP. The Mg^{2+}-dependent cooperative increase in neuronal activity, needs to be match by the promptness of the cooperative responsiveness of Hb deoxygenation, adjusting the supply of O_2 to that of G in CSF. This allows the sudden 10 times increment of the rate of aerobic glycolysis, at the activated neuronal circuits.

- Hence, a metabolic-ionic feedback for metabolic integration rests in a hemoglobin sensor role.

The aforementioned results were examined for reductionist modeling circumscribed to the participation of a core energy transduction molecular mechanism, within a generalized perspective of neuronal function. However, a wider more specific perspective of long-term memory depends of very significant studies interrelating molecular, physiological, learning, behavioral and pathological levels, briefly treated under discussion.

Not only noradrenalin, but numerous neurotransmitter are released into the extracellular space in the brain and their biological actions are mediated by specific receptors, which interconnect or link with the adenylyl cyclase (AC) 2nd messenger system.

The nerve impulse breakdown of MgATP by the Na^+/K^+ ATPase, releases ADP^{3-} and Pi^{2-} modifing ionic equilibrium, by releasing free Mg^{2+}. The divalent metal activates responsiveness to l-noradreanline stimulated and basal AC. During the depolarization phase of the action potential, synchronized changes in the concentrations of ATP^{4-} and Ca^{2+} within the cytoplasmic side of the membrane generate CaATP. Completing a neurotransmitter induced nerve impulse cycle, by a turn-off of the resposiveness of AC to neurotransmitter activation.

The generated cAMP up-regulates the Na^+/K^+ ATPase and induces protein kinase A (PKA) phosphorylation of the adrenergic receptor coupled with a stimulatory G-protein (Gs). A Pi^{2-} increment in Cerebral spinal fluid (CSF), allows to augment the uptake of Pi^{2-} and glucose by the anaerobic glycolysis incrementing sugar phosphates in red cells. At pH=7.4 maximal production of 2,3-DPG in the Hb-red cell system results in O_2-release, which in the neuron enhance turnover of the electron transport chain.

The metabolically, stimulated neurons up-regulate Krebs cycle, stepping-up mitochondrial concentrations of $citrate^{3-}$ and $dicarboxylic^{2-}$ acids. These ones chelate Ca^{2+} and allow its influx into mitochondria, limiting a Ca^{2+}-dependent inhibition of basal AC activity.

Hence, the findings interrelates the neuronal electrochemical impulse with neurotransmitter and basal activation of AC. Thus, at the required stages, a persistent flow of cAMP initiates long-term memory, involving activation of protein kinases and the cAMP response element binding proteins. Hence, the code and decode of the information flow, became sustainable by aerobic glycolysis at the activated brain areas.

Acknowledgments

The contributions of my PhD. Students: S. Brydon-Golz, H. Ohanian, R.H. Harris, V.A. De Bari, P.P. Vicario, R. Cruz and K. Borhanian, from special projects of the College of Medicine and Dentistry of New Jersey (CMDNJ). S. De Farias on leave from the Instituto Butantan Sao Paulo, Brazil and under graduate project N. Seidler, plus the assistances of E.M. De Pahn, N.A. Novak and N. Bennun, are all gratefully recognized.

References

[1] L. Cahill and J.L. McGaugh, "Mechanisms of emotional arousal and lasting declarative memory", *Trends Neurosci.*, 21, 294-9 (1998).

[2] M. Costa-Mattioli, N. Sonenberg, "Translational control of gene expression: a molecular switch for memory storage", *Prog. Brain Res.*, 169, 81-95 (2008).

[3] T. Sadeh, D. Shohamy, D.R. Levy, N. Reggev and A. Maril, "Cooperation between the hippocampus and the striatum during episodic encoding", *J. Cogn. Neurosci.*, 23 (7), 1597-608 (2010).

[4] E. Tulving and H.J. Markowitsch, "Episodic and declarative memory: role of the hippocampus", *Hippocampus*, 8 (3), 198-204 (1998).

[5] A. Martin and L.L. Chao, "Semantic memory and the brain: structures and processes", *Current Opinion in Neurobiology*, 11 (2), 194-201(2001).

[6] B. Szepietowska, W. Zhu, O. Chan, A. Horblitt, J. Dziura and R.S. Sherwin, "Modulation of β-Adrenergic Receptors in the Ventromedial Hypothalamus Influences Counterregulatory Responses to Hypoglycemia", *Diabetes.*, 60 (12), 3154-8 (2011).

[7] S.S. Kety, "The biogenic amines in the central nervous system: their possible roles in arousal, emotion and learning", *In the neurosciences: second study program*, 324-335 (1970).

[8] S. Brydon-Golz, H. Ohanian and A. Bennun, "Effects of noradrenaline on the activation and the stability of brain adenylate cyclase", *Biochem. J.*, 166, 473-483 (1977).

[9] A. Bennun, "Characterization of the norepinephrine-activation of adenylate cyclase suggests a role in memory affirmation pathways. Overexposure to epinephrine inactivates adenylate cyclase, a casual pathway for stress-pathologies", *BioSystems*, 100, 87-93(2010).

[10] H. Ohanian, K. Borhanian and A. Bennun, "The effect of manganese on the regulation of brain adenylate cyclase by magnesium and adenosine triphosphate", *Biochem Soc Trans.*, 6(6), 1179-82 (1978).

[11] H. Ohanian, K. Borhanian, S. De Farias and A. Bennun, "A model for the regulation of brain adenylate cyclase by ionic equilibria", *Journal of Bienergetics and biomembranes*, 13, Nos. 5/6 (1981).

[12] K. Tully, V.Y. Bolshakov, "Emotional enhancement of memory: how norepinephrine enables synaptic plasticity", *Mol Brain*, 13(3), 15 (2010).

[13] J.R. McReynolds, K. Donowho, A. Abdi, J.L. McGaugh, B. Roozendaal, C.K. McIntyre, "Memory-enhancing corticosterone treatment increases amygdala norepinephrine and Arc protein expression in hippocampal synaptic fractions", *Neurobiol Learn Mem.*, 93(3), 312-21 (2010).

[14] D. Mueller and S.P., "Cahill Noradrenergic modulation of extinction learning and exposure therapy", *Behav Brain Res.*, 208(1), 1-11 (2010).

[15] D.L. Korol and P.E. Gold, "Epinephrine converts long-term potentiation from transient to durable form in awake rats", *Hippocampus.*, 18(1), 81-91 (2008).

[16] H. Mitoma and S. Konishi, "Monoaminergic long-term facilitation of GABA-mediated inhibitory transmission at cerebellar synapses", *Neuroscience.*, 88(3), 871-83 (1999).

[17] J.F. López-Téllez, M.F. López-Aranda, I. Navarro-Lobato, M. Masmudi-Martín, E.M. Montañez, E.B. Calvo and Z.U. Khan. "Prefrontal inositol triphosphate is molecular

correlate of working memory in nonhuman primates", *J. Neurosci.*, 30(8), 3067-71 (2010).
[18] T.J. Gould and P.C. Bickford. "Age-related deficits in the cerebellar beta adrenergic signal transduction cascade in Fischer 344 rats", *J. Pharmacol Exp. Ther.*, 281(2), 965-71 (1997).
[19] P.J. Scarpace and I.B. Abrass, "Alpha- and beta-adrenergic receptor function in the brain during senescence", *Neurobiology of Aging*, 9, 53-58 (1988)
[20] P. Valverius, S. Borg, M.R. Valverius, P.L. Hoffman and B. Tabakoff. "Beta-adrenergic receptor binding in brain of alcoholics", *Experimental Neurology*, 105, Issue: 3, Pages: 280-286 (1989)
[21] G.N. Pandey. and J.M. Davis. "Treatment with antidepressants and down regulation of beta-adrenergic receptors", *Drug Development Research, 3, Issue 5*, 393-406 (1983).
[22] D. León, C.A. Castillo, M.A. Ruiz, J.L. Albasanz and M. Martín., "Metabotropic glutamate receptor/phospholipase C pathway is increased in rat brain at the end of pregnancy", *Neurochem Int.*, 50(5), 681-8 (2007)
[23] J.F. López-Téllez, M.F. López-Aranda, I. Navarro-Lobato, M. Masmudi-Martín, E. Martín Montañez, E. Blanco Calvo and Z.U. Khan. "Prefrontal Inositol Triphosphate Is Molecular Correlate of Working Memory in Nonhuman Primates", *The Journal of Neuroscience*, 30(8), 3067-3071 (2010).
[24] R. Bourtchuladze, B. Frenguelli, J. Blendy, D. Cioffi, G. Schutz and A.J. Silva. "Deficient long-term memory in mice with a targeted mutation of the cAMP-responsive element-binding protein", *Cell.,* 79(1), 59-68 (1994)
[25] Y. Liu and J. Zhang., "Recent development in NMDA receptors", *Chin. Med. J.*, 113 (10), 948-56 (2000).
[26] A. Bennun, "The unitary hypothesis on the coupling of energy transduction and its relevance to the modeling of mechanisms", *Annals of the New York Academy of Sciences*, 227, 116-145 (1974).
[27] S. Brydon-Golz and A. Bennun, "Postsynthetic stabilized modification of adenylate cyclase by metabolites", *Biochemical Society Transactions*, 3, 721-724 (1975).
[28] A. Bennun, "Hypothesis for coupling energy transduction with ATP synthesis or ATP hydrolysis", *Nature New Biology*, 233 (35), 5-8 (1971).
[29] Bennun, "Hypothesis on the role of liganded states of proteins in energy transducing systems", *BioSytems, 7*, 230-244 (1975).
[30] J. Glowinski and L.C. Iverson, *J. Neurochem.*, 13, 655-669 (1966).
[31] A.G. Gilman, *Proc. Natl. Acad. Sci. USA*, 67, 305-312 (1970).
[32] B.L. Brown, R.P. Ekis and J.D.M. Albano, "In advances in Cyclic Nucleotide", *Research*, 2, 25-40 (1972).
[33] R.H. Harris, R. Cruz, A. Bennun, "The effect of hormones on metal and metal-ATP interactions with fat cell adenylate cyclase", *BioSystems*, 11, 29-46 (1979).
[34] A. Bennun and E. Racker, "Partial resolution of the enzymes catalyzing photophospharylation IV. Interaction of coupling factor I from chloroplast with components of the chloroplast membrane", *J. Biol. Chem.*, 244, 1325-1331 (1969).
[35] R. Harris and A. Bennun, "Hormonal control of fat cells adenylate cyclase", *Mol. Cell. Biochem.*, 13, 141-146 (1976).
[36] G.L. Collingridge, S. Peineau, J.G. Howland and Y.T. Wang, "Long-term depression in the CNS", *Nat Rev Neurosci.*, 11(7), 459-73 (2010).

[37] M.M. Taqui-Khan, A.E. Martell, *J. Phys. Chem.*, 66, 10-15 (1962).
[38] M.M. Taqui-Khan, A.E. Martell, *J. Am. Chem. Soc.*, 88, 668-671 (1966).
[39] D.E. Atkinson, *Biochemistry*, 7, 4030-4034 (1968).
[40] C.H. Bailey, E.R. Kandel, "Synaptic remodeling, synaptic growth and the storage of long-term memory in Aplysia", *Prog Brain Res.*, 169, 179-98 (2008).
[41] G.F. Reis, M.B. Lee, A.S. Huang, K.D. Parfitt, "Adenylate cyclase-mediated forms of neuronal plasticity in hippocampal area CA1 are reduced with aging", *J. Neurophysiol.*, 93(6), 3381-9. (2005)
[42] D. Veloso, R.W. Guynn, M. Oskarsson and R.L. Veech, *J. Biol. Chem.*, 248, 4811-4819 (1973).
[43] M.S. Seeling, "In magnesium deficiency in the pathogenesis of disease", *Plenum Press*, New York (1980).
[44] H. Rasmussen, *Science*, 170, 404-412 (1970).
[45] H. Rasmussen, D.B.P. Goodman, *Ann. N.Y. Acad. Sci.*, 253, 789-796 (1975).
[46] A. Chanutin and R.R. Curnish, "Effect of organic and inorganic phosphates on the oxygen equilibrium of human erythrocytes", *Arch Biochem Biophys.*, 121(1), 96-102 (1967).
[47] Arnone, "X-ray diffraction study of binding of 2,3-diphosphoglycerate to human deoxyhaemoglobin", *Nature*, 237, 146-9 (1972).
[48] M.F. Perutz, "Stereochemistry of cooperative effects in heamoglobin", *Nature*, 228, 726-739 (1970).
[49] M.F. Perutz, A.J. Wilkinson, M. Paoli, G.G. Dodson, "The stereochemical mechanism of the cooperative effects in hemoglobin revisited", *Annu Rev Biophys Biomol Struct*; 27, 1-34 (1998).
[50] H.F. Bunn, B.J. Ransil, A. Chao, "The interaction between erythrocyte organic phosphates, magnesium ion, and hemoglobin", *J. Biol Chem.*, 246(17), 5273-9 (1971).
[51] R. Benesch and R.E. Benesch, "The effect of organic phosphates from the human erythrocyte on the allosteric properties of hemoglobin", *Biochem Biophys Res Commun*, 26, 162-174 (1967).
[52] E. Antonini and M. Brunori, *J. Biol. Chem.*, 244, 3909-3912 (1969).
[53] R. Benesch and R.E. Benesch, *Nature (London)*, 221, 618-622 (1969).
[54] R.E. Benesch, "The stability of the heme-globin linkage: measurement of heme exchange", Methods Enzymol., 231, 496-502 (1994).
[55] A. Bennun, N. Seidler and V.A. De Bari, "A model for the regulation of haemoglobin affinity for oxygen", *Biochemical Society Transactions*, 13, 364-366 (1985).
[56] P.C. Churchil and M.C. Churchil, *Arch. Int. Pharmacodyn. Ther.*, 258, 300-312 (1982).
[57] M.F. Perutz, M. Paoli, A.M. Lesk, "Fix L, a haemoglobin that acts as an oxygen sensor: signalling mechanism and structural basis of its homology with PAS domains", *Chem Biol.*, 6, 291-7 (1999).
[58] P.P. Vicario and A. Bennun, "Separate effects of Mg^{2+}, MgATP, and ATP^{4-} on the kinetic mechanism for insulin receptor tyrosine kinase", *Archives of Biochemistry and Biophysics*, 278, 99-105 (1990)
[59] A. Bennun, N. Seidler and V.A. De Bari, "Divalent Metals in the Regulation of Hemoglobin Affinity for Oxygen", *Ann. N.Y. Acad. Sci.*, 463, 76-79 (1986).
[60] G. Gerber, H. Berger, G.R. Janig and S.M. Rapoport, *Eur. J. Biochem.*, 38, 563-571 (1973).

[61] G. Fermi, M.F. Perutz, in: Atlas of molecular structures in Biology, 2 Haemoglobin & Myoglobin D.C., Phillips and Richards F.M. (Eds.) Clarendon Press, Oxford (1982).
[62] J.M. Rifkind, J.M. Heim, *Biochemistry*, 16, 4438-4443 (1977).
[63] A.O.M. Stoppani, A. Bennun and E.M. De Pahn, "Energy requirement for the anaerobic oxidation of acetate in baker's yeast", *Biochimica et Biophysica*, 92, 176-178 (1964).
[64] A. Bennun, E.M. De Pahn and A.O.M. Stoppani, "Some properties of particle-bound intracellular ATPase from baker's yeast", *Biochimica et Biophysica*, 89, 532 539 (1964).
[65] A.O.M. Stoppani, A. Bennun and E.M. De Pahn, "Effect of 2,4-dinitrophenol on krebs cycle and phosphate metabolism in baker's yeast", *Archives of Biochemistry and Biophysics*, 108(2), 258-265 (1964).
[66] Deuticke, J. Duhm and R. Dierkesmann, "Maximal elevation of 2,3-diphosphoglycerate concentrations in human erythrocytes: Influence on glycolytic metabolism and intracellular Ph", *Pflügers Archiv European Journal of Physiology*, 326 (1), 15-34 (1971).
[67] Brewer, G.J. General red cell metabolism. In: Surgenor D. ed. The red blood cell. New York: Academic Press, 387-433, (1974)
[68] E.L. Thomas, L.E. King, M. Morrison, "The uptake of cyclic AMP by human erythrocytes and its effect of membrane phosphorylation", *Archs. Biochem. Biophys.* 196, 459-464 (1979).
[69] V.A. DeBari and A. Bennun, "Cyclic GMP in the human erythrocyte. Intracellular levels and transport in normal subjects and chronic hemodialysis patients", *Clin Biochem.*, 15, 219-221 (1982).
[70] V.A. DeBari, N.A. Novak, A. Bennun, "Cyclic Nucleotide Metabolism in the Human Erythrocyte", *Clin. Physiol. Biochem.*, 2, 227-238 (1984).
[71] P. Novembre, J. Nicotra, V.A. DeBari, M.A. Needle and A. Bennun, "Erythrocyte transport of cyclic nucleotides", *Annals of the New York Academy of Sciences*, 435, 190-194 (1984).
[72] J. Marie, L. Tichonicky, J.C. Dreyfus, A. Kahn, "Endogenous, cyclic 3',5' AMP-dependent phosphorylation of Human red cell pyruvate kinase", *Bichem. Biophys. Res. Commun.*, 87, 862-867 (1979).
[73] P.P., Vicario and A. Bennun, "Interaction of MnATP and peptide substrate with insulin receptor tyrosine kinase", *Biochem Soc Trans.* 17(6), 1108-9 (1989).
[74] V.A. De Bari and A. Bennun, "Cyclic GMP in the human erythrocyte. Intracellular levels and transport in normal subjects and chronic hemodialysis patients", *Clinical Biochemistry*, 15(4), 219-221 (1982).
[75] V.A. De Bari, N. Novak and A. Bennun, "Cyclic nucleotide metabolism in the human erythrocyte", *Clinical Physiology and Biochemistry*, 2, 227-238 (1984).
[76] P.P., Vicario and A. Bennun, "Regulation of insulin receptor tyrosine kinase by metabolic intermediates", *Biochem Soc Trans.* 17(6), 1110-1 (1989).
[77] A. Bennun, M.A. Needle, V.A. De Bari, "Stimulation of the hexose monophosphate pathway in the human erythrocyte by Mn^{2+}: Evidence for a Mn^{2+}-dependent NADPH peroxidase activity", *Biochemical Medicine*, 33, 17-21 (1985).
[78] S. Ramón y Cajal, "The Croonian Lecture: La Fine Structure des Centres Nerveux", *Proceedings of the Royal Society of London*, 55, 444-468 (1894)

[79] A. Mørk, "Actions of Lithium on the Cyclic AMP Signalling System in Various Regions of the Brain - Possible Relations to its Psychotropic Actions -: A Study on the Adenylate Cyclase in Rat Cerebral Cortex, Corpus Striatum and Hippocampus", *Pharmacology & Toxicology*, 73 (1993)

[80] T. Lømo, "The discovery of long-term potentiation", *Phil. Trans. Roy. Soc. London B (Biological Sciences)*, 358(1432), 617-620 (2003).

[81] K. Doya, "Complementary roles of basal ganglia and cerebellum in learning and motor control", *Current Opinion in Neurobiology*, 10 (6), 732-739 (2000).

[82] J.H. Medina, J.I. Rossato, L.R.M. Bevilaqua, I. Izquierdo, M. Cammarota, "Dopamine controls persistence of long term memory storage", *Science*, 21, 1017-1020 (2009).

[83] A. Bennun, "A coupling mechanism to inter-relate regulatory with haem-haem interactions of haemoglobin", *Biomed. Biochim.*, 46, 314-319 (1987).

[84] R. Malenka and M. Bear, "LTP and LTD: an embarrassment of riches", *Neuron* 44 (1), 5-21 (2004).

[85] C.E. Curtis, M. D'Esposito, "Persistent activity in the prefrontal cortex during working memory", *Trends in Cognitive Sciences*, 7(9): 415-423 (2003).

[86] A.G. Gilman., "G Proteins: Transducers of Receptor-Generated Signals", *Annual Review of Biochemistry*, 56, 615-649 (1987).

[87] R.C. Malenka, M.F. Bear, "LTP and LTD: an embarrassment of riches", *Neuron*, 44, 5-21 (2004).

[88] N. Wettschureck, S. Offermanns, "Mammalian G proteins and their cell type specific functions", *Physiol. Rev.*, 85(4), 1159-204 (2005).

[89] S. Brydon-Golz and A. Bennun, "Postsynthetic stabilized modification of adenylate cyclase by metabolites", *Biochem. Soc. Trans.*, 3, 721-724 (1975).

[90] Von Hungen and S. Roberts, "Adenylate-cyclase receptors for adrenergic neurotransmitters in rat cerebral cortex", *Eur. J Biochem.* 36(2), 391-401 (1973).

[91] S. Kakiuchi, T.W. Rall, "Studies on adenosine 3,5-phosphate in rabbit cerebral cortex", *Mol Pharmacol.*, 4(4), 379-388 (1968).

[92] K. Tanaka, "Alteration of second messengers during acute cerebral ischemia - adenylate cyclase, cyclic AMP-dependent protein kinase, and cyclic AMP response element binding protein", *Prog. Neurobiol.*, 65(2), 173-207 (2001)

[93] Karaçay, G. Li, N.J. Pantazis and D.J. Bonthius, "Stimulation of the cAMP pathway protects cultured cerebellar granule neurons against alcohol-induced cell death by activating the neuronal nitric oxide synthase (nNOS) gene", *Brain Res.*, 1143, 34-45 (2007).

[94] A. Mørk, "Calmodulin-dependent adenylate cyclase activity in rat cerebral cortex : effects of divalent cations, forskolin and isoprenaline", *Archives of Physiology and Biochemistry*, 97(3) (1989)

[95] Mørk, "Mode of Action of Lithium on the Catalytic Unit of Adenylate Cyclase from Rat Brain", *Pharmacology & Toxicology*, 60(4) (1987).

[96] D.K. Vassilatis, J.G. Hohmann, H. Zeng, F. Li and J.E. Ranchalis., "The G protein-coupled receptor repertoires of human and mouse", *Proc Natl Acad Sci USA*, 100 (8), 4903-4908 (2003).

[97] N. Del Campo, S.R. Chamberlain, B.J. Sahakian and T.W. Robbins, "The roles of dopamine and noradrenaline in the pathophysiology and treatment of attention-deficit/hyperactivity disorder", Biol Psychiatry., 69(12), 145-57 (2011).

[98] J. Prince, "Catecholamine dysfunction in attention-deficit/hyperactivity disorder: an update", *J. Clin. Psychopharmacol.* 28, 39-45 (2008).

[99] A. Mørk, "Effects of Lithium on Calmodulin-stimulated Adenylate Cyclase Activity in Cortical Membranes from Rat Brain", *Pharmacology & Toxicology*, 60(1) (1987).

[100] S. Cull-Candy, S. Brickley and M. Farrant, "NMDA receptor subunits: diversity, development and disease", *Curr. Opin. Neurobiol.*, 11(3), 327-35 (2001).

[101] P. Paoletti, J. Neyton ., "NMDA receptor subunits: function and pharmacology", *Curr. Opin. Pharmacol.*, 7(1), 39-47 (2007).

[102] R. Dingledine, K. Borges, D. Bowie and S.F. Traynelis, "The glutamate receptor ion channels", *Pharmacol. Rev.*, 51(1), 7-61 (1999).

[103] D.O. Levitsky, D.A. Nicoll and K.D., "Philipson Identification of the high affinity Ca^{2+}-binding domain of the cardiac Na^{+}-Ca^{2+} exchanger", *J. Biol. Chem.*, 269, 22847-22852 (1994).

[104] L. Annunziato, G. Pignataro and G.F. Di Renzo, "Pharmacology of brain Na+/Ca2+ exchanger: from molecular biology to therapeutic perspectives", *Pharmacol Rev.*, 56(4), 633-54 (2004).

[105] K.D. Philipson and D.A. Nicoll, "Sodium-calcium *Annu Rev Physiol.*, 62, 111-33 (2000).

[106] P. Caroni and E. Carafoli, "The regulation of the Na^{+}-Ca^{2+} exchanger of heart sarcolemma", *Eur J Biochem*, 132, 451-460 (1983).

[107] S. Asano, T. Matsuda, K. Takuma, H.S. Kim, T. Sato, T. Nishikawa and A. Baba, "Nitroprusside and cyclic GMP stimulate Na^{+}/Ca^{2+} exchange activity in neuronal preparations and cultured rat astrocytes", *J Neurochem*, 64, 2437-2441 (1995).

[108] W. Harley, C. Floyd, T. Dunn, X.D. Zhang, T.Y. Chen, M. Hegde, H. Palandoken, M.H. Nantz, L. Leon, K.L. Carraway, B. Lyeth and F.A. Gorin, "Dual inhibition of sodium-mediated proton and calcium efflux triggers non-apoptotic cell death in malignant gliomas", *Brain Res.*, 2(1363), 159-69 (2010).

[109] M.P. Blaustein and E.M. Santiago, "Effects of internal and external cations and ATP on sodium-calcium exchange and calcium-calcium exchange in squid axons", *Biophys J.*, 20, 79-111 (1977).

[110] Linck, Z. Qiu, Z. He, Q. Tong, D.W. Hilgemann and K.D. Philipson, "Functional comparison of the three isoforms of the Na^{+}/Ca^{2+} exchanger (NCX1, NCX2, NCX3)", *Am J Physiol.*, 274, C415-C423 (1998).

[111] A.J. Silva, J.H. Kogan, P.W. Frankland and S. Kida, "Creb and memory", *Annu. Rev. Neurosci.*, 21, 127-48 (1998).

[112] Purves, J. A. George, D. Fitzpatrick, W.C. Hall, A. LaMantia, J.O. McNamara and L.E. White. *Neuroscience. 4th ed.*. 170-6 (2008).

[113] Y. Wang, A. Ghezzi, J.C. Yin and N.S. Atkinson, "CREB regulation of BK channel gene expression underlies rapid drug tolerance", *Genes, Brain, and Behavior*, 8(4), 369-76 (2009).

[114] A. Barco, C. Bailey and E. Kandel, "Common molecular mechanisms in explicit and implicit memory", *J. Neurochem.*, 97(6), 1520-33 (2006).

[115] B. Mayr and M. Montminy, "Transcriptional regulation by the phosphorylation-dependent factor CREB", *Nat. Rev. Mol. Cell Biol.* 2(8), 599-609 (2001).

[116] J. Mackliff, "Bilateral Electro Coagulation of Adrenal Medulla a Pioneer Surgery for Disappearence of Parkinson Symptons", 2nd world annual congress of Neurotalk 2011, Dalian, China (2011) http://www.drjosemackliff.com.

[117] A. Delorenzi, M.E. Pedreira, A. Romano, S.I. Garcia, C.J. Pirola, V.E. Nahmod and H. Maldonado, "Angiotensin II enhances long-term memory in the crab Chasmagnathus", Brain Research Bulletin Volume: 41, Issue: 4, Pages: 211-220 (1996)

[118] J.M. Saavedra, "Brain angiotensin II: new developments, unanswered questions and therapeutic opportunities", *Cell Mol Neurobiol.* 25(3-4), 485-512 (2005).

[119] J.W. Wright and J.W. Harding; "Brain renin-angiotensin--a new look at an old system", *Prog Neurobiol.* 95(1), 49-67 (2011).

In: Horizons in Neuroscience Research. Volume 15
Editors: Andres Costa and Eugenio Villalba
ISBN: 978-1-63321-523-8

Chapter 4

Reorganization of Elementary Functions (REF) after Brain Injury and in the Intact Brain: A Novel Understanding of Neurocognitive Organization and Reorganization

Jesper Mogensen[*]
The Unit for Cognitive Neuroscience, Department of Psychology,
University of Copenhagen, Denmark

Abstract

The REF (Reorganization of Elementary Functions) model describes the dynamic neural and cognitive reorganizations mediating behavioural and conscious flexibility. These mechanisms have phylogenetically evolved to enable such flexibility in intact individuals. It is, however, the same mechanisms, which mediate the functional recovery after brain injury. Thus, the REF-model provides a theoretical framework for the understanding of normally occurring cognitive flexibility as well as the posttraumatic reorganization of the injured brain and for the development of improved therapeutic procedures in the cognitive rehabilitation of brain injured patients. A key feature of the model is that the "reprogramming" of the injured brain depends crucially on the feedback to the patient in the specific training situation. The resultant cognitive strategies may lead to proficient task solutions in the training situation (e.g. in the clinical setting) while not being able to generalize to the daily life of the patient. It is therefore important to develop more "ecologically valid" training methods in the therapeutic settings and to continue the rehabilitative training into the daily life of the patient. The use of advanced technologies is one of the promising ways to "bring reality into the institution" and to "bring training into the real world". Frequently, however, traditional methods of training and testing

[*] Correspondence: Prof. Jesper Mogensen, The Unit for Cognitive Neuroscience, Department of Psychology, University of Copenhagen, Oester Farimagsgade 2A, DK-1353 Copenhagen K, Denmark, e-mail: *jesper.mogensen@psy.ku.dk*.

cannot easily be translated into advanced technology applications. Such efforts have to be undertaken as part of a thorough research process. This research process, on the other hand, provides a deeper insight into previously overlooked neural and cognitive mechanisms of the normal and injured brain. It is also possible to achieve an improved generalization from clinical training to reallife situations by specifically aiming the training at the development of what is in the REF-model called Algorithmic Modules (AMs). An AM participates in many different cognitive programmes (each of which is related to a specific task). Rehabilitative training aimed at improving AMs can therefore give the patient basic mechanisms, which can be applied broadly. The REF-model stresses that the prognosis regarding posttraumatic cognitive rehabilitation depends significantly on the pretraumatic experience of the patient. And that the rehabilitative training has to be tailored to the "cognitive profile" of the patient. The REF-model reconciles the concepts of "functional localization" and "functional recovery" by describing a connectionist network (with a high flexibility and potential for reprogramming) within which the basic units are strictly localized basic information processing units. This theoretical reconceptualisation highlights important challenges to neurorehabilitation and points the way towards an improved clinical practise. The model also represents an alternative to "modular" as well as "connectionist" models of the neurocognitive organization of the brain.

Introduction

Brain injury due to vascular incidents (stroke) or mechanical impact leading to traumatic brain injury (TBI) is almost always associated with a certain level of posttraumatic impairments within the motor, sensory and/or cognitive domains. A growing body of basic and clinical research is devoted to scrutinizing the mechanisms of such injury and developing therapeutic interventions, which can diminish the initial level of posttraumatic impairments and promote the posttraumatic functional rehabilitation of brain injured patients. These efforts have, however, for the most part been devoted to understanding and treating the motor and sensory symptoms – as opposed to the cognitive processes such as attention, learning, memory, problem solving and language. This state of affairs reflects the fact that conceptually as well as methodologically research and clinical interventions are relatively easier to conduct within domains situated closer to the input and output of the brain – as opposed to the more complicated and less understood mechanisms involved in cognitive information processing. In the long run, however, cognitive impairments are likely to have the greatest impact in terms of quality of life, ability to return to independent living, and potentially to return to work (e.g. Moore and Stambrook, 1995). While the need for further research and conceptual clarifications (leading to improved therapeutic methods) are clearly also needed within the sensory and motor domains (e.g. Levin et al., 2009), such needs are even more compelling within the area of posttraumatic cognitive rehabilitation.

The need for better models of posttraumatic cognitive recovery and cognitive rehabilitation of brain injured patients has been stressed repeatedly (e.g. Robertson, 1999; Wilson, 2002). As emphasized by Wilson (2002), a comprehensive theoretical framework for clinical rehabilitation of brain injured patients has to include theories from a significant number of domains, each of which contributes to the outcome of the rehabilitative therapeutic endeavour. There can be no doubt that such a holistic approach is important and may even be essential for any attempt to provide a complete theoretical basis for clinical neurorehabilitation. However, there may be good reason to expect that more limited neurocognitive models of posttraumatic functional recovery and neural plasticity can provide information of clinical relevance (e.g. Mogensen, 2011a, 2011c, 2012; Robertson, 1999). Such models may ultimately also be included in more holistic and comprehensive theoretical frameworks for clinical rehabilitation (e.g. Wilson, 2002). The topic of the present chapter is such a neurocognitive model, the REF (Reorganization of Elementary Functions) model (Mogensen, 2011a, 2011c, 2012, 2014; Mogensen and Malá, 2009). In addition to a description of the REF-model the theoretical and clinical implications of the dynamic neural and cognitive posttraumatic reorganizations of the model will be discussed.

Conceptualisations of Posttraumatic Functional Recovery

As has been stressed elsewhere (e.g. Mogensen, 2011a, 2011c, 2012, 2014; Mogensen and Malá, 2009; Overgaard and Mogensen, 2011, 2014) and as will be discussed later in the present chapter, references to posttraumatic functional recovery can be confusing and even misleading if not related to a particular level of analysis. For example, in this chapter, the term "cognitive recovery" is used to refer to the process by which the trauma-associated cognitive impairments, as manifested in the performance of tasks, are diminished or eliminated – allowing the patient or experimental animal to achieve a task solution of a higher level of proficiency than what is observed immediately posttraumatically. This use of the term "recovery" is different from that used, for instance by Wilson (e.g. Wilson, 2002; Wilson et al., 2000) who makes a distinction between effects of treatment and effects of recovery. In the latter context, recovery is conceptualised at the neural level and appears to refer to neural processes that more or less restore the pretraumatic neural state. One of the purposes of the REF-model is to provide a framework within which to conceptualise the levels of analysis at which various kinds of posttraumatic cognitive and neural plastic events can be understood (e.g. Mogensen, 2011a, 2011c, 2014; Mogensen and Malá, 2009; Overgaard and Mogensen, 2011). With regard to the posttraumatic re-establishment of motor functions, Levin et al. (2009) have argued along lines that are somewhat similar to those presently presented for the cognitive processes. As discussed in more detail below and elsewhere (Mogensen, 2014), the distinctions made by Levin et al. (2009) between posttraumatic "recovery" and "compensation" may be less meaningful within the cognitive domains.

The principles of posttraumatic structural as well as functional changes may differ somewhat between the more cognitive domains and the domains associated with the direct input or output of the brain. For instance, when an injury destroys the input or output pathways of the brain (e.g. the optic nerve or the major descending motor pathways) this may lead to a rather chronic loss of sensory input or ability to execute motor action, respectively. In those cases, the degree of posttraumatic functional recovery (as defined at the behavioural level) may remain limited, although processes of a clearly "compensational" nature may allow the individual to achieve at least some degree of "recovery". In, for example, the case of complete loss of sensory input within one modality (e.g. vision), intermodal plasticity may allow input via other modalities (such as touch or hearing) to substitute (at least to an extent) for the lost input (e.g. Bach-y-Rita et al., 1969, 1998; Kaczmarek et al., 1991; Ptito et al.,

2005). In the domain of direct motor output, spared components of the motor pathways may allow the patient to achieve at least some level of mobility (e.g. Levin et al., 2009). In contrast to these somewhat limited potentials of a functional recovery observed at the behavioural level, posttraumatic functional recovery and clinical rehabilitation appear to be more the rule than the exception for the cognitive domains (e.g. Mogensen, 2011a, 2012).

If posttraumatic cognitive recovery is defined as a more or less complete return to the pretraumatic level of behavioural/cognitive proficiency of task performance and/or conscious representation, cognitive recovery has been documented in numerous studies carried out in both animal models and patients (e.g. Buller and Hardcastle, 2000; Carney et al., 1999; Leon-Carrion and Machuca-Murga, 2001; Mogensen, 2011a, 2011b, 2011c; Mogensen and Malá, 2009; Mogensen et al., 2004a, 2007; Overgaard and Mogensen, 2011; Panksepp and Panksepp, 2000; Ramachandran and Blakeslee, 1998; Rohling et al., 2009). Most of the cases of posttraumatic cognitive recovery have been demonstrated in association with formalized rehabilitative training. There are, however, also instances of what is termed "spontaneous recovery" (e.g. Leon-Carrion and Machuca-Murga, 2001). For the most part, a recovery process is defined as being "spontaneous" if the subjects (patients or experimental animals) have not been subjected to a specific posttraumatic training procedure. It is frequently implied that the "spontaneous" recovery is the result of one or another type of experience-independent process. It may, however, be inappropriate to assume that functional recovery in the absence of formalized training is necessarily experience-independent. Even when formalized training is absent, patients and experimental animals are constantly exposed to the challenges of daily activities. Under almost all circumstances, daily living, such as the more or less successful coordination of movements, communication (or attempted communication), feeding, and other basic activities, constitute in themselves informal "training". Therefore, it is in most such cases difficult to discriminate between experience-dependent and experience-independent types of recovery processes. There can, however, be no doubt that truly experience-independent processes do occur. An example of experience-independent processes is the recovery accompanying the disappearance of an injury-associated "penumbra". Briefly, the penumbra phenomenon is a situation in which injury to one part of the brain results in a reduced level of blood supply to other parts of the brain. Although the reduced blood supply within the penumbra region is sufficient for the survival of the neurons, normal levels of functionality are not possible. Consequently, the symptoms observed during the presence of the penumbra are a combination of the consequences of the actually lost tissue (within the region of the primary trauma) and the functional impairments within the brain regions affected by the penumbra. Penumbras mostly disappear spontaneously and when that happens a normal level of functional performance is restored within the part of the brain that initially lies in the penumbra (e.g. Choi et al., 2007).

In humans as well as in animal models, the degree to which the cognitive recovery manages to result in the elimination of posttraumatic symptoms varies substantially. In some instances, extensive rehabilitative training achieve only a limited degree of functional recovery, while in other instances the recovery may be "complete" – defined as the acquisition of a posttraumatic proficiency equal to that seen in the absence of any brain injury (e.g. Mogensen et al., 2004a). In animal models utilizing comparisons to well-established pretraumatic functional baselines (e.g. Mogensen, 2011b), such a "complete" functional recovery can be demonstrated under circumstances that involve the complete removal of the brain structure in question. From a theoretical point of view, such a functional recovery poses a severe challenge to some of the traditionally rather well-established concepts of cognitive neuroscience. In particular, one has to face the apparent contradiction between the concept of "cognitive functional recovery" and the well-established concept of "functional localization".

Functional Localization and Recovery

There is a widespread consensus that a regional specialization exists within the brain. Accordingly, the structures and substructures are functionally specialized and it appears appropriate to speak about the "localization" of various "functions" (e.g. Coltheart, 2001; Kringelbach and Rolls, 2004; Monakow, 1914; Selnes, 2001). The idea of a functional specialization within the specific structures of the brain is mainly supported by data from two sources: (1) neuroimaging techniques (e.g. PET and fMRI) reveal rather consistent patterns of regional cerebral activation when subjects are stimulated in particular ways, perform particular tasks etc., and (2) "lesion experiments" (be it clinical examinations of brain injured patients or controlled experiments in animal models) demonstrate that lesions or regional inactivations within particular brain structures are associated with specific patterns of behavioural/cognitive symptoms.

The latter source of evidence, however, is directly related to the apparent contradiction between "functional localization" and posttraumatic "functional recovery" mentioned above. While it seems logical that loss (lesion) of a specialized brain structure leads to a predictable type of impairment, it would be illogical to expect the "recovery" of the lost "function", since regrowth of the missing brain region appears not to occur. Nevertheless, localization as well as posttraumatic recovery of cognitive functions are highly documented empirical facts (see above and e.g., Mogensen, 2011a, 2011c; Mogensen and Malá, 2009). To resolve this apparent contradiction, it is necessary to understand the functional organization and reorganization of the brain. But it is also an important issue in the context of the above-mentioned goal of a better theoretical basis for clinical efforts aiming at improving the rehabilitation of cognitive processes.

The aphasias are perhaps the best-studied clinical conditions when it comes to mapping the neural substrate of functional recovery. Aphasias are primarily the result of injury to the left hemisphere and ipsilateral contributions to the mediation of reacquired linguistic functions have been documented by for instance Szaflarski et al. (2011), Perani et al. (2003), Specht et al. (2009), and Meinzer et al. (2008). Meinzer et al. (2008) demonstrated treatment-induced reintegration of various perilesional areas. However, a dominating issue within studies addressing the posttraumatic reacquisition of language is, whether the contralateral (right) hemisphere contributes significantly to the mediation of posttraumatic recovery. Numerous studies have found the apparent involvement of structures within the right hemisphere in the mediation of posttraumatic recovery of language (e.g. Ansaldo and Arguin, 2003; Ansaldo et al., 2002; Baumgaertner et al., 2005; Perani et al., 2003; Specht et al., 2009; Thomas et al., 1997; Thulborn et al., 1999). There are indications (e.g. Thomas et al., 1997) that the pattern of shift towards right hemisphere mediation of linguistic functions may differ between types of aphasia. Additionally, changes in the direction of right hemisphere mediation of language may be accompanied by internal reorganizations within the left hemisphere (shifts to mediation by ipsilateral uninjured regions). Reorganizations such as these may lead to a more bilateral representation of language – due to the concurrent shift of linguistic mediation within the left hemisphere and to the contralateral, right hemisphere (e.g. Thompson et al., 2010). The recovery-associated shift towards right hemisphere mediation of linguistic functions seem mostly to occur without rehabilitative training specifically aimed at such a shift (rather, rehabilitation has been aimed in a more general way towards recovery of linguistic abilities). In some cases, however, aphasic patients have – somewhat successfully – been subjected to training aimed at achieving a higher degree of right hemisphere mediation of linguistic tasks (e.g. Crosson et al., 2009). Crosson et al. (2009) used a manipulation task performed with the patient's left hand to initiate naming trials. In this procedure the manipulation task achieved an independent right hemisphere activation timed to occur concurrently with the naming task. This combination of an independent right hemisphere activation and the performance of a naming task is believed to facilitate an interhemispheric shift of linguistic task mediation. In many studies of recovering or recovered aphasic patients, fMRI is utilized in order to determine the changes in neural activity associated with such a recovery process. It may, however, be questioned whether all the changes in regional activation demonstrated by for instance fMRI during posttraumatic rehabilitation in aphasic patients are necessarily associated with the reacquisition of language. Posttraumatic changes in activity within a given structure may be the consequence of any trauma-related process – e.g., "disinhibition" due to lack of input from the injured brain region or any other process, which is not directly related to the recovery of linguistic abilities. There are, however, cases in which specific regional activations can with a higher level of certainty be related to the reacquisition of linguistic abilities. For instance, Meinzer et al. (2007b) studied the recovery seen in a bilingual aphasic patient. In this case, activation of part of the superior temporal lobe of the right hemisphere was exclusively associated with the use of the trained language while no such activation was observed associated with the untrained language.

Mechanisms of Posttraumatic Functional Recovery

As mentioned, in many instances there may be technical reservations regarding studies addressing the functional reorganizations in brain injured patients undergoing cognitive rehabilitation. It is clear, however, that the overall picture from for example recovering aphasic patients is that there is a degree of "shift" of mediation of the originally lost or impaired cognitive processes to the preserved structures of the brain – and presumably cases of "vicariation" (the phenomenon that brain areas with different functions can assume or "take over" the function of an injured brain region (e.g. Finger and Stein, 1982; Slavin et al., 1988)). These plastic properties of the brain have within philosophy of mind been taken as evidence for multiple realizations and as an argument in favour of functionalism (e.g. Block and Fodor, 1972). As argued by Overgaard and Mogensen (2011), conclusions regarding multiple realizations do, however, require a much deeper and detailed analysis in order to utilize the results of studies mapping posttraumatic task mediation. Only by addressing more thoroughly the detailed mechanisms of posttraumatic reorganization of the brain can one achieve a comprehensive understanding of the degree to which functions are truly "relocalized" (e.g. Mogensen, 2011a, 2011c, 2012, 2014; Mogensen and Malá, 2009).

In the context of trying to clarify whether or not the information processing normally performed by the region of the brain lost to injury is re-established elsewhere in the brain, a very basic issue is to scrutinize the likelihood that the basic circuitry of the injured brain region is (re)established within other parts of the brain. This topic is discussed in detail elsewhere (Mogensen, 2011a, 2011c). Although the adult brain possesses an impressive level of plasticity, it still seems unlikely that the posttraumatic processes include an actual recreation of the circuitry of the lost region of the brain. Mature neurons are in a number of important respects different from the immature neurons, which originally formed the local circuitry of the brain (e.g. Chen et al., 1995; Fawcett et al., 1989; Goldberg et al., 2002). In the adult brain there is, however, an ongoing neurogenesis and this neurogenesis is potentiated by brain injury (e.g. Arvidsson et al., 2002; Chen et al., 2004; Magavi et al., 2000; Nakatomi et al., 2002; Sharff et al., 2000). As opposed to mature neurons, such newly formed neurons may be better equipped to recreate a particular local circuitry of the brain. And they may receive support in reaching the relevant brain regions since, upon injury, mature astrocytes are able to transform themselves into radial glial cells similar to those guiding the neuromigration during development (Leavitt et al., 1999; Rakic, 1971, 1985). Such observations may create a level of optimism regarding the potential for recreation of lost circuitry. There is, however, a number of crucial factors, which appears to prevent the adult brain (injured as well as uninjured) from recreating the basic, local networks. From the final stages of the original ontogenic development (and thereby the formation of the basic circuitry of the brain) a number of factors associated with glial cells and myelin appear to prevent restructuring and presumably recreation of such basic circuitry (e.g. Berry, 1982; Schäfer et al., 2008; Schwab and Thoenen, 1985). Of special importance in this context are the astrocyte-produced chondroitin sulphate proteoglycans (CSPGs). The CSPGs play an important role in termination of the developmentally "critical periods" and they are believed to "consolidate" the originally formed circuitry in its "final" form (e.g. Berardi et al., 2004; Del Rio and Soriano, 2007; McGee et al., 2005; Pizzorusso et al., 2002; Schäfer et al., 2008). It seems that the presence of these CSPGs block the possibility of an adult recreation of traumatically lost networks (e.g. Del Rio and Soriano, 2007; Schäfer et al., 2008). Consequently, the recreation of the basic circuitry in response to brain injury is unlikely to occur – at least if such local networks are completely lost to injury. The situation and the chances of a partial re-establishment of the lost circuitry may be somewhat different in case of subtotal lesions of a particular local circuitry (e.g. Robertson and Murre, 1999). Since, however, the majority of cases of major brain injury will deprive the individual of the entire local circuitry of particular brain regions, there is a clear need for a better understanding of the reorganizations and recovery-associated processes of the injured brain.

Clinical studies (such as those that address the neural substrate of reacquisition of language in aphasic patients – see above) are clearly of significant importance to the understanding of posttraumatic functional reorganizations. It is, however, only in well-controlled animal models (e.g. Mogensen, 2011b) that one can avoid some of the short-comings of clinical studies – for instance the occurrence of subtotal lesions of brain structures and the fact that in most brain injured patients, multiple brain regions are simultaneously affected. The use of animal models also allows a degree of experimental manipulation, which is impossible to achieve in human studies. An extensive animal model-based research programme (e.g. Mogensen et al., 2002, 2003, 2004a, 2005, 2007) has scrutinized the neural and cognitive mechanisms of posttraumatic recovery of cognitive processes. By including a spectrum of organic and behavioural/cognitive "challenge" methods (e.g. Mogensen, 2011b; Mogensen and Malá 2009) these studies have provided insights into the neural and cognitive processes that mediate the posttraumatic functional recovery of various cognitive processes.

A number of principles of posttraumatic functional recovery have emerged from the above-mentioned studies and others (for reviews see: Mogensen, 2011a, 2011c; Mogensen and Malá, 2009). Three general principles are especially important and describe the situation after a successful posttraumatic rehabilitation within a cognitive domain:

1. *Modification of the degree of contribution to task mediation by individual brain structures*:
Some structures within the spared regions of the brain exhibit an increased or decreased level of contribution to mediation of a particular task.

2. *Task-dependent and dissimilar neural substrates:*
After a given lesion, the functional recovery of various cognitive tasks is mediated by unique and dissimilar neural substrates.

3. *Application of new cognitive strategies:*
The fully recovered individuals solve a given task by applying new strategies that are dissimilar to those applied pretraumatically.

Point 2 and 3 of these principles emphasize that the lost information processing (i.e. the destroyed circuitry) appears not to have returned even though the individual has obtained a full functional recovery, thereby supporting the above-mentioned conclusion that posttraumatic recreation of the lost basic circuitry is unlikely. If, within any part of the brain, the rehabilitative training had been accompanied by the creation of a circuitry similar to that available in the pretraumatic situation, one would expect all cognitive domains affected by the lesion posttraumatically to receive equal contributions to functional recovery from the brain region within which the circuitry had been recreated. Such a situation is contradicted by Principle 2. Principle 3 contradicts the re-establishment of the information processing lost to injury. If posttraumatic processes had led to the re-establishment of the information processing of the injured brain structure, one would expect not only task solutions of a proficiency similar to that seen preoperatively – which is, indeed, seen in some instances (e.g. Mogensen et al., 2004a) – but also that such a task solution would employ similar strategies to those seen in the pretraumatic situation.

Jointly, the clinical and especially animal model-based data (some of which has been reviewed above and elsewhere – see for instance Mogensen, 2011a, 2011c, 2012, 2014; Mogensen and Malá, 2009) indicate that although the brain is posttraumatically capable of a high degree of behaviourally defined functional recovery, this is mainly (potentially exclusively) achieved without recreation of the basic circuitry which has been lost to injury. To account for this situation, Mogensen (2011a, 2011c, 2012, 2014) and Mogensen and Malá (2009) have proposed the REF-model.

The REF-Model

While the REF-model has been developed primarily on the basis of research that focuses on posttraumatic functional recovery within the cognitive domains, it is believed to account for neural and cognitive processes, which have evolved in order to mediate behavioural and cognitive flexibility (including problem solving) in the intact brain (see Mogensen (2011a, 2011c) for further discussions).

Elementary Functions

Within the REF-model, three levels of analysis are important for the understanding of the mechanisms of posttraumatic functional recovery (see Table 1).

Table 1. The three levels of analysis of the REF-model – including some of the characteristics of Algorithmic Strategies (ASs) and Elementary Functions (EFs). For further details: see the present text

SURFACE PHENOMENA (Mental/Behavioural manifestation level). Final products in terms of mental states (potentially conscious) and overt behavior. Realized by individual Algorithmic Strategies (ASs).
ALGORITHMIC STRATEGIES (ASs) Consists of numerous interacting Elementary Functions (EFs). Mostly a result of experience and learning. *Neural substrate:* The substrate of the constituent EFs plus the interconnections between these EFs – that is: most ASs are distributed across many regions of the brain.
ELEMENTARY FUNCTIONS (EFs) Truly localized. Perform basic information processing. *Neural substrate:* Substructures/local circuits – within a given brain structure – that is: EFs are fully localized.

At the lowest of these three levels are the basic information processing modules, referred to as Elementary Functions (EFs). Each EF contributes a “modular” type of information processing and the EFs are truly localized. It is at the level of the EFs that the true “functional localization” is found – in the sense that each EF is mediated by a specific substructure of a brain region. Traditionally defined brain structures (e.g. the hippocampus, the prefrontal cortex, or the neostriatum) contain the neural substrate of numerous EFs. When a brain structure is lost to injury, all EFs mediated by the subregions within that structure are irreversibly lost. Presently, the functional properties of individual EFs are poorly characterized. There is a significant conceptual distance between the EFs and what is traditionally defined as cognitive or “psychological” functions (e.g. allocentric spatial orientation, egocentric orientation, episodic memory, facial recognition, focussed attention etc.). When the characterization of an EF becomes possible, it is likely that the functional properties of EFs may best be expressed in mathematical terms rather than in the vocabulary of cognitive psychology. For purely illustrative purposes – this is not the place for an evaluation of that particular approach to visual perception – one may say that while the entire "zero-crossing" algorithm proposed by Marr and Hildreth (1980) to explain visual edge detection cannot be considered an EF, the individual calculation steps within that algorithm may be at the level of EFs.

At the top of the three levels of the REF-model are the surface phenomena of behavioural and/or mental manifestations – including conscious awareness. It is at the level of these surface phenomena that the symptoms upon brain injury are normally characterized (in terms of behavioural and/or cognitive impairments). It is also at the level of surface phenomena that posttraumatic functional recovery is normally observed (be it clinically or in most cases of animals model-based experiments).

Algorithmic Strategies

To bridge the gab between the non-recovering EFs and the surface phenomena, the level of Algorithmic Strategies (ASs) has been inserted. An AS consists of numerous interacting EFs. The ASs are primarily established as a consequence of experience and learning – some may, however, be genetically preprogrammed. The neural substrate of an AS consists of the neural substrates of all its constituent EFs plus the interconnections between the neural substrates of these EFs. Consequently, while an EF is strictly localized to a particular subregion of a brain structure, most ASs are distributed across many regions of the brain. The information processing of an AS is the mechanism mediating a specific surface phenomenon – e.g., a specific solution of a specific task. Most surface phenomena can be realized via the activity of a multitude of ASs (although only by one at a time). The task solutions achieved via activation of various ASs may provide outcomes that are equally proficient and unless special analytical techniques are employed, it is normally not possible at the surface level to discriminate between behavioural phenomena reflecting two related but different ASs. When injury destroys the neural substrate of one or more of the constituent EFs within an AS, that AS is lost and posttraumatically the surface phenomenon relying on that AS is impaired.

Activation of ASs is performed by a selector/evaluator mechanism, which besides the selection and activation of ASs also performs a continuous check – via feedback regarding the success or failure of the resultant behaviour – of the appropriateness of the AS in question in the current situation. Under normal circumstances most situations are already associated (via information stored by the selector/evaluator mechanism) with activation of a successful AS – either because the current situation is identical to previous situations in which such an association to an AS has been established or because there is such a degree of similarity between the current situation and previous problem solving situations that a generalization can be readily achieved. The selector/evaluator mechanism controlling the activation and evaluation of ASs resembles (but is not identical to) the "Supervisory Attentional System" (SAS) of Norman and Shallice (1986). The selector/evaluator mechanism is not a unitary system. Rather, it is a number of interacting ASs – including the SAS (Situational Algorithmic Strategy) of Overgaard and Mogensen (2014).

An individual may encounter a task for which there is no established procedure available in at least two different contexts. The situation may be of a novel nature in that it presents the intact individual with a new task to which there is no possibility to generalize from previous experience. Or the situation may be known from previous experience, but brain injury has robbed the individual of the possibility of utilizing previously established procedures. Whether for one or the other of these reasons, the situation is not a priori associated with activation of a specific AS. In such situations, a process of activation of individual ASs as mediators of the behaviour is initiated (see Figure 1). The quality of the resultant behaviour and/or conscious representation is evaluated by the selector/evaluator mechanism and in case of success a future association between that situation and the tested AS is established.

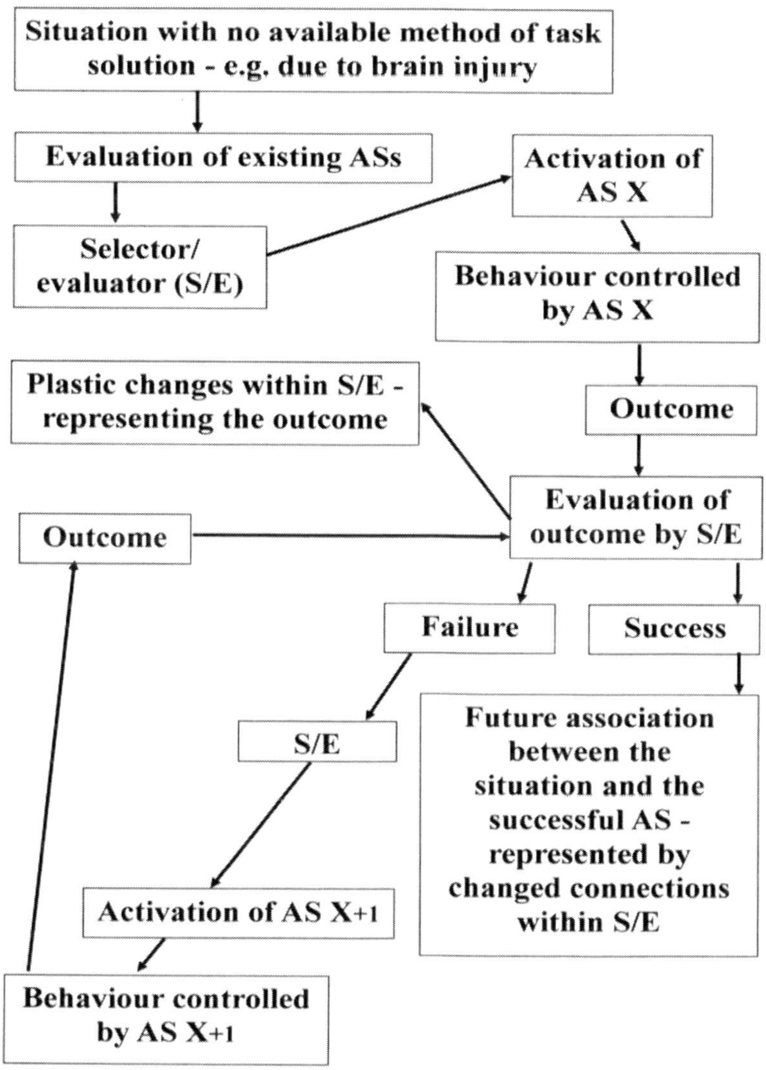

Figure 1. Flow diagram depicting the sequence of events, which (according to the REF-model) leads to a successful functional recovery after brain injury – in case that a pre-existing AS can achieve a successful task solution. For further details: see the present text as well as Figures 3, 4, and 5 in Mogensen and Malá (2009), Figure 1 in Mogensen (2011c) and Figure 2 in Mogensen (2011a).

In case of failure, an alternative AS is activated and evaluated – continuing a process which potentially includes the evaluation of numerous pre-existing ASs (for further descriptions: see Mogensen, 2011a, 2011c; Mogensen and Malá, 2009). This process bears some resemblence to what was described as the "hypothesis" evaluation of Krechevsky (e.g. 1932, 1933). If a pre-existing AS is eventually found to successfully give rise to the required surface phenomenon, the situation will be associated with activation of that AS in the future. Every search for novel associations between an AS and a particular situation/task is accompanied by neuroplasticity modifying the synaptic connections within the selector/evaluator mechanism. This plasticity mediates the future association between the situation in question and the selection and activation of the successfully utilized AS.

Reorganization of Elementary Functions (REF)

In case the evaluation of existing ASs does not lead to a successful solution of the task, a novel AS will have to be created and associated with the situation/task in question. The creation of a novel AS involves the reorganization of the functional interactions between EFs.

This is the "Reorganization of Elementary Functions" (REF) process. This reorganization of the connectivity between the neural substrates of EFs is based on feedback regarding the outcome of the processes of activating various ASs (and/or potential ASs) – including a process resembling the backpropagation algorithm (e.g. Bryson and Ho, 1969; Parker, 1986; Rumelhart and McClelland, 1986; Werbos, 1974, 1994). A highly simplified version of this process is illustrated in Figure 2. Due to the backpropagation process, a set of EFs, which previously did not in themselves constitute an interacting entity, is functionally united to form a novel AS. Although the backpropagation mechanism is presently stressed as being of crucial importance to the situation in which none of the pre-existing ASs can lead to a successful task solution, it should be stressed that backpropagation is not exclusively active in such situations. Backpropagation mechanisms constantly modify the connectivity between EFs as a consequence of the relative degrees of success or failure of behaviour/cognitive representation in various situations. The outcome of an attempted task solution (be it successful or not) leads to a backpropagation process modifying the connectivity – and thereby functional interaction – between individual EFs. In parallel to this, feedback to the selector/evaluator mechanism regarding the degree of success of an attempted task solution also modifies connectivity within the selector/evaluator mechanism itself (which, as mentioned, is also a special class of ASs) (for further descriptions: see Mogensen, 2011a, 2011c, 2012; Mogensen and Malá, 2009).

It should be emphasized that it is the connectivity between the neural substrates of EFs rather than any aspect of the internal circuitry of the EFs that is modified as a result of the backpropagation mechanism. According to the REF-model, the internal circuitry of individual EFs (and thereby their information processing) remains unchanged by these processes. The EFs do, however, perform that information processing in a novel context and on information from novel sources. Such a situation may most easily be illustrated within domains of sensory and perceptual analysis. In individuals in whom one of the hands has been amputated, the region of the somatosensory cortex, which used to represent the now absent hand, does not remain "vacant". Rather, the neighbouring somatosensory regions – representing the arm and face, respectively – incroaches on the "vacant" area in such a way that the original hand area is eventually taken over by inputs from arm and face, respectively (e.g. Karl et al., 2001; Weiss et al., 2000; Yang et al., 1994). Additonally, training restricted to part of the body may be associated with related shifts within the somatosensory representations (e.g. Elbert et al., 1995; Merzenich and Jenkins, 1993; Münte et al., 2002; Xerri et al., 1996). Somewhat similar processes are found within the auditory system where the tonotopic representations at various levels may be reorganized as a result of both partial loss of input – for instance due to restricted cochlear lesions – and specific learning experiences in intact individuals (e.g. Irvine, 2007; Recanzone et al., 1993; Robertson and Irvine, 1989; Scheich, 1991; Thai-Van et al., 2007). In the cases of both the somatotopic and tonotopic representations these reorganizations represent situations in which neural "modules" which originally performed their information processing (and thereby contributions to sensory and perceptual processes) on information from one part of the body or aspects of the tonotopic spectrum, respectively, received a modified input – leading to a situation in which the "modules" most likely continued to perform more or less the same type of information processing on information from another part of the body or tonotopic spectrum, respectively (further discussions can be found in Mogensen, 2011a, 2011c; Overgaard and Mogensen, 2011, 2014).

A somewhat similar although slightly more complicated situation is found in some cases of intermodal plasticity (e.g. Bach-y-Rita et al., 1969, 1998; Kaczmarek et al., 1991; Ptito et al., 2005). For instance, Ptito et al. (2005) have demonstrated that in "early blind" individuals spatial orientation discrimination performed on somatosensory information is partly mediated by a cortical region that in individuals with intact vision is associated with performance of visual tasks in which the spatial orientation of figures is to be determined.

It is likely (e.g. Mogensen, 2011a, 2011c) that the cortical region in question contributes the same type of information processing in sighted and blind individuals, respectively – but receiving the relevant inputs for this information processing from visual and somatosensory sources, respectively.

In the contexts of the above-mentioned examples, it should be noticed that the "modules" found within the somatotopic representation of the primary somatosensory cortex and the tonotopic representations at various levels of the auditory system may actually be the neural substrates of EFs contributing specific information processing on inputs within these modalities. In contrast, the cortical region which appears to be able to contribute to spatial orientation tasks on the basis of both visual and somatosensory sources of input (Ptito et al., 2005) constitutes an entity that is much larger than the neural substrate of any EF. A neural region of this size contains the neural substrates of numerous EFs. There may, however, be benefits from conceptualising such an interacting collection of EFs, which do not in themselves constitute an entire AS as an entity. Such a "partial AS" is termed an "Algorithmic Module" (AM).

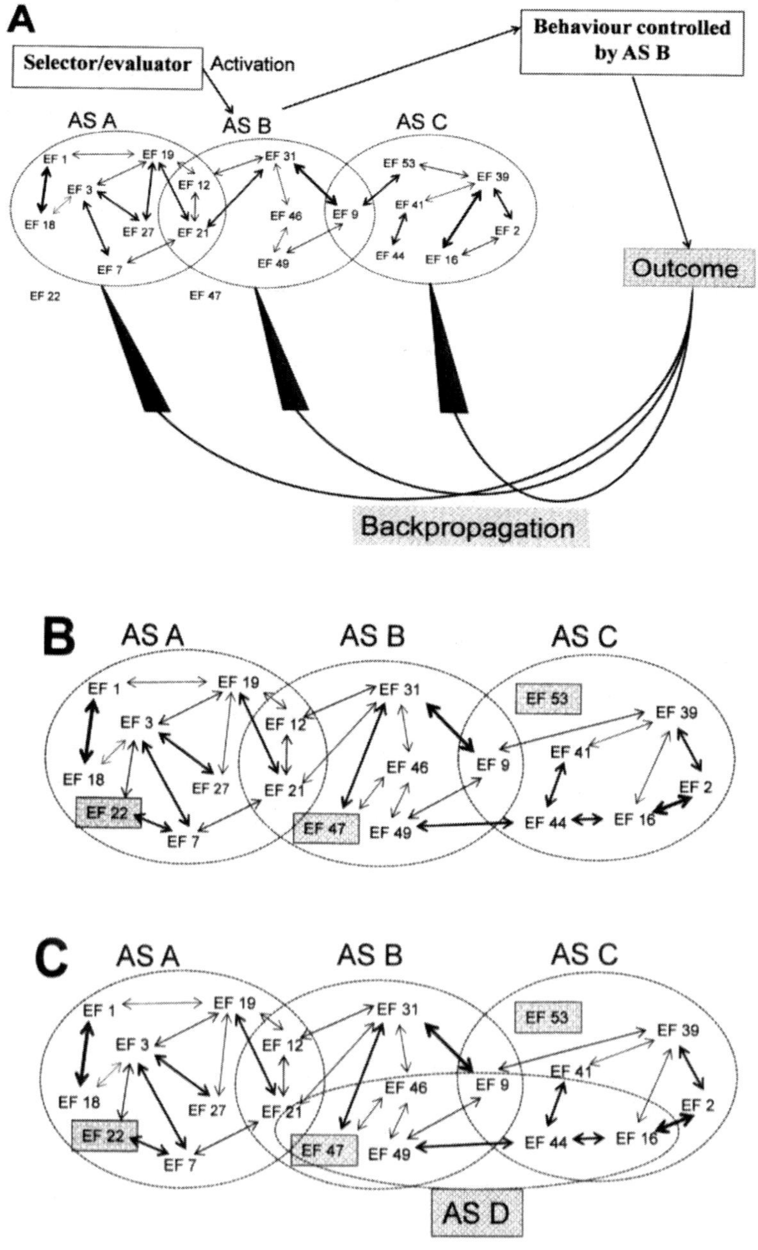

Figure 2. Schematic and simplified representation of the experience-associated reorganizations of connectivity between the neural substrates of EFs. Panel A illustrates a situation in which AS B is selected as the mediator of behavioural performance and the outcome of the specific attempted task solution leads to feedback in the form of backpropagation – modifying the connectivity within AS A, AS B and AS C. Panel B shows the thus modified connectivity of the three ASs including the situation that EF 22 and EF 47 are now constituent parts of AS A and AS B, respectively, while EF 53 is no longer part of AS C. Panel C illustrates the fact that reorganizations such as those shown in Panels A and B can lead to the establishment of a novel AS (AS D) – as well as the already mentioned modifications of the pre-existing ASs. For further details: see the present text as well as Figure 5 in Mogensen and Malá (2009).

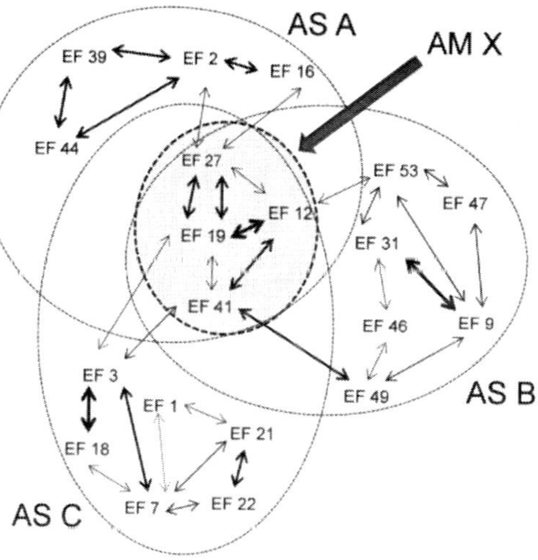

Figure 3. Schematic and highly simplified illustration of the relationship between three ASs (AS A, AS B and AS C) and an AM (AM X), which participates in all of the three (otherwise independent) ASs. For further details: see the present text.

Algorithmic Modules

Being "partial ASs", the AMs are best understood in terms of their similarities with and differences from ASs.

An AM is similar to an AS with respect to:

1. Consisting of a number of EFs and a pattern of interconnections between these EFs – defining their interaction.
2. Having a neural substrate, which consists of the neural substrates of the constituent EFs and the neural connections mediating the interactions between the neural substrates of the constituent EFs.
3. Having been formed primarily as the result of experience-associated (mostly backpropagation) mechanisms – but in some instances potentially having a structure, which is purely genetically predetermined.

An AM differs from an AS with respect to:

1. Not being able in itself to mediate a task solution and/or conscious representation.
2. Not being independently activated by the selector/evaluator mechanism – and consequently never being independently evaluated and modified via backpropagation mechanisms.

The latter of the above mentioned differences between AMs and ASs should not be taken to indicate that AMs are not being evaluated and potentially modified via backpropagation. On the contrary, while not being subjected to such processes as an independent entity, the AMs are typically a constituent subdivision within many ASs and thereby as part of these selected and potentially undergoing backpropagation-mediated modifications. A highly simplified illustration of the relationship between an AM and a number of ASs is given in Figure 3.

In the description of EFs above it was stressed that the entire “zero-crossing” algorithm of Marr and Hildreth (1980) cannot be considered an EF (while individual calculation steps of this algorithm may have such a status). One could, however, consider the “zero-crossing” algorithm to be an example of a potential AM. The entire algorithm is achieved by an interaction between individual steps accomplished by what could be EFs and the entire “zero-crossing” algorithm achieves a standardized information processing which in itself does not lead to a task solution, behavioural manifestation or complete conscious representation. However, as previously stressed this is stated for purely illustrative purposes and without any attempt to enter into a debate regarding the validity of this particular algorithm as part of the mediation of visual perception, the “zero-crossing” algorithm may participate as a crucial step in the mediation of numerous perceptual and for that matter problem solving processes. In the terminology of the REF-model, this would make the “zero-crossing” algorithm an AM, which constitutes a subdivision – a more or less standardized algorithm “module” – within numerous ASs each of which mediates a particular perceptual process and/or problem solving behaviour.

Another potential example of AMs may be the networks leading to what has been identified as place cells and grid cells, respectively (e.g. Fyhn et al., 2004; Hafting et al., 2005; Henze et al., 2000; Moser et al., 2008; O’Keefe and Dostrovsky, 1971; Rolls et al., 2006; Solstad et al., 2006). The individual place or grid cells are obviously able to represent their respective spatial representations due to a number of previous computational steps – each of which might be conceptualised as constituting an EF. Jointly the functional assembly leading to what is registered in the firing patterns of a place or grid cell may be adequately conceptualised as constituting an AM.

A special class of AMs – the Perceptual Algorithmic Modules (PAMs) – plays an important role in the analysis of sensory information and potentially conscious awareness of such information (e.g. Overgaard and Mogensen, 2014).

Neural, Behavioural and Conscious Flexibility in the Normal and Brain Injured Individual

The processes described by the REF-model have most likely evolved in the context of problem solving and behavioural as well as conscious flexibility in the intact individual (Mogensen, 2011a, 2011c). Having phylogenetically evolved as an aspect of the flexibility and plasticity of the intact individual, they have automatically also become the mechanisms of posttraumatic functional recovery when brain injury robs the individual of the EFs and ASs, thereby eliminating mechanisms, which pretraumatically allowed a particular task to be solved. What was pretraumatically associated with an efficient task solution (i.e. activation of the appropriate AS) is posttraumatically the equivalent to a “novel problem solving situation”. Consequently, that situation now calls for the above-described mechanism of search for an adequate AS and potentially the creation of a novel AS.

The behaviourally defined “complete functional recovery” upon brain injury is in the context of the REF-model to be seen as a situation in which the posttraumatic task solution is accomplished via an AS that allows behavioural and/or conscious manifestations which cannot be readily distinguished from the behaviour occurring pretraumatically.

The apparent contradiction between "functional localization" and "functional recovery" discussed above is in reality the result of a "confusion of levels". The term "function" is used in two different contexts. What is truly localized is the information processing of the individual EFs. This information processing is permanently lost when the neural substrate of those EFs are destroyed by injury. On the other hand the "functional recovery" is observed and defined according to the surface phenomenon which can be achieved posttraumatically at a more or less similar proficiency to that seen pretraumatically via the activation of alternative ASs (for further discussion: see Mogensen, 2011a, 2011c; Mogensen and Malá, 2009).

Within the uninjured brain, the life-long reorganizations of ASs – reorganizations occurring whenever the individual is faced with a novel situation – imply that there is a high individuality associated with the neurocognitive organization of the brain. Many tasks and behavioural challenges are common to practically all human beings. And many of these tasks can most easily be solved by application of ASs including certain populations of EFs (at least under normal circumstances). Consequently, it is not surprising to find a relatively high similarity of neurocognitive organizations when comparing intact individuals. But when such comparisons include groups of subjects with a more specialized type of experiences, it may become more obvious that the neural and functional organization is extensively influenced by the experience of the individual. Studies of the Broca area within the neocortex of the left frontal lobe provide one such example. The Broca area is in practically all individuals associated with mediation of expressive language (e.g. Alexander et al., 1990). In professional musicians, however, this brain region also – in contrast to what is the case in musically naïve individuals – participates in the mediation of mental rotation (Sluming et al., 2007). Apparently, aspects of extensive musical experience (potentially related to sight-reading of music) reorganize the neural and cognitive organization associated with mental rotation. And do so in such a way that modules within the Broca area become involved in such ASs – in contrast to what is seen in musically naïve individuals.

Implications of the REF-Model for Posttraumatic Rehabilitative Training

The neuroplastic changes associated with the posttraumatic rehabilitative training fall (Mogensen, 2011a) into two main categories: (1) the changes within the selector/evaluator mechanisms mediating the association between a successful AS and a particular problem solving situation and (2) the reorganization of the connectivity between the neural substrates of EFs – mediating the novel configuration of ASs. These two categories of plasticity are "instructed" by feedback regarding the outcome of the processes in which the behavioural and/or conscious manifestations of a particular AS meet the current environment: the feedback to the selector/evaluator mechanism and the backpropagation mechanism, respectively. As stressed by Mogensen (2011a) it can be argued that it is at the level of the surface phenomena that the primary causation of plastic changes is to be found. It is when the functional manifestations at the surface level interact with the specific environment that the feedback and backpropagation mechanisms lead to a "downward causation". And it is according to this "causation" that selection of ASs and potentially creation of novel ASs is achieved. This situation has important implications for the clinical practise in terms of rehabilitative training.

Generalization of Institutionalised Training to Reallife Situations

It is the outcome of the specific interaction between the individual and the current environment which is the source of both the plastic modifications within the selector/evaluator mechanism and the modified connectivity between individual EFs. Therefore, the future nature, as well as selection of ASs related to a particular task, depends crucially on the situation in which the training has occurred. One may, in case of posttraumatic cognitive rehabilitative training, expect a potentially worrying degree of situational dependence on the outcome of the rehabilitative training (e.g. Mogensen, 2011a, 2011c, 2012; Mogensen and Malá, 2009; Wilms and Mogensen, 2011). A rehabilitative training programme may lead to highly proficient task solutions for the particular tasks utilized in the training programme and in the context of the training environment. However, there may be little or no generalization to the everyday life situation of the patient. In the terminology of the REF-model: while proficient ASs have been developed and selected within the clinical setting, those ASs are not able to mediate proficient task solutions in the different setting of the everyday life of the patient. A patient may appear fully "recovered" when subjected to formal testing in the hospital or other clinical setting and still subsequently demonstrate severe residual problems in non-institutionalized contexts (e.g. Mogensen, 2011a, 2012; Wilms and Mogensen, 2011).

To counteract these kinds of shortcomings, the rehabilitative training should be conducted under circumstances that can be generalized to the everyday environment of the patient. Within clinical institutions, training should ideally be organized in manners as similar as possible to the everyday challenges of the patient's home and potentially workplace. Training should also be continued as seemlessly as possible into the daily environment to which the patient eventually returns after leaving the hospital and/or other institutions. One way to partly achieve such goals is through the adequate use of advanced technology.

Utilization of Advanced Technology in Neurorehabilitation

The use of advanced technology is a growing and important field within neurorehabilitation. The use of these technologies may, indeed, be one of the best ways in the future to achieve "lifelike" training situations in the institution and the possibility of continuing rehabilitative training in an "out-of-institution" setting. In the institution much more lifelike training situations may be achieved by using virtual reality (e.g. Rizzo et al., 1997). The utilization of various hand-held microcomputers and similar devices may also allow a relatively formalized but highly flexible "training" to continue into the daily life of the patient. An example of the latter may be the success of the "NeuroPage" project (e.g. Wilson et al., 1997) in which hand-held devices assist amnestic patients. While these devices may be considered a "cognitive prosthesis", it has been demonstrated also to produce training effects – thereby making it part of the actual rehabilitative training process (e.g. Wilson et al., 2001).

There are significant promises associated with the utilization of computers and other types of advanced technology in the rehabilitative training. A more "ecologically valid" cognitive training may be achieved by using for instance virtual reality in order to "bring reality into the institution" and by utilizing devices such as the NeuroPage system in order to "bring training into the real world" (for further discussions: see Wilms and Mogensen, 2011). In parallel with the possibilities of increasing the "ecologic validity" of training situations, the use of advanced technologies also holds at least a number of already realized and not the least potential benefits. Rehabilitative training may become more intensive by supplementing the face-to-face therapeutic sessions with a therapist with training sessions in which the patient exclusively interacts with technological devices (e.g. Katz, 2009; Rizzo et al., 2004; Tsirlin et al., 2009). The value of a high intensity of training has repeatedly been demonstrated – in animal models (e.g. Malá et al., 2012) as well as in clinical studies (e.g. Cicerone et al., 2004; Cifu et al., 2003; Meinzer et al., 2007a; Pulvermüller et al., 2001; Slade et al., 2002). A high intensity of posttraumatic training may, however, not always be advantageous (e.g. Bland et al., 2001; DeBow et al., 2004; Humm et al., 1998; Kozlowski et al., 1996) and beneficial effects of a high intensity of training may only be achieved within certain posttraumatic time-frames (e.g. Malá et al., 2012). Utilization of computer-based systems using artificial intelligence (e.g. Wilms, 2011) can also allow advanced technology-based systems to adapt in highly flexible and dynamic manners to many aspects of the progress of the patient. The pattern of progressions in task performance during cognitive recovery after brain injury differs considerably among patients – and from day to day within the same patient. A dynamic and adjustable approach is therefore required in order to provide the optimal training parameters (e.g. Robertson and Murre, 1999). Effects of training may also be measured and utilized in the guidance of the progression of training by inclusion of the novel field of brain-computer interactions (e.g. Coyle et al., 2003). These techniques may allow the training situation (including the demands and feedback to the patient) to be regulated by direct measurements of neural activity (e.g. Coyle et al., 2003; Daly and Wolpaw, 2008; Sitaram et al., 2009). Utilization of such techniques will, however, obviously only be optimal when a thorough – and presently only partly existing – knowledge of the neural processes mediating the desired functional rehabilitative processes has been obtained.

While the utilization of advanced technologies in postraumatic cognitive rehabilitation holds a significant degree of promise, there is also a need for caution, as well as extensive testing of methods, when traditional types of training are translated to a computer-based setting. As stressed elsewhere (Wilms and Mogensen, 2011) naïve "translations" from for example a "paper-and-pencil" version of a test or training procedure to a computer-based version may create unforseen discrepancies in the results obtained. This situation poses a significant challenge to the clinical use of advanced technologies, though it also provides a "window" through which cognitive neuroscience may gain an improved understanding of some of the cognitive mechanisms mediated by both the intact and the injured brain. An example of this can be seen in the results of a study by Wilms and Malá (2010). The so-called Prism Adaptation Therapy (PAT) (e.g. Frassinetti et al., 2002; Rossetti et al., 1998) may be used successfully in patients suffering hemispatial neglect (e.g. Rossetti et al., 1998). In PAT, the patients are trained in a task requiring them to point – without being able to follow their arm visually during the pointing movement – to targets defined by the therapist. While doing so, the patient is wearing prism goggles, which diverts the visual field ten degrees to the right (the patients are exhibiting a hemispatiel neglect of the left hemispace). In the traditional version of PAT, the feedback to the patient is the sight of the pointing finger at the moment when the pointing movement has been terminated. In most cases, the patient adapts gradually to the perceptual shift imposed by the prism goggles and eventually shows an after-effect in the form of a relative shift of the pointing movement. This shift persists after the removal of the goggles – which means that the procedure constitutes an at least partial therapeutic intervention regarding the neglect of the left hemispace. An essential element of the procedure is the feedback regarding the precision of the pointing movements during the training period (e.g. Frassinetti et al., 2002; Sarri et al., 2008; Serino et al., 2006, 2007). The study of Wilms and Malá (2010) included both the traditional version of the PAT-procedure and a procedure in which the patients pointed to a touch-sensitive computer screen and feedback was provided graphically (in the form of an X) on the screen rather than via the direct view of the pointing finger. Although the two versions of the test (i.e. the traditionally administered and the computerized version) apparently were similar regarding all formal demands, both injured and uninjured subjects demonstrated significantly dissimilar outcomes when subjected to the two PAT-procedures. The version in which an icon on the computer screen provided the feedback resulted in significantly less after-effects than did the traditional version. The almost complete absence of an after-effect in the computerized version of the test was unexpected. The results, however, emphasized, as is stressed by the REF-model, that the exact setting of the training procedure, as well as the nature of the feedback provided to the patient, are crucial factors for the outcome of neurorehabilitative training. This is stressed by the fact that while feedback in the form of an X failed to produce a therapeutic outcome, a positive effect was obtained, when the X was replaced by a virtual finger (e.g. Wilms and Mogensen, 2011). Additionally, such findings do provide (as emphasized above) a novel window into the understanding of important neural and cognitive mechanisms in brain injured as well as intact individuals.

Training Aimed at Developing AMs

Significant improvements in the degree of generalization from specific rehabilitative training to the reallife situations of the patient may be achieved by various steps towards more naturalistic settings in the clinical situation and continuation of the training process into the reallife situations of the patient (for both types of measures: see above). However, the situational specificity of an ASs and the potential lack of generalization accompanying this feature of rehabilitative training may lead to a level of pessimism. To counteract this pessimism, one may look for additional ways of achieving an improved level of generalization from specific training to reallife situations. While the development of ASs and the association between an AS and a specific training situation may have less potential for generalizing, more positive results may be achieved if the training can ensure the strengthening and further development of AMs. The AMs are, as described above, not in themselves recruited as the mediators of behaviour and/or conscious experience. They do, however, constitute significant elements within a multitude of ASs. By developing improved or even novel AMs rehabilitative training may provide the patient with crucial elements/modules which will allow the subsequent task solutions of reallife situations to develop on an improved and more effective basis.

One of the challenges currently facing basic research as well as clinical practise is to develop an improved understanding of the ways by which the posttraumatic rehabilitative training can achieve an improved strengthening and/or de novo development of AMs. Our present level of understanding does not allow clear predictions and/or recommendations regarding procedures able to achieve such a goal. One could, however, speculate that examples of training methods able to improve AMs are some of the techniques successfully utilized in improving various parameters regarding working memory (e.g. Bell et al., 2003; Buschkuehl et al., 2008; Günther et al., 2003; Westerberg et al. 2007). Some such procedures have even been demonstrated to generalize to the task performance of stroke patients in related but not identical test-procedures (e.g. Westerberg et al., 2007).

Implications of the REF-Model for the Prognosis Regarding Posttraumatic Cognitive Performance and Rehabilitation

The therapeutic setting and the degree to which rehabilitative training can be improved in terms of the patient's ability to generalize to daily life situations are crucial factors with respect to the eventual outcome of cognitive recovery in brain injured patients. Other crucial elements that influence such an outcome spring from the pretraumatic experience and cognitive profile of the patient.

The Importance of the Pretraumatic Experience of the Patient

The spectrum of ASs available during posttraumatic functional recovery is crucial to the efficacy of training. This is not limited to cases in which activation of pre-existing ASs may achieve successful task solution. Even if no spared AS is able to mediate a successful task solution, the interconnectivity between the neural substrates of EFs – and not the least adequate AMs – are essential "building blocks" in the process of shaping new and successful ASs. Therefore, one can expect the speed, efficacy and for that matter eventual outcome of rehabilitative training to be dependent on the pretraumatic cognitive status of the patient. If the brain injured individual posttraumatically still possesses varied and efficient ASs as well as adequate AMs within a number of cognitive domains related to the area in which the symptoms are seen, one will expect a more proficient and potentially quicker process of recovery.

Not only the efficacy of rehabilitative training but also the cognitive profile of the outcome achieved via such training may depend on pretraumatic cognitive factors – i.e., the nature of available ASs and/or AMs. An example of this can be seen in a series of fascinating observations related to the metacontrast masking experimental setup (e.g. Schwiedrzik et al., 2009) in which training-induced increases in the utilization as well as subjective awareness of originally not consciously perceived stimuli can be demonstrated. These experiments are especially interesting on the background that both clinical and experimental data point to the fact that brain injured individuals exhibiting "blindsight" can improve their task performance via training. Improved performance in the forced choice procedure can be seen in both monkeys suffering bilateral ablation of the primary visual cortex (V1) (Dineen and Keating, 1981; Humphrey, 1974) and in patients demonstrating blindsight (e.g. Bridgeman and Staggs, 1982; Chokron et al., 2008; Henriksson et al., 2007; Raninen et al., 2007; Stoerig, 2006; Zihl, 1980; Zihl and Werth, 1984). Improvements such as these are often not accompanied by any change in subjective awareness of the stimulus. In general, patients exhibit blindsight by behavioural demonstrations of a "perceptual processing" of the stimulus without demonstrating any conscious awareness of the stimulus. However, there is now a growing body of evidence indicating that perceptual training can also increase the reported perceptual awareness of stimuli in blindsight patients (e.g. Sahraie et al., 2006). Additionally, studies in intact subjects exposed to subliminal presentation of visual stimuli (mostly utilizing the above-mentioned metacontrast masking in a "stimulus onset asynchrony" paradigm) have demonstrated that in individuals without brain injury, training can also increase the degree of perceptual awareness of a stimulus which was originally unavailable to consciousness (e.g. Albrecht et al., 2010; Schwiedrzik et al., 2009). In the context of such metacontrast masking experiments (e.g. Schwiedrzik et al., 2009) it appears that individual differences between normal subjects reflect different solution strategies (i.e. selection of dissimilar ASs) (Albrecht et al., 2010). When individuals displaying such differences at the outset of training are subjected to identical training procedures, it turns out that the solution strategies become even more dissimilar during the period of training. This demonstrates a potentiation rather than an elimination of these differences in strategies (Albrecht et al., 2010). These results emphasize that subjecting individuals with pre-existing differences in the available ASs, as well as the potential differences in the "biases" of the selector/evaluator mechanism, to identical rehabilitative training may not necessarily shape the cognitive processes in question in a particular direction. Rather, it may produce different outcomes depending on the pretraumatic condition of the patients.

It is important to notice that such biases in the strategy selection and "cognitive styles" do not in themselves constitute a division between a "poor" and a "good" prognosis. Whether or not a particular cognitive bias will manifest itself in a positive or negative manner during posttraumatic rehabilitative training will depend on many factors – not the least whether or not the therapeutic procedures are tailored towards the individual needs and the unique style of progression of the patient in question. If utilized correctly, even vastly dissimilar cognitive biases may be turned into an advantage during the rehabilitative training. A somewhat related phenomenon may be found within the research area dealing with the "cognitively enhancing" effects of an upbringing in an "enriched" (varied and stimulating) environment (e.g. Rosenzweig, 1971). The majority of the initial research in this area (e.g. Bennett et al., 1964, 1969; Renner and Rosenzweig, 1987; Rosenzweig, 1984; Rosenzweig et al., 1961) focused mainly on biochemical and anatomical effects on the brain. It was more or less assumed that such changes at the neural level would at the behavioural level be accompanied by an improved performance of cognitive tasks. It has, however, turned out that such environments may not always be "cognitively enhancing" in a more global sense. Rather, the consequences may be a changed tendency to select particular solution strategies (ASs in the terminology of the REF-model) (e.g. Mogensen, 1991). Whether or not the environmental enrichment has an enhancing effect on cognition (potentially it might be reflected in a cognitively impoverished performance) will depend on the subsequent regime of training and testing (Mogensen, 1991).

This emphasizes that potentially cognitively enhancing biases still need to be harvested in regimes of training and testing which take into account the biases and abilities of the individual.

Brain Reserve and Cognitive Reserve

Individual differences and their consequences regarding the outcome in case of pathological and traumatic processes in the brain are dealt with intensively in the literature on the issues of "brain reserve" (e.g. Luk et al., 2011; Satz, 1993; Scheibel et al., 2009; Schmand et al., 1997; Valenzuela and Sachdev, 2006) and "cognitive reserve" (e.g. Alexander et al., 1997; Craik et al., 2010; Fuentes et al., 2010; Kesler et al., 2003; Ropacki and Elias, 2003; Schweizer et al., 2012; Stern, 2002, 2003, 2006). These concepts are primarily discussed in the context of Alzheimer's dementia but also with respect to various types of vascular and/or traumatic brain injury (e.g. Kesler et al., 2003; Satz, 1993; Stern, 2002, 2003). It is, however, obvious that even the literature dealing with "brain reserve" and/or "cognitive reserve" in the context of Alzheimer's dementia has implications for the cognitive prognosis of patients suffering various types of brain injury. Both "brain reserve" and "cognitive reserve" refer to situations in which the pretraumatic condition influences the degree to which patients posttraumatically (or during degenerative neural processes, e.g., Alzheimer's dementia) are able to "compensate" for the neural and cognitive loss. "Brain reserve" primarily emphasizes structural aspects of the brain (e.g. density of synaptic contacts), while the emphasis in "cognitive reserve" is on the availability of cognitive processes and strategies. As will be discussed below, a brain may in certain respects hold a "reserve" which is in itself rather cognitively independent but can promote aspects of cognitive rehabilitation and/or "compensation". In most instances, however, the neural features considered in "brain reserve" may reflect the neural aspects of what is at another level of analysis considered the "cognitive reserve". Number of neurons, density of synaptic connections, level of cerebral metabolism or grey matter volume within a cortical region may for instance all reflect the neural substrate of what is at the cognitive level seen as a higher richness of available cognitive strategies. In the context of the REF-model, "brain reserve" may in a particular study be seen as the analysis of the degree of connectivity between the neural substrate of individual EFs, while "cognitive reserve" may refer to the same situation – but analysed at the cognitive level of the availability of ASs and/or AMs. In "brain reserve" there may be a tendency to (over-)stress the quantitative aspects of synaptic connectivity, while the idea of "cognitive reserve" puts a higher emphasis on the quality (shape) of the pattern of synaptic connections – conceptualised as available cognitive strategies.

In the context of the REF-model, "reserve" (be it conceptualised as "brain" or "cognitive") must primarily reflect an enhanced capability to select and establish adequate ASs. Although many factors may influence such an ability, three such factors may be of primary importance:

A. The availability of adequate ASs and/or AMs.
B. The efficacy of the selector/evaluator mechanism.
C. The efficacy of the cellular and biochemical mechanisms involved in the synaptic modifications necessary to achieve the neural reorganizations mediating the

association between a particular AS and a given task situation as well as the modified connectivity between the neural substrates of EFs necessary for the establishment of novel ASs.

Regarding the first of these points (A), it is interesting to notice that the level of premorbid education (e.g. Kesler et al., 2003; Schneider et al., 2014) and the presence of premorbid "complex mental activity" (e.g. Valenzuela and Sachdev, 2006) are amongst the factors associated with improved cognitive outcome in case of traumatic brain injury and dementia. In general, a relatively low level of education appears to be a risk factor regarding the development of dementia (e.g. Gatz et al., 2006, 2007; Roe et al., 2008). Since ASs and (at least to an extent) AMs are established and optimized as a result of situations in which the individual is confronted with more or less novel problem solving situations, it is expected that a richer library of both ASs and AMs is available to individuals who have experienced a higher level of formal or informal education as well as a life-style dominated by more complex cognitive activities. Although brain injury may rob such an individual of the neural substrate of the ASs which pretraumatically formed the basis of important cognitive activities, the more extensive library of ASs may allow a rather prompt selection of alternative ASs – allowing an optimal posttraumatic rehabilitative process. In some instances, the posttraumatic selection of alternative ASs may even permit an "immediate recovery" in which no period of posttraumatic symptoms can be noticed (e.g. Mogensen et al., 2002; Wörtwein et al., 1995). In such an instance, the brain injury in question – or for that matter the level of neurodegeneration associated with a progressive process of dementia – may remain asymptomatic and clinically unnoticed. However, even in the absence of appropriate ASs, the de novo establishment of ASs may be significantly eased by a higher availability of AMs which can be recruited as essential components to novel ASs.

Although this issue has not been discussed extensively in the previous descriptions of the REF-model (e.g. Mogensen, 2011a, 2011c, 2012; Mogensen and Malá, 2009), the posttraumatic processes of establishing novel ASs and associating new or pre-existing ASs with particular situations may differ significantly in their efficacy with varied efficacy of the evaluator/selector mechanism. It is variations such as these in selector/evaluator efficacy that is referred to under B. One factor which may contribute to a more efficient evaluator/selector mechanism is the premorbid "training" of such a system. The "training" of the selector/evaluator mechanism is in itself associated with a reorganization of connectivity between the neural substrates of and interactions between EFs – since the selector/evaluator mechanism is formed by a number of interacting ASs (see above). In general, the above-mentioned effects of level of education and presence of complex cognitive activities may, in themselves, represent different levels of challenges to and thereby training of the evaluator/selector mechanism.

Another way in which the selector/evaluator mechanism may be exposed to an increased level of activity and "training" is via the demands posed by bilingualism. An individual faced with the parallel use of two languages (as opposed to only one) will have to perform an ongoing selection and monitoring of the cognitive and neural mechanisms associated with one or the other language, respectively. In this context, it is interesting to notice that bilingualism has been demonstrated to delay the onset of Alzheimer's dementia (e.g. Bialystok et al., 2007; Chertkow et al., 2010; Craik et al., 2010; Schweizer et al., 2012). Additionally, Luk et al. (2011) found bilingualism to be associated with a higher level of white matter integrity in older individuals. Bialystok (2009) has emphasized that bilingualism is associated with both beneficial and detrimental effects on various aspects of cognition. For instance, memory tasks based on verbal recall are performed more poorly by bilinguals. However – and in agreement with what would be expected on the basis of improved performance of the selector/evaluator mechanism – memory tasks based primarily on executive control are performed better by bilinguals (Bialystok, 2009).

Numerous factors (genetic as well as environmental) affect the cellular and biochemical processes referred to above under point C. One of the prerequisites for optimal neural functioning – including plastic abilities – is an adequate supply of neurotrophic factors (including the neurotrophin Brain Derived Neurotrophic Factor (BDNF)) (e.g. Kafitz et al., 1999; Levi-Moltalcini, 1982; Lewin and Barde, 1996). BDNF is produced by glial cells as well as neurons and the supply to a neuron, in the form of synaptic take-up followed by retrograde axonal transport, is mostly achieved through synaptic interactions within an efficient functional network (e.g. Hamburger, 1934, 1975; Hollyday and Hamburger, 1976; Levi-Moltalcini, 1982; Levi-Moltalcini and Levi, 1942). Ultimately, the lack of BDNF may lead to the death of a neuron. One of the many (e.g. Siesjo and Siesjo, 1996) mechanisms of secondary and tertiary neural degeneration after brain injury is that neurons within brain regions, which are originally spared by brain injury, but have lost the possibility to interact (and receive for instance BDNF as part of that interaction) with neurons lost to the primary injury are deprived of neurotrophins and die (e.g. Sofroniew et al., 1993). Situations in which higher levels of BDNF are available may therefore lead to improved posttraumatic cognitive performance and rehabilitation via at least two mechanisms: (I) reduced secondary and tertiary degeneration may better preserve brain regions originally spared by the primary trauma – making a richer library of ASs and AMs available during the posttraumatic reorganizational processes – and (II) the improved plastic abilities associated with higher levels of BDNF (e.g. Kafitz et al., 1999; Levi-Moltalcini, 1982; Lewin and Barde, 1996) can contribute to the relative ease of the neuroplastic processes referred to under C. In the central nervous system, the hormone erythropoietin (EPO) has a broad range of effects (e.g. Mammis et al., 2009), including stimulation of the production of BDNF (e.g. Mahmood et al., 2007; Viviani et al., 2005; Zhang et al., 2006). Administration of EPO (and consequently increased production of BDNF) has been found to be associated with improved posttraumatic symptomatology as well as rehabilitation – including improved cognitive performance and rehabilitation – after vascular as well as traumatic brain injury (e.g. Alafaci et al., 2000; Brines et al., 2000; Buemi et al., 2000; Calapai ct al., 2000; Grasso, 2001; Lu et al., 2005; Malá et al., 2005, 2007; Mogensen et al., 2004b, 2008a, 2008b; Siren et al., 2001; Springborg et al., 2002; Wang et al., 2006). The degree to which these therapeutic effects are mediated via stimulation of BDNF remains unknown, and the relative contribution of reduced secundary/tertiary degeneration and improved neural plastic abilities, respectively, is also poorly understood at the present moment. It is, however, interesting to note that posttraumatic functional recovery and rehabilitation after various types of brain injury can also be improved by increased physical activity in the form of exercise (e.g. Arida et al., 2009; Devine and Zafonte, 2009; Griesbach et al., 2004a, 2004b, 2009; Hayes et al., 2008; Luo et al., 2007; Malá et al., 2008; Seo et al., 2010). The relevance of such findings in the present context is not the least that physical activation and exercise are known to increase the production of BDNF (e.g. Griesbach et al., 2004a, 2004b, 2009; Moltini et al., 2004; Vaynman et al., 2003). Under certain circumstances, the exercise-induced improvement in cognitive abilities after brain injury have been found to depend upon stimulation of the production of BDNF (Griesbach et al., 2009).

To the extent that a more direct "brain reserve" can be distinguished from a "cognitive reserve", the former, a "pure brain reserve", may primarily be associated with an improved plastic potential and maybe even a stronger resistance to secondary and tertiary brain injury. Such "reserve factors" may, for instance, be associated with situations of pretraumatically increased production of BDNF. An example of such a situation may be seen in the results obtained by Malá et al. (2008) where pretraumatic physical activation was associated with an improved posttraumatic cognitive recovery after mechanical brain injury. It may also be of interest to notice that a higher level of physical exercise appears to be associated with reduced risk of Alzheimer's dementia (e.g. Baker et al., 2010; Teri et al., 2003).

Localization and Recovery of Cognitive Functions – Rehabilitation and Reconceptualisation

The reorganizational neural and cognitive processes upon brain injury are often described as being divided into two phases: initially a relearning of compromized and/or lost "functions", later followed by compensational processes which support the behavioural abilities without re-establishing what has been lost to injury (e.g. Stein and Hoffman, 2003). Viewed within the framework of the REF-model, such a description may be somewhat artificial and potentially misleading with regard to the possibility of re-establishing lost functions – if "function" is considered at the level of the EFs. As mentioned above and discussed extensively elsewhere (Mogensen, 2011a), the lost neural substrate of EFs is unlikely to be re-established elsewhere in the brain. However, in case of subtotal lesions of various structures (or more likely substructures) of the brain, mechanisms such as those suggested by Robertson and Murre (1999) may allow a degree of "re-establishment" of the original substrate of task mediation (see also Mogensen and Malá, 2009). Additionally, an actual "relearning" of the original "function" may be possible in case the trauma-associated impairment is caused by loss of EFs which fall within one of the categories where a large number of essentially similar EFs are available. This may potentially be the situation within some of the "modules" of primary sensory and motor areas of for instance the neocortex. Potentially, sensory representations such as the somatotopic map of somatosensory representations and the tonotopic map of auditory frequency representations may contain the neural substrates of a large number of essentially identical EFs, which derive their different "functional qualities" (representations of a particular body part or tonal frequency, respectively) exclusively from their input and not from their internal information processing. Loss of such EFs may eventually lead to a reorganization in which the input to the now missing EFs is redirected to the neural substrate of the preserved EFs within the same domain – EFs which are able to perform an information processing similar to what has been lost. Potentially, these conditions could exist even within functional domains further from the input and output pathways of the brain. If this is the case, the posttraumatic reorganization may, indeed, be able to replace the lost EFs with EFs which are able to perform an identical information processing to that seen pretraumatically. However, as has been stressed above, such a situation is likely to be an exception rather than anything which can normally be expected to occur.

Even with respect to the second phase of the posttraumatic processes described by Stein and Hoffman (2003), the compensational processes, there may be a need for a degree of reconceptualisation. Although the mechanism of selection and potentially even creation of alternative ASs can in a strict sense be seen as "compensational processes" (since the pretraumatic information processing is never re-established), there is a need for a word of caution regarding what is to be understood by the term "compensation". As stressed by Levin et al. (2009) within the motor domain, there is a pressing need for better clarification and conceptual rigor with regard to the use of the terms "recovery" and "compensation". This definitely holds for the motor systems addressed by Levin et al. (2009), but even more so for the posttraumatic cognitive processes presently addressed (e.g. Mogensen, 2014). At the neural level and at the level of the strictly localized, and in case of injury permanently lost, EFs, the posttraumatic processes will according to the REF-model not include the re-establishment of the lost circuitry and can therefore by some definitions be claimed to be exclusively "compensational". It is, however, essential to remember that this type of "compensation" is likely to manifest itself at the level of the surface phenomena in the form of behaviour/cognitive representations which may appear completely identical to what was seen pretraumatically (unless challenged by detailed analytical methods). At the level of the surface phenomena, such a "compensation" differs radically from what is mostly understood by such a term. In an extensive and thorough review, Wilson (2000) deals with the concept and methods of compensation within neurorehabilitation upon brain injury. When addressing compensational methods in case of the amnesias, Wilson (2000) emphasizes the support of memory by diaries, tape-recorders, and alarm-watches, while compensation for speech and language deficits include voice-synthesizers, alternative communication systems as well as mime and gesture. There can be no doubt that rehabilitative cognitive training upon brain injury can greatly benefit from the utilization of many such compensational strategies. It is, however, to be noted that the type of "compensation" emphasized by the REF-model will typically at the level of the surface phenomena result in behavioural and conscious manifestations which mostly will be judged clinically as a return to more or less complete pretraumatic proficiency and thereby be considered "re-establishment of function". When, for instance, the so-called constraint-induced rehabilitative training of aphasic patients deliberately strives to eliminate all "compensational" processes (e.g. various types of non-verbal communication) and achieves a successful re-establishment of language (e.g. Cherney et al., 2008; Meinzer et al., 2005; Pulvermüller et al., 2001) the re-established linguistic abilities are still (according to the REF-model) the result of "compensational" processes at the neural level as well as at the level of the EFs – having been established via alternative and potentially newly created ASs. These reconceptualizations of "compensation" are discussed in detail by Mogensen (2014).

At the theoretical level, the REF-model has provided a framework within which the concepts of localization of function and functional recovery can coexist. Additionally, it has provided a structure within which connectionist networks (e.g. McClelland et al., 1986; McLeod et al., 1998; Rumelhart and McClelland, 1986) can coexist with "modularity" (e.g. Barrett and Kurzban, 2006; Fodor, 2000; Pinker, 1999). The functionally highly specialized EFs place the REF-model within the framework of what is called "Massive Modularity" (e.g. Barrett and Kurzban, 2006) and it consequently embrases the idea of "rules of computation" associated with such modularity-based models. However, at the level of the ASs, the REF-model must be considered connectionist and distributed – and the ASs are established and modified according to backpropagation mechanisms. Therefore, the basis for an AS is more or less a connectionist network of interconnected EFs. Via this combination, the REF-model accomodates both the modularity-based predictability of for instance lesion effects and the connectionist flexibility and potential for dynamic reorganizations which allow behavioural and cognitive flexibility in the intact individual as well as the posttraumatic cognitive rehabilitation of the brain injured patient.

The REF-model thus provides a theoretical framework linking the mechanisms of neural rehabilitation to both the dynamic processes mediating behavioural and cognitive flexibility in the intact brain and to a number of theoretical concepts utilized in the attempts to understand the organization and reorganization of the brain. However, the REF-model is clearly not the type of comprehensive theoretical framework for clinical rehabilitation which Wilson (2002) convincingly argues is lacking. That said, the implications provided by the REF-model point to the need for improvements regarding both clinical practise and future research. Fortunately, the exceptionally dynamic nature of our brains enables a degree of posttraumatic functional recovery to occur in most cases of brain injury. Often, the level of such a recovery is impressively high. It is, however, a challenge to both basic research and clinical practise to develop even better methods to harvest the results of such a neural and cognitive flexibility. Developing more ecologically valid rehabilitative training programmes, utilizing advanced technologies in better ways, developing rehabilitative training techniques which can more adequately ensure a comprehensive repertoire of AMs to be available to the patient during posttraumatic life are but a few of these challenges. More than anything else basic neuroscience as well as clinical rehabilitative practises need to exhibit a conceptual and methodological flexibility which can in the future better match the increadible plasticity and flexibility of the normal and injured brain.

Conclusion

The REF-model resolves the apparent contradiction between "functional localization" and "functional recovery" after brain injury. It does so by combining two organizational principles, which are often seen as opposed to each other. A flexible connectionist network, which is constantly modified by backpropagation associated with situational feedback. And a modular organization based on the strictly localized (and in case of injury permanently lost) EFs. The EFs are the basic elements within the connectionist network. By this combination, the model provides a conceptual framework within which to understand the neural and cognitive mechanisms of behavioural and conscious flexibility in intact and in brain injured individuals. Thereby the REF-model offers a theoretical basis on which to understand and therapeutically support the posttraumatic recovery of cognitive functions. The model stresses a number of challenges to and shortcomings of neurorehabilitation – not the least a frequent lack of generalization of training results from clinical settings to the reallife situations of the patient. Such problems have to be solved, partly by the use of advanced technologies such as virtual reality and portable microcomputers. Also, there is a need for rehabilitative training methods, which improve the AMs – and by doing so gives the patient a better basis for the future challenges of daily life. Finally, it is clear that the prognosis regarding short- and long-term cognitive outcome after brain injury depends on the pretraumatic experiences (e.g. education) of the patient. And that it is crucial that the rehabilitative training is specifically tailored to the background and "cognitive profile" of the individual patient.

Acknowledgment

The present study was supported by a grant from the Danish Council for Independent Research.

References

Alafaci, C., Salpietro, F., Grasso, G., Sfacteria, A., Passalacqua, M., Morabito, A., Tripodo, E., Calapai, G., Buemi, M. and Tomasello, F. (2000). Effect of recombinant human erythropoietin on cerebral ischemia following experimental subarachnoid hemorrhage. *European Journal of Pharmacology*, 406: 219-225.

Albrecht, T., Klapötke, S. and Mattler, U. (2010). Individual differences in metacontrast masking are enhanced by perceptual learning. *Consciousness and Cognition*, 19: 656-666.

Alexander, G.E., Furey, M.L., Grady, C.L., Pietrini, P., Brady, D.R., Mentis, M.J. and Schapiro, M.B. (1997). Association of premorbid intellectual function with cerebral metabolism in Alzheimer's disease: implications for the cognitive reserve hypothesis. *American Journal of Psychiatry,* 154: 165-172.

Alexander, M.P., Naeser, M.A. and Palumbo, C. (1990). Broca's area aphasias: aphasia after lesions including the frontal operculum. *Neurology*, 40: 353-362.

Ansaldo, A.I. and Arguin, M. (2003). The recovery from aphasia depends on both the left and right hemispheres: three longitudinal case studies on the dynamics of language function after aphasia. *Brain and Language,* 87: 177-178.

Ansaldo, A.I., Arguin, M. and Lecours, A.R. (2002). The contribution of the right cerebral hemisphere to the recovery from aphasia: a single longitudinal case study. *Brain and Language,* 82: 206-222.

Arida, R.M., Scorza, F.A., Scorza, C.A. and Cavalheiro, E.A. (2009). Is physical activity beneficial for recovery in temporal lobe epilepsy? Evidences from animal studies. *Neuroscience and Biobehavioral Reviews,* 33: 422-431.

Arvidsson, A., Collin, T., Kirik, D., Kokaia, Z. and Lindvall, O. (2002). Neuronal replacement from endogenous precursors in the adult brain after stroke. *Nature Medicine,* 8: 963-970.

Bach-y-Rita, P., Collins, C.C., Saunders, F.A., White, B. and Scadden, L. (1969). Vision substitution by tactile image projection. *Nature,* 221: 963-964.

Bach-y-Rita, P., Kaczmarek, K.A., Tyler, M.E. and Garcia-Lara, J. (1998). Form perception with a 49-point electrotactile stimulus array on the tongue: A technical note. *Journal of Rehabilitation Research and Development,* 35: 1-7.

Baker, L.D., Frank, L.L., Foster-Schubert, K., Green, P.S., Wilkinson, C.W., McTiernan, A., Plymate, S.R., Fishel, M.A., Watson, S., Cholerton, B.A., Duncan, G.E., Mehta, P.D. and Craft, S. (2010). Effects of aerobic exercise on mild cognitive impairment. *Archives of Neurology,* 67: 71-79.

Barrett, H.C. and Kurzban, R. (2006). Modularity in cognition: Framing the debate. *Psychological Review*, 113: 628-647.

Baumgaertner, A., Schraknepper, V. and Saur, D. (2005). Differential recovery of aphasia and apraxia of speech in an adolescent after infarction of the left frontal lobe: longitudinal behavioral and fMRI data. *Brain and Language,* 95: 211-212.

Bell, M., Bryson, G. and Wexler, B.E. (2003). Cognitive remediation of working memory deficits: durability of training effects in severely impaired and less severely impaired schizophrenia. *Acta Psychiatrica Scandinavica,* 108: 101-109.

Bennett, E.L., Diamond, M.L., Krech, D. and Rosenzweig, M.R. (1964). Chemical and anatomical plasticity of brain. *Science*, 146: 610-619.

Bennett, E.L., Rosenzweig, M.R. and Diamond, M.C. (1969). Rat brain: Effects of anvironmental enrichment on wet and dry weights. *Science*, 163: 825-826.

Berardi, N., Pizzorusso, T. and Maffei, L. (2004). Extracellular matrix and visual cortical plasticity: Freeing the synapse. *Neuron*, 44: 905-908.

Berry, M. (1982). Post-injury myelin-breakdown products inhibit axonal growth: An hypothesis to explain the failure of axonal regeneration in the mammalian central nervous system. *Bibliotheca Anatomica*, 23: 1-11.

Bialystok, E. (2009). Bilingualism: the good, the bad, and the indifferent. *Bilingualism: Language and Cognition,* 12: 3-11.

Bialystok, E., Craik, F.I.M. and Freedman, M. (2007). Bilingualism as a protection against the onset of symptoms of dementia. *Neuropsychologia,* 45: 459-464.

Bland, S.T., Pillai, R.N., Aronowski, J., Grotta, J.C. and Schallert, T. (2001). Early overuse and disuse of the affected forelimb after moderately severe intraluminal suture occlusion of the middle cerebral artery in rats. *Behavioral Brain Research,* 126: 33-41.

Block, N. and Fodor, J. (1972). What psychological states are not. *Philosophical Review*, 81: 159-181.

Bridgeman, B. and Staggs, D. (1982). Plasticity in human blindsight. *Vision Research*, 22: 1199-1203.

Brines, M.L., Ghezzi, P., Keenan, S., Agnello, D., de Lanerolle, N.C., Cerami, C., Itri, L.M. and Cerami, A. (2000). Erythropoietin crosses the blood-brain barrier to protect against experimental brain injury. *Proceedings of the National Academy of Sciences, US,* 97: 10526-10531.

Bryson, A.E. and Ho, Y-C. (1969). *Applied Optimal Control*. New York: Blaisdell.

Buemi, M., Grasso, G., Corica, F., Calapai, G., Salpietro, F.M., Casuscelli, T., Sfacteria, A., Aloisi, C., Alafaci, C., Sturiale, A., Frisina, N. and Tomasello, F. (2000). In vivo evidence that erythropoietin has a neuroprotective effect during subarachnoid hemorrhage. *European Journal of Pharmacology,* 392: 31-34.

Buller, D.J. and Hardcastle, V.G. (2000). Evolutionary psychology, meet developmental neurobiology: against promiscuous modularity. *Brain and Mind,* 1: 307-325.

Buschkuehl, M., Jaeggi, S.M., Hutchison, S., Perrig-Chiello, P., Däpp, C., Müller, M., Breil, F., Hoppeler, H. and Perrig, W.J. (2008). Impact of working memory training on memory performance in old-old adults. *Psychology and Aging,* 23: 743-753.

Calapai, G., Marciano, M.C., Corica, F., Allegra, A., Parisi, A., Frisina, N., Caputi, A.P. and Buemi, M. (2000). Erythropoietin protects against brain ischemic injury by inhibition of nitric oxide formation. *European Journal of Pharmacology*, 401: 349-356.

Carney, N., Chesnut, R.M., Maynard, H., Mann, N.C., Patterson, P. and Helfand, M. (1999). Effect of cognitive rehabilitation on outcomes for persons with traumatic brain injury: A systematic review. *Journal of Head Trauma Rehabilitation*, 14: 277-307.

Chen, D.F., Jhaveri, S. and Schneider, G.E. (1995). Intrinsic changes in developing retinal neurons result in regenerative failure of their axons. *Proceedings of the National Academy of Sciences US*, 92: 7287-7291.

Chen, J., Magavi, S.S. and Macklis, J.D. (2004). Neurogenesis of corticospinal motor neurons extending spinal projections in adult mice. *Proceedings of the National Academy of Sciences US*, 101: 16357-16362.

Cherney, L.R., Patterson, J.P., Raymer, A., Frymark, T. and Schooling, T. (2008). Evidence-based systematic review: effects of intensity of treatment and constraint-induced language therapy for individuals with stroke-induced aphasia. *Journal of Speech, Language, and Hearing Research,* 51: 1282-1299.

Chertkow, H., Whitehead, V., Phillips, N., Wolfson, C., Atherton, J. and Bergman, H. (2010). Multilingualism (but not always bilingualism) delays the onset of Alzheimer disease: evidence from a bilingual community. *Alzheimer Disease and Associated Disorders,* 24: 118-125.

Choi, J.Y., Lee, K.H., Na, D.L., Byun, H.S., Lee, S.J., Kim, H., Kwon, M., Lee, K-H. and Kim, B-T. (2007). Subcortical aphasia after striatocapsular infarction: Quantitative analysis of brain perfusion SPECT using statistical parametric mapping and a statistical probabilistic anatomic map. *Journal of Nuclear Medicine*, 48: 194-200.

Chokron, S., Perez, C., Obadia, M., Gaudry, I., Laloum, L. and Gout, O. (2008). From blindsight to sight: cognitive rehabilitation of visual field defects. *Restorative Neurology and Neuroscience*, 26: 305-320.

Cicerone, K.D., Mott, T., Azulay, J. and Friel, J.C. (2004). Community integration and satisfaction with functioning after intensive cognitive rehabilitation for traumatic brain injury. *Archives of Physical Medicine and Rehabilitation,* 85: 943-950.

Cifu, D.X., Kreutzer, J.S., Kolakowsky-Hayner, S.A., Marwitz, J.H. and Englander, J. (2003). The relationship between therapy intensity and rehabilitative outcomes after traumatic brain injury: a multicenter analysis. *Archives of Physical Medicine and Rehabilitation,* 84: 1441-1448.

Coltheart, M. (2001). Assumptions and methods in cognitive neuropsychology. In: Rapp, B. (Ed.), *The Handbook of Cognitive Neuropsychology.* Philadelphia, PA: Psychology Press; pp. 3-21.

Coyle, S., Ward, T. and Markhan, C. (2003). Brain-computer interfaces: a review. *Inter-disciplinary Science Reviews,* 28: 112-118.

Craik, F.I.M., Bialystok, E. and Freedman, M. (2010). Delaying the onset of Alzheimer disease. Bilingualism as a form of cognitive reserve. *Neurology,* 75: 1726-1729.

Crosson, B., Moore, A.B., McGregor, K.M., Chang, Y-L., Benjamin, M., Gopinath, K., Sherod, M.E., Wierenga, C.E., Peck, K.K., Briggs, R.W., Rothi, L.J.G. and White, K.D. (2009). Regional changes in word-production laterality after a naming treatment designed to produce a rightward shift in frontal activity. *Brain and Language,* 111: 73-85.

Daly, J.J. and Wolpaw, J.R. (2008). Brain-computer interfaces in neurological rehabilitation. *The Lancet Neurology,* 7: 1032-1043.

DeBow, S.B., McKenna, J.E., Kolb, B. and Colbourne, F. (2004). Immediate constraint-induced movement therapy causes local hyperthermia that exacerbates cerebral cortical injury in rats. *Canadian Journal of Physiology and Pharmacology,* 82: 231-237.

Del Rio, J.A. and Soriano, E. (2007). Overcoming chondroitin sulphate proteoglycan inhibition of axon growth in the injured brain: Lessons from chondroitinase ABC. *Current Pharmacological Design*, 13: 2485-2492.

Devine, J.M. and Zafonte, R.D. (2009). Physical exercise and cognitive recovery in acquired brain injury: a review of the literature. *Physical Medicine and Rehabilitation,* 1: 560-575.

Dineen, J. and Keating, E.G. (1981). The primate visual system after bilateral removal of striate cortex. Survival of complex pattern vision. *Experimental Brain Research*, 41: 338-345.

Elbert, T., Pantev, C., Weinbruch, C., Rockstroh, B. and Taub, E. (1995). Increased cortical representation of the fingers of the left hand in string players. *Science,* 270: 305-307.

Fawcett, J.W., Housden, E., Smith-Thomas, L. and Meyer, R.L. (1989). The growth of axons in three-dimensional astrocyte cultures. *Developmental Biology*, 135: 449-458.

Finger, S. and Stein, D.G. (1982). Vicariation theory and radical reorganization of function. In: Finger, S. and Stein, D.G. (Eds.), *Brain Damage and Recovery: Research and Clinical Perspectives.* New York: Academic Press; pp. 287-302.

Fodor, J. (2000*). The Mind Doesn't Work That Way: The Scope and Limits of Computational Psychology.* Cambridge, MA: MIT Press.

Frassinetti, F., Angeli, V., Meneghello, F., Avanzi, S. and Ladavas, E. (2002). Long-lasting amelioration of visuospatial neglect by prism adaptation. *Brain,* 125: 608-623.

Fuentes, A., McKay, C. and Hay, C. (2010). Cognitive reserve in paediatric traumatic brain injury: relationship with neuropsychological outcome. *Brain Injury,* 24: 995-1002.

Fyhn, M., Molden, S., Witter, M.P., Moser, E.I. and Moser, M-B. (2004). Spatial representation in the entorhinal cortex. *Science,* 305: 1258-1264.

Gatz, M., Mortimer, J.A., Fratiglioni, L., Johansson, B., Berg, S., Reynolds, C.A. and Pedersen, N.L. (2006). Potentially modifiable risk factors for dementia in identical twins. *Alzhemier's and Dementia,* 2: 110-117.

Gatz, M., Mortimer, J.A., Fratiglioni, L., Johansson, B., Berg, S., Andel, R., Crowe, M., Fiske, A., Reynolds, C.A. and Pedersen, N.L. (2007). Accounting for the relationship between low education and dementia: a twin study. *Physiology and Behavior,* 92: 232-237.

Goldberg, J.L., Klassen, M.P., Hua, Y. and Barres, B.A. (2002). Amacrine-signaled loss of intrinsic axon growth ability by retinal ganglion cells. *Science*, 296: 1860-1864.

Grasso, G. (2001). Neuroprotective effect of recombinant human erythropoietin in experimental subarachnoid hemorrhage. *Journal of Neurosurgical Sciences,* 45: 7-14.

Griesbach, G.S., Gomez-Pinilla, F. and Hovda, D.A. (2004a). The upregulation of plasticity-related proteins following TBI is disrupted with acute voluntary exercise. *Brain Research,* 1016: 154-162.

Griesbach, G.S., Hovda, D.A., Molteni, R., Wu, A. and Gomez-Pinilla, F. (2004b). Voluntary exercise following traumatic brain injury: brain-derived neurotrophic factor upregulation and recovery of function. *Neuroscience,* 125: 129-139.

Griesbach, G.S., Hovda, D.A. and Gomez-Pinilla, F. (2009). Exercise-induced improvement in cognitive performance after traumatic brain injury in rats is dependent on BDNF activation. *Brain Research,* 1288: 105-115.

Günther, V.K., Schäfer, P., Holzner, B.J. and Kemmler, G.W. (2003). Long-term improvements in cognitive performance through computer-assisted cognitive training: a pilot study in a residential home for older people. *Aging and Mental Health,* 7: 200-206.

Hafting, T., Fyhn, M., Molden, S., Moser, M-B. and Moser, E.I. (2005). Microstructure of a spatial map in the entorhinal cortex. *Nature,* 436: 801-806.

Hamburger, V. (1934). The effects of wing bug extirpation in chick embryos on the development of the central nervous system. *Journal of Experimental Zoology*, 68: 449-494.

Hamburger, V. (1975). Cell death in the development of the lateral motor column of the chick embryol. *Journal of Comparative Neurology*, 160: 535-546.

Hayes, K., Sprague, S., Guo, M., Davis, W., Friedman, A., Kumar, A., Jimenez, D.F. and Ding, Y. (2008). Forcecd, not voluntary, exercise effectively induces neuroprotection in stroke. *Acta Neuropathologica,* 115: 289-296.

Henriksson, L., Raninen, A., Näsänen, R., Hyvärinen, L. and Vanni, S. (2007). Training-induced cortical representation of a hemianopic hemifield. *Journal of Neurology, Neurosurgery and Psychiatry*, 78: 74-81.

Henze, D.A., Borhegyi, Z., Csicsvari, J., Mamiya, A., Harris, K.D. and Buzsaki, G. (2000). Intracellular features predicted by extracellular recordings in the hippocampus in vivo. *Journal of Neurophysiology,* 84: 390-400.

Hollyday, M. and Hamburger, V. (1976). Reduction of the naturally occurring motor neuron loss by enlargement of the periphery. *Journal of Comparative Neurology*, 170: 311-320.

Humm, J.L., Kozlowski, D.A., James, D.C., Gotts, J.E. and Schallert, T. (1998). Use-dependent exacerbation of brain damage occurs during an early post-lesion vulnerable period. *Brain Research,* 783: 286-292.

Humphrey, N.K. (1974). Vision in a monkey without striate cortex: a case study. *Perception,* 3: 241-255.

Irvine, D.R.F. (2007). Auditory cortical plasticity: does it provide evidence for cognitive processing in the auditory cortex? *Hearing Research*, 229: 158-170.

Kaczmarek, K.A., Webster, J.G., Bach-y-Rita, P. and Tompkins, W.J. (1991). Electrotactile and vibrotactile displays for sensory substitutionsystems. *Biomedical Engineering*, 38: 1-16.

Kafitz, K.W., Rose, C.R., Thoenen, H. and Konnerth, A. (1999). Neurotrophin-evoked rapid excitation through TrkB receptors. *Nature*, 401: 918-921.

Karl, A., Birbaumer, N., Lutzenberger, W., Cohen, L.G. and Flor, H. (2001). Reorganization of motor and somatosensory cortex in upper extremity amputees with phantom limb pain. *Journal of Neuroscience*, 15: 3609-3618.

Katz, R.C. (2009). Application of computers to the treatment of US veterans with aphasia. *Aphasiology,* 23: 1116-1126.

Kesler, S.R., Adams, H.F., Blasey, C.M. and Bigler, E.D. (2003). Premorbid intellectual functioning, education, and brain size in traumatic brain injury: an investigation of the cognitive reserve hypothesis. *Applied Neuropsychology,* 10: 153-162.

Kozlowski, D.A., James, D.C. and Schallert, T. (1996). Use-dependent exaggeration of neuronal injury after unilateral sensorimotor cortex lesions. *Journal of Neuroscience,* 16: 4776-4786.

Krechevsky, I. (1932). "Hypotheses" in rats. *Psychological Reviews,* 39: 516-532.

Krechevsky, I. (1933). Hereditary nature of "hypotheses". *Journal of Comparative Psychology*, 16: 99-116.

Kringelbach, M.L. and Rolls, E.T. (2004). The functional neuroanatomy of the human orbitofrontal cortex: evidence from neuroimaging and neuropsychology. *Progress in Neurobiology,* 7: 341-372.

Leavitt, B.R., Hernit-Grant, C.S. and Macklis, J.D. (1999). Mature astrocytes transform into transitional radial glia within adult mouse neocortex that supports directed migration of transplanted immature neurons. *Experimental Neurology*, 157: 43-57.

León-Carrión, J. and Machuca-Murga, F. (2001). Spontaneous recovery of cognitive functions after severe brain injury: When are neurocognitive sequelae established? *Revista Española de Neuropsicologia*, 3: 58-67.

Levi-Montalcini, R. (1982). Developmental neurobiology and the natural history of nerve growth factor. *Annual Review of Neuroscience*, 5: 341-362.

Levi-Montalcini, R. and Levi, G. (1942). Les conséquences de la destruction d'un territoire d'innervation périphérique sur le développement des centres nerveux correspondants dans l'embryon de poulet. *Archives of Biology*, 53: 537-545.

Levin, M.F., Kleim, J.A. and Wolf, S.L. (2009). What do motor "recovery" and "compensation" mean in patients following stroke? *Neurorehabilitation and Neural Repair,* 23: 313-319.

Lewin, G.R. and Barde, Y.A. (1996). Physiology of the neurotrophins. *Annual Review of Neuroscience*, 19: 289-317.

Lu, D., Mahmood, A., Qu, C., Goussev, A., Schallert, T. and Chopp, M. (2005). Erythropoietin enhances neurogenesis and restores spatial memory in rats after traumatic brain injury. *Journal of Neurotrauma,* 22: 1011-1017.

Luk, G., Bialystok, E., Craik, F.I.M. and Grady, C.L. (2011). Lifelong bilingualism maintains white matter integrity in older adults. *Journal of Neuroscience,* 31: 16808-16813.

Luo, C.X., Jiang, J., Zhou, Q.G., Zhu, X.J., Wang, W., Zhang, Z.J., Han, X. and Zhu, D.Y. (2007). Voluntary exercise-induced neurogenesis in the postischemic dentate gyrus is associated with spatial memory recovery from stroke. *Journal of Neuroscience Research,* 85: 1637-1646.

Magavi, S.S., Leavitt, B.R. and Macklis, J.D. (2000). Induction of neurogenesis in the neocortex of adult mice. *Nature*, 405: 951-955.

Mahmood, A., Lu, D., Qu, C., Goussev, A., Zhang, Z.G., Lu, C. and Chopp, M. (2007). Treatment of traumatic brain injury in rats with erythropoietin and carbamylated erythropoietin. *Journal of Neurosurgery,* 107: 392-397.

Malá, H., Alsina, C.G., Madsen, K.S., la Cour Sibbesen, E., Stick, H. and Mogensen, J. (2005). Erythropoietin improves place learning in an 8-arm radial maze in fimbria-fornix transected rats. *Neural Plasticity***,** 12: 329-340.

Malá, H., Castro, M.R., Jørgensen, K.D. and Mogensen, J. (2007). Effects of erythropoietin on posttraumatic place learning in fimbria-fornix transected rats after a 30-day postoperative pause. *Journal of Neurotrauma*, 24: 1647-1657.

Malá, H., Castro, M.R., Knippel, J., Køhler, P.J., Lassen, P. and Mogensen, J. (2008). Therapeutic effects of a restraint procedure on posttraumatic place learning in fimbria-fornix transected rats. *Brain Research,* 1217: 221-231.

Malá, H., Castro, M.R., Pearce, H., Kingod, S.C., Nedergaard, S.K., Scharff, Z., Zandersen, M. and Mogensen, J. (2012). Delayed intensive acquisition training alleviates the lesion-induced place learning deficits after fimbria-fornix transection in the rat. *Brain Research,* 1445: 40-51.

Mammis, A., McIntosh, T.K. and Maniker, A.H. (2009). Erythropoietin as a neuroprotective agent in traumatic brain injury. Review. *Surgical Neurology,* 71: 527-531.

Marr, D. and Hildreth, E. (1980). Theory of edge detection. *Proceedings of the Royal Society of London B*, 207: 187-217.

McClelland, J.L., Rumelhart, D.E. and The PDP Research Group (1986). *Parallel Distributed Processing: Vol. 2. Psychological and Biological Models.* Cambridge, MA: MIT Press.

McGee, A.W., Yang, Y., Fischer, Q.S., Daw, N.W. and Strittmatter, S.M. (2005). Experience-driven plasticity of visual cortex limited by myelin and Nogo receptor. *Science*, 309: 2222-2226.

McLeod, P., Plunkett, K. and Rolls, E.T. (1998). *Introduction to Connectionist Modelling of Cognitive Processes.* Oxford: Oxford University Press.

Meinzer, M., Djundja, D., Barthel, G., Elbert, T. and Rockstroh, B. (2005). Long-term stability of improved language functions in chronic aphasia after constraint-induced aphasia therapy. *Stroke,* 36: 1462-1466.

Meinzer, M., Elbert, T., Djundja, D., Taub, E. and Rockstroh, B. (2007a). Extending the Constraint-Induced Movement Therapy (CIMT) approach to cognitive functions: Constraint-Induced Aphasia Therapy (CIAT) of chronic aphasia. *NeuroRehabilitation,* 22: 311-318.

Meinzer, M., Obleser, J., Flaisch, T., Eulitz, C. and Rockstroh, B. (2007b). Recovery from aphasia as a function of language therapy in an early bilingual patient demonstrated by fMRI. *Neuropsychologia,* 45: 1247-1256.

Meinzer, M., Flaisch, T., Breitenstein, C., Wienbruch, C., Elbert, T. and Rockstroh, B. (2008). Functional re-recruitment of dysfunctional brain areas predicts language recovery in chronic aphasia. *NeuroImage,* 39: 2038-2046.

Merzenich, M.M. and Jenkins, W.M. (1993). Reorganization of cortical representations of the hand following alterations of skin inputs induced by nerve injury, skin island transfers, and experience. *Journal of Hand Therapy*, 6: 89-104.

Mogensen, J. (1991). Influences of the rearing conditions on functional properties of the rat's prefrontal system. *Behavioural Brain Research*, 42: 135-142.

Mogensen, J. (2011a). Almost unlimited potentials of a limited neural plasticity: levels of plasticity in development and reorganization of the injured brain. *Journal of Consciousness Studies,* 18: 13-45.

Mogensen, J. (2011b). Animal models in neuroscience, In: Hau, J. and Schapiro, S.J. (Eds.), *Handbook of Laboratory Animal Science*, Third Edition, Volume II. Animal Models. Boca Raton, FL: CRC Press LLC; pp. 47-73.

Mogensen, J. (2011c). Reorganization in the injured brain: implications for studies of the neural substrate of cognition. *Frontiers in Psychology,* 2:7: 1-10.

Mogensen, J. (2012). Cognitive recovery and rehabilitation after brain injury: mechanisms, challenges and support. In: Agrawal, A. (Ed.), *Brain Injury – Functional Aspects, Rehabilitation and Prevention.* Rijeka, Croatia: InTech; pp. 121-150.

Mogensen, J. (2014). Recovery, compensation and reorganization in neuropathology – levels of conceptual and methodological challenges. In: Tracy, J. (Ed.), *Plasticity of Cognition in Neurologic Disorders.* New York: Oxford University Press; pp. in press.

Mogensen, J. and Malá, H. (2009). Post-traumatic functional recovery and reorganization in animal models. A theoretical and methodological challenge. *Scandinavian Journal of Psychology*, 50: 561-573.

Mogensen, J., Christensen, L.H., Johansson, A., Wörtwein, G., Bang, L.E. and Holm, S. (2002). Place learning in scopolamine treated rats: the roles of distal cues and catecholaminergic mediation. *Neurobiology of Learning and Memory*, 78: 139-166.

Mogensen, J., Wörtwein, G., Plenge, P. and Mellerup, E.T. (2003). Serotonin, locomotion, exploration, and place recall in the rat. *Pharmacology, Biochemistry and Behavior*, 75: 381-395.

Mogensen, J., Lauritsen, K.T., Elvertorp, S., Hasman, A., Moustgaard, A. and Wörtwein, G. (2004a). Place learning and object recognition by rats subjected to transection of the fimbria-fornix and/or ablation of the prefrontal cortex. *Brain Research Bulletin*, 63: 217-236.

Mogensen, J., Miskowiak, K., Sørensen, T.A., Lind, C.T., Olsen, N.V., Springborg, J.B. and Malá, H. (2004b). Erythropoietin improves place learning in fimbria-fornix transected rats and modifies the search pattern of normal rats. *Pharmacology, Biochemistry and Behavior*, 77: 381-390.

Mogensen, J., Moustgaard, A., Khan, U., Wörtwein, G. and Nielsen, K.S. (2005). Egocentric spatial orientation in a water maze by rats subjected to transection of the fimbria-fornix and/or ablation of the prefrontal cortex. *Brain Research Bulletin*, 65: 41-58.

Mogensen, J., Hjortkjær, J., Ibervang, K.L., Stedal, K. and Malá, H. (2007). Prefrontal cortex and hippocampus in posttraumatic functional recovery: Spatial delayed alternation by rats subjected to transection of the fimbria-fornix and/or ablation of the prefrontal cortex. *Brain Research Bulletin*, 73: 86-95.

Mogensen, J., Boyd, M.H., Nielsen, M.D., Kristensen, R.S. and Malá, H. (2008a). Erythropoietin improves spatial delayed alternation in a T-maze in rats subjected to ablation of the prefrontal cortex. *Brain Research Bulletin,* 77: 1-7.

Mogensen, J., Jensen, C., Kingod, S.C., Hansen, A., Larsen, J.A.R. and Malá, H. (2008b). Erythropoietin improves spatial delayed alternation in a T-maze in fimbria-fornix transected rats. *Behavioural Brain Research*, 186: 215-221.

Molteni, R., Wu, A., Vaynman, S., Ying, Z., Bernard, R.J. and Gomez-Pinella, F. (2004). Exercise reverses the harmful effects of consumption of a high-fat diet on synaptic and behavioral plasticity associated to the action of brain-derived neurotrophic factor. *Neuroscience,* 123: 429-440.

Monakow, C.V. (1914). *Die Lokalisation im Grosshirn und der Abbau der Funktion durch Kortikale Herde.* Wiesbaden: Bergmann.

Moore, A.D. and Stambrook, M. (1995). Cognitive moderators of outcome following traumatic brain injury: a conceptual model and implications for rehabilitation. *Brain Injury*, 9: 109-130.

Moser, E.I., Kropff, E., and Moser, M-B. (2008). Place cells, grid cells, and the brain's spatial representation system. *Annual Review of Neuroscience,* 31: 69-89.

Münte, T.F., Altenmüller, E. and Jäncke, L. (2002). The musician's brain as a model of neuroplasticity. *Nature Reviews. Neuroscience*, 3: 473-478.

Nakatomi, H., Kuriu, T., Okabe, S., Yamamoto, S-C., Hatano, O., Kawahara, N., Tamura, A., Kirino, T. and Nakafuku, M. (2002). Regeneration of hipppocampal pyramidal neurons after ischemic brain injury by recruitment of endogenous neural progenitors. *Cell*, 110: 429-441.

Norman, D.A. and Shallice, T. (1986). Attention to action: Willed and automatic control of behavior. In: Davidson, R.J., Schwartz, G.E. and Shapiro, D. (Eds.), *Consciousness and Self-Regulation* (Vol. 4). New York: Plenum Press; pp. 1-18.

O'Keefe, J. and Dostrovsky, J. (1971). The hippocampus as a spatial map. Preliminary evidence from unit activity in the freely-moving rat. *Brain Research,* 34: 171-175.

Overgaard, M. and Mogensen, J. (2011). A framework for the study of multiple realizations: the importance of levels of analysis. *Frontiers in Psychology*, 2:79: 1-10.

Overgaard, M. and Mogensen, J. (2014). Visual perception from the perspective of a representational, non-reductionistic, level-dependent account of perception and conscious awareness. *Philosophical Transactions of the Royal Society of London, Series B: Biological Sciences,* 369: 20130209.

Panksepp, J. and Panksepp, J.B. (2000). The seven sins of evolutionary psychology. *Evolution and Cognition,* 6: 108-131.

Parker, D.B. (1986). A comparison of algorithms for neuron-like cells. In: Denker, J. (Ed.), *Proceedings of the Second Annual Conference on Neural Networks for Computing, Proceedings* Vol. 151. New York: American Institute of Physics; pp. 327-332.

Perani, D., Cappa, S.F., Tettamanti, M., Rosa, M., Scifo, P., Miozzo, A., Basso, A. and Fazio, F. (2003). A fMRI study of word retrieval in aphasia. *Brain and Language,* 85: 357-368.

Pinker, S. (1999). *How the Mind Works.* London: Penguin Books.

Pizzorusso, T., Medini, P., Berardi, N., Chierzi, S., Fawcett, J.W. and Maffei, L. (2002). Reactivation of ocular dominance plasticity in the adult visual cortex. *Science*, 298: 1248-1251.

Ptito, M., Moesgaard, S.M., Gjedde, A. and Kupers, R. (2005). Cross-modal plasticity revealed by electrotactile stimulation of the tongue in the congenitally blind. *Brain,* 128: 606-614.

Pulvermüller, F., Neininger, B., Elbert, T., Mohr, B., Rockstroh, B., Koebbel, P. and Taub, E. (2001). Constraint-induced therapy of chronic aphasia after stroke. *Stroke,* 32: 1621-1626.

Rakic, P. (1971). Guidance of neurons migrating to the fetal monkey neocortex. *Brain Research*, 33: 471-476.

Rakic, P. (1985). Mechanisms of neuronal migration in developing cerebellar cortex. In: Edelman, G.M., Cowan, W.M. and Gull, E. (Eds.), *Molecular Basis of Neural Development.* New York: Wiley; pp. 139-160.

Ramachandran, V.S. and Blakeslee, S. (1998). *Phantoms in the Brain: Probing the Mysteries of the Human Mind*. William Morrow, New York.

Raninen, A., Vanni, S., Hyvärinen, L. and Näsänen, R. (2007). Temporal sensitivity in a hemianopic visual field can be improved by long-term training using flicker stimulation. *Journal of Neurology, Neurosurgery and Psychiatry*, 78: 66-73.

Recanzone, G.H., Schreiner, C.E. and Merzenich, M.M. (1993). Plasticity in the frequency representation of primary auditory cortex following discrimination training in adult owl monkeys. *Journal of Neuroscience*, 13: 87-103.

Renner, M.J. and Rosenzweig, M.R. (1987). *Enriched and Impoverished Environments: Effects on Brain and Behavior*. New York: Springer.

Rizzo, A.A., Buckwalter, J.G. and Neumann, U. (1997). Virtual reality and cognitive rehabilitation: a brief review of the future. *Journal of Head Trauma Rehabilitation,* 12: 1-15.

Rizzo, A.A., Schultheis, M., Kerns, K.A. and Mateer, C. (2004). Analysis of assets for virtual reality applications in neuropsychology. *Neuropsychological Rehabilitation,* 14: 207-239.

Robertson, D. and Irvine, D.R.F. (1989). Plasticity of frequency organization in auditory cortex of guinea pigs with partial unilateral deafness. *Journal of Comparative Neurology*, 282: 456-471.

Robertson, I.H. (1999). Cognitive rehabilitation: attention and neclect. *Trends in Cognitive Sciences,* 3: 385-393.

Robertson, I.H. and Murre, J.M.J. (1999). Rehabilitation of brain damage: brain plasticity and principles of guided recovery. *Psychological Bulletin,* 125: 544-575.

Roe, C.M., Xiong, C., Miller, J.P., Cairns, N.J. and Morris, J.C. (2008). Interaction of neuritic plaques and education predicts dementia. *Alzheimer Disease and Associated Disorders,* 22: 188-193.

Rohling, M.L., Faust, M.E., Beverly, B. and Demakis, G. (2009). Effectiveness of cognitive rehabilitation following acquired brain injury: A meta-analytic re-examination of Cicerone et al.'s (2000, 2005) systematic reviews. *Neuropsychology*, 23: 20-39.

Rolls, E.T., Stringer, S.M. and Elliot, T. (2006). Entorhinal cortex grid cells can map to hippocampal place cells by competitive learning. *Network: Computation in Neural Systems,* 447: 447-465.

Ropacki, M.T. and Elias, J.W. (2003). Preliminary examination of cognitive reserve theory in closed head injury. *Archives of Clinical Neuropsychology,* 18: 643-654.

Rosenzweig, M.R. (1971). Effects of environment of development of brain and behavior. In: Tobach, E., Aronson, L.R. and Shaw, E. (Eds.), *The Biopsychology of Development*, New York: Academic Press: pp. 303-342.

Rosenzweig, M.R. (1984). Experience, memory, and the brain. *American Psychologist*, 39: 365-376.

Rosenzweig, M.R., Krech, D. and Bennett, E.L. (1961). Heredity, environment, brain biochemistry, and learning. In: *Current Trends in Psychological Theory*. Pittsburgh, PA: University of Pittsburg Press; pp. 87-110.

Rossetti, Y., Rode, G., Pisella, L., Farne, A., Li, L., Boisson, D. and Perenin, M-T. (1998). Prism adaptation to a rightward optical deviation rehabilitates left hemispatial neglect. *Nature,* 395: 166-169.

Rumelhart, D. and McClelland, J. (1986). *Parallel Distributed Processing*. Cambridge, MA: MIT Press.

Sahraie, A., Trevethan, C.T., MacLeod, M.J., Murray, A.D., Olson, J.A. and Weiskrantz, L. (2006). Increased sensitivity after repeated stimulation of residual spatial channels in blindsight. *Proceedings of the National Academy of Sciences US*, 103: 14971-14976.

Sarri, M., Greenwood, R., Kalra, L., Papps, B., Husain, M. and Driver, J. (2008). Prism adaptation aftereffects in stroke patients with spatial neglect: pathological effects on subjective straight ahead but not visual open-loop pointing. *Neuropsychologia,* 46: 1069-1080.

Satz, P. (1993). Brain reserve capacity on symptom onset after brain injury: a formulation and review of evidence for threshold theory. *Neuropsychology,* 7: 273-295.

Schäfer, R., Dehn, D., Burbach, G.J. and Deller, T. (2008). Differential regulation of chondroitin sulfate proteoglycan mRNAs in the denervated rat fascia dentata after unilateral entorhinal cortex lesion. *Neuroscience Letters,* 439: 61-69.

Scharff, C., Kirn, J.R., Grossman, M., Macklis, J.D. and Nottebohm, G. (2000). Targeted neuronal death affects neuronal replacement and vocal behavior in adult songbirds. *Neuron,* 25: 481-492.

Scheibel, R.S., Newsome, M.R., Troyanskaya, M., Steinberg, J.L., Goldstein, F.C., Mao, H. and Levin, H.S. (2009). Effects of severity of traumatic brain injury and brain reserve on cognitive-control related brain activation. *Journal of Neurotrauma,* 26: 1447-1461.

Scheich, H. (1991). Auditory cortex: comparative aspects of maps and plasticity. *Current Opinion in Neurobiology,* 1: 236-247.

Schmand, B., Smit, J.H., Geerlings, M.I. and Lindeboom, J. (1997). The effects of intelligence and education on the development of dementia. A test of the brain reserve hypothesis. *Psychological Medicine,* 27: 1337-1344.

Schneider, E.B., Sur, S., Raymont, V., Duckworth, J., Kowalski, R.G., Efron, D.T., Hui, X., Selvarajah, S., Hambridge, H.L. and Stevens, R.D. (2014). Functional recovery after moderate/severe traumatic brain injury. *Neurology,* doi: 10.1212/WNL. 0000000000000379.

Schwab, M.E. and Thoenen, H. (1985). Dissociated neurons regenerate into sciatic but not optic nerve explants in culture irrespective of neurotrophic factors. *Journal of Neuroscience,* 5: 2415-2423.

Schweizer, T.A., Ware, J., Fischer, C.E., Craik, F.I.M. and Bialystok, E. (2012). Bilingualism as a contributor to cognitive reserve: evidence from brain atrophy in Alzheimer's disease. *Cortex,* 48: 991-996.

Schwiedrzik, C.M., Singer, W. and Melloni, L. (2009). Sensitivity and perceptual awareness increase with practice in metacontrast masking. *Journal of Vision,* 9: 1-18.

Selnes, O.A. (2001). A historical overview of contributions from the study of deficits. In: Rapp, B. (Ed.), *The Handbook of Cognitive Neuropsychology.* Philadelphia PA: Psychology Press; pp. 23-41.

Seo, T-B., Kim, B-K., Ko, I-G., Kim, D-H., Shin, M-S., Kim, C-J., Yoon, J-H. and Kim, H. (2010). Effect of treadmill exercise on Purkinje cell loss and astrocytic reaction in the cerebellum after traumatic brain injury. *Neuroscience Letters,* 481: 178-182.

Serino, A., Angeli, V., Frassinetti, F. and Ladavas, E. (2006). Mechanisms underlying neglect recovery after prism adaptation. *Neuropsychologia,* 44: 1068-1078.

Serino, A., Bonifazi, S., Pierfederici, L. and Ladavas, E. (2007). Neglect treatment by prism adaptation: what recovers and for how long. *Neuropsychological Rehabilitation,* 17: 657-687.

Siesjo, B.K. and Siesjo, P. (1996). Mechanisms of secondary brain injury. *European Journal of Anaesthesiology,* 13: 247-268.

Siren, A.L., Fratelli, M., Brines, M., Goemans, C., Casagrande, S., Lewczuk, P., Keenan, S., Gleiter, C., Pasquali, C., Capobianco, A., Mennini, T., Heumann, R., Cerami, A., Ehrenreich, H. and Ghezzi, P. (2001). Erythropoietin prevents neuronal apoptosis after cerebral ischemia and metabolic stress. *Proceedings of the National Academy of Sciences US,* 98: 4044-4049.

Sitaram, R., Caria, A. and Birbaumer, N. (2009). Hemodynamic brain-computer interfaces for communication and rehabilitation. *Neural Networks,* 22: 1320-1328.

Slade, A., Tennant, A. and Chamberlain, M.A. (2002). A randomised controlled trial to determine the effect of intensity of therapy upon length of stay in a neurological rehabilitation setting. *Journal of Rehabilitation Medicine,* 34: 260-266.

Slavin, M.D., Laurence, S. and Stein, D.G. (1988). Another look at vicariation. In: Finger, S., LeVere, T.E., Almli, C.R. and Stein, D.G. (Eds.), *Brain Injury and Recovery: Theoretical and Controversial Issues.* New York: Plenum Press; pp. 165-178.

Sluming, V., Brooks, J., Howard, M., Downes, J.J. and Roberts, N. (2007). Broca's area supports enhanced visuospatial cognition in orchestral musicians. *Journal of Neuroscience,* 27: 3799-3806.

Sofroniew, M.V., Cooper, J.D., Svendsen, C.N., Crossman, P., Ip, N.Y., Lindsay, R.M., Zafra, F. and Lindholm, D. (1993). Atrophy but not death of adult septal cholinergic neurons after ablation of target capacity to produce mRNAs for NGF, BDNF, and NT3. *Journal of Neuroscience,* 13: 5263-5276.

Solstad, T., Moser, E.I. and Einevoll, G.T. (2006). From grid cells to place cells: a mathematical model. *Hippocampus,* 16: 1026-1031.

Specht, K., Zahn, R., Willmes, K., Weis, S., Holtel, C., Krause, B.J., Herzog, H. and Huber, W. (2009). Joint independent component analysis of structural and functional images reveals complex patterns of functional reorganisation in stroke aphasia. *NeuroImage,* 47: 2057-2063.

Springborg, J.B., Ma, X.D., Rochat, P., Knudsen, G.M., Amtorp, O., Paulson, O.B., Juhler, M. and Olsen, N.V. (2002). A single subcutaneous bolus of erythropoietin normalizes cerebral blood flow autoregulation after subarachnoid haemorrhage in rats. *British Journal of Pharmacology,* 135: 823-829.

Stein, D.G. and Hoffman, S.W. (2003). Concepts of CNS plasticity in the context of brain damage and repair. *Journal of Head Trauma Rehabilitation,* 18: 317-341.

Stern, Y. (2002). What is cognitive reserve? Theory and research application of the reserve concept. *Journal of the International Neuropsychological Society,* 8: 448-460.

Stern, Y. (2003). The concept of cognitive reserve: a catalyst for research. *Journal of Clinical and Experimental Neuropsychology,* 25: 589-593.

Stern, Y. (2006). Cognitive reserve and Alzheimer disease. *Alzheimer Disease and Associated Disorders,* 20: 112-117.

Stoerig, P. (2006). Blindsight, conscious vision, and the role of primary visual cortex. *Progress in Brain Research,* 155: 217-234.

Szaflarski, J.P., Eaton, K., Ball, A.L., Banks, C., Vannest, J., Allendorfer, J.B., Page, S. and Holland, S.K. (2011). Poststroke aphasia recovery assessed with functional magnetic resonance imaging and a picture identification task. *Journal of Stroke and Cerebrovascular Diseases,* 20: 336-345.

Teri, L., Gibbons, L.E., McCurry, S.M., Logsdon, R.G., Buchner, D.M., Barlow, W.E., Kukull, W.A., LaCroix, A.Z., McCormick, W. and Larson, E.B. (2003). Exercise plus behavioral management in patients with Alzheimer Disease: a randomized controlled trial. *JAMA,* 290: 2015-2022.

Thai-Van, H., Micheyl, C., Norena, A., Veuillet, E., Gabriel, D. and Collet, L. (2007). Enhanced frequency discrimination in hearing-impaired individuals: a review of perceptual correlates of central neural plasticity induced by cochlear damage. *Hearing Research*, 233: 14-22.

Thomas, C., Altenmüller, E., Marckmann, G., Kahrs, J. and Dichgans, J. (1997). Language processing in aphasia: changes in lateralization patterns during recovery reflect cerebral plasticity in adults. *Electroencephalography and Clinical Neurophysiology*, 102: 86 97.

Thompson, C.K., den Ouden, D-B., Bonakdarpour, B., Garibaldi, K. and Parrish, T.B. (2010). Neural plasticity and treatment-induced recovery of sentence processing in agrammatism. *Neuropsychologia*, 48: 3211-3227.

Thulborn, K.R., Carpenter, P.A. and Just, M.A. (1999). Plasticity of language-related brain function during recovery from stroke. *Stroke*, 30: 749-754.

Tsirlin, I., Dupierrix, E., Chokron, S., Coquillart, S. and Ohlmann, T. (2009). Uses of virtual reality for diagnosis, rehabilitation and study of unilateral spatial neglect: review and analysis. *CyberPsychology and Behavior*, 12: 175-181.

Valenzuela, M.J. and Sachdev, P. (2006). Brain reserve and dementia: a systematic review. *Psychological Medicine*, 36: 441-454.

Vaynman, S., Ying, Z. and Gomez-Pinilla, F. (2003). Interplay between brain-derived neurotrophic factor and signal transduction modulators in the regulation of the effects of exercise on synaptic-plasticity. *Neuroscience*, 122: 647-657.

Viviani, B., Bartesaghi, S., Corsini, E., Villa, P., Ghezzi, P., Garau, A., Galli, C.L. and Marinovich, M. (2005). Erythropoietin protects primary hippocampal neurons increasing the expression of brain-derived neurotrophic factor. *Journal of Neurochemistry*, 93: 412-421.

Wang, K.W., Larner, S.F., Robinson, G. and Hayes, R.L. (2006). Neuroprotection targets after traumatic brain injury. *Current Opinion in Neurology*, 19: 514-519.

Weiss, T., Miltner, W.H.R., Huonker, R., Friedel, R., Schmidt, I. and Taub, E. (2000). Rapid functional plasticity of the somatosensory cortex after finger amputation. *Experimental Brain Research*, 134: 199-203.

Werbos, P.J. (1974). *Beyond Regression: New Tools for Prediction and Analysis in the Behavioral Sciences*. Harvard University: Applied Mathematics.

Werbos, P.J. (1994). *The Roots of Backpropagation: From Ordered Derivatives to Neural Networks and Political Forecasting*. New York: John Wiley and Sons.

Westerberg, H., Jacobaeus, H., Hirvikoski, T., Clevberger, P., Östensson, M-L., Bartfai, A. and Klingberg, T. (2007). Computerized working memory training after stroke – a pilot study. *Brain Injury*, 21: 21-29.

Wilms, I. (2011). Using artificial intelligence to control and adapt level of difficulty in computer based, cognitive therapy – an explorative study. *Journal of Cybertherapy and Rehabilitation*, 4: 387-397.

Wilms, I. and Malá, H. (2010). Indirect versus direct feedback in computer-based Prism Adaptation Therapy. *Neuropsychological Rehabilitation*, 20: 830-853.

Wilms, I. and Mogensen, J. (2011). Dissimilar outcomes of apparently similar procedures as a challenge to clinical neurorehabilitation and basic research – when the same is not the same. *NeuroRehabilitation*, 29: 221-227.

Wilson, B.A. (2000). Compensating for cognitive deficits following brain injury. *Neuropsychology Review*, 10: 233-243.

Wilson, B.A. (2002). Towards a comprehensive model of cognitive rehabilitation. *Neuropsychological Rehabilitation,* 12: 97-110.

Wilson, B.A., Evans, J.J., Emslie, H. and Malinek, V. (1997). Evaluation of NeuroPage: a new memory aid. *Journal of Neurology, Neurosurgery, and Psychiatry,* 63: 113-115.

Wilson, B.A., Watson, P.C., Baddeley, A.D., Emslie, H. and Evans, J.J. (2000). Improvement or simply practice? The effects of 20 repeated assessments on people with and without brain injury. *Journal of the International Neuropsychological Society*, 6: 469-479.

Wilson, B.A., Emslie, H.C., Quirk, K. and Evans, J.J. (2001). Reducing everyday memory and planning problems by means of a paging system: a randomised control crossover study. *Journal of Neurology, Neurosurgery, and Psychiatry,* 70: 477-482.

Wörtwein, G., Saerup, L.H., Charlottenfeld-Starpov, D. and Mogensen, J. (1995). Place learning by fimbria-fornix transected rats in a modified water maze. *International Journal of Neuroscience*, 82: 71-81.

Xerri, C., Coq, J., Merzenich, M. and Jenkins, W. (1996). Experience-induced plasticity of cutaneous maps in the primary somatosensory cortex of adult monkeys and rats. *Journal of Physiology*, 90: 277-287.

Yang, T.T., Gallen, C.C., Ramachandran, V.S., Cobb, S., Schwartz, B.J. and Bloom, F.E. (1994). Noninvasive detection of cerebral plasticity in adult human somatosensory cortex. *Neuroreport*, 5: 701-704.

Zhang, F., Signore, A.P., Zhou, Z., Wang, S., Cao, G. and Chen, J. (2006). Erythropoietin protects CA1 neurons against global cerebral ischemia in rat: potential signalling mechanisms. *Journal of Neuroscience Research,* 83: 1241-1251.

Zihl, J. (1980). "Blindsight": improvement of visually guided eye movements by systematic practice in patients with cerebral blindness. *Neuropsychologia*, 18: 71-77.

Zihl, J. and Werth, R. (1984). Contributions to the study of "blindsight" – II. The role of specific practice for saccadic localization in patients with postgeniculate visual field defects. *Neuropsychologia*, 22: 13-22.

In: Horizons in Neuroscience Research. Volume 15
Editors: Andres Costa and Eugenio Villalba
ISBN: 978-1-63321-523-8

Chapter 5

The Effects of Acoustic Stimulation on Comatose Patients

***João F. F. L. Simões*[1], *Luis M. T. Jesus*[2,*], *David Voegeli*[3], *Celso Martins*[4], *Andreia Hall*[5] *and David Simpson*[6]**

[1]School of Health Sciences (ESSUA), University of Aveiro, Aveiro, Portugal
[2]School of Health Sciences (ESSUA) and Institute of Electronics and Informatics Engineering of Aveiro (IEETA), University of Aveiro, Aveiro, Portugal
[3]School of Health Sciences, University of Southampton, Southampton, UK
[4]Hospital de Santo António, Centro Hospitalar do Porto, Porto, Portugal
[5]Department of Mathematics and Institute of Electronics and Informatics Engineering of Aveiro (IEETA), University of Aveiro, Aveiro, Portugal
[6]Institute of Sound and Vibration Research, University of Southampton, Southampton, UK

Abstract

One of the most important clinical challenges in patients with severe brain damage is to estimate different aspects of their actual state of consciousness and, in particular, their residual conscious perception of the environment. Different methods have been developed to quantify the preservation of responses in the ascending sensory systems and the projection areas, e.g., bedside behavioural assessment and analyses of short-latency event-related potentials.

This study includes a detailed, objective and systematic analysis of the effects of acoustic stimulation on comatose patients using a standardised stimulus message. The curves of electrocardiography, blood pressure and pulse plethysmography and the pulse frequency, systolic, diastolic and mean arterial pressure, oxygen saturation level and peripheral body temperature, were monitored before, during and after the auditory stimulation of comatose patients with an unknown voice and a familiar voice. This study

* Corresponding author: Luis M. T. Jesus, University of Aveiro, 3810-193 Aveiro, Portugal; Tel.: + 351 234372458; Fax: + 351 234401597; e-mail: lmtj@ua.pt

contributes new evidence to improve the quality of care and help in establishing the prognosis.

1. Introduction

Most of current research on coma focuses on how to distinguish irreversible from potentially reversible coma but the question of how health professionals should address comatose patients has escaped critical appraisal. Not talking to comatose patients may promote the notion that they are dead or nearly dead, and may influence health professionals to inappropriately withhold or withdraw therapy (LaPuma et al., 1990). Extensive research (LaPuma et al., 1990) on the auditory brainstem function of comatose patients shows that the majority have normal brainstem auditory evoked responses. Verbal communication is a way of orientating and providing meaningful sensory input to an unconscious patient. Familiar voices or simply using the patient's name and individualising care may provide better sensory input (Elliot and Wright, 1999; Holeckova et al., 2006; Perrin et al., 2006; Boly et al., 2011).

Communication amongst healthcare providers is one of the most important factors associated with quality of care and patient safety (Hoonakker et al., 2008), especially in Intensive Care Units (ICUs). However, relatively little is known about the different aspects of communication in health care, specifically in ICUs, and the relation between communication and outcome measures.

Despite its importance, there is evidence (Alasad and Almad, 2005) that communication in Intensive Care Unit (ICU) settings is not common practice. Inadequate health professional-patient communication results in increased levels of stress and anxiety. Thus, communication within the ICU environment is of paramount importance, and effective communication is an essential aspect of health professionals' care.

One of the most important and challenging tasks for healthcare professionals caring for patients who survive brain injury is establishing a prognosis for long-term functional outcome, as soon as possible after injury, even while the patient is unconscious. Severe brain injury changes many lives instantaneously and prognostication during coma recovery can help with early counselling and adjustment of the patient and family, as well as guide the planning for present and future rehabilitation.

The inconsistencies in the research findings (Johnson, Omery and Nikas, 1989; Doman et al., 1993; Jones et al., 1994; Walker, Eakes and Siebelink, 1998; Holeckova et al., 2006; Perrin et al., 2006; Puggina, 2006) point to the need for further detailed investigation of the effects of voice stimulation on comatose patients.

Based on a literature review (Simões, Jesus and Voegeli, 2013) we developed a study that includes a detailed, objective and systematic analysis of the effects of acoustic stimulation on comatose patients using a standardised stimulus message. The curves of electrocardiography (ECG), blood pressure and pulse plethysmography and the parameters of pulse frequency, systolic, diastolic and mean arterial pressure, oxygen saturation level and peripheral body temperature, were monitored before, during and after the auditory stimulation of the comatose patient with an unknown voice and a familiar voice.

This study contributes to the reflection on the practice of communication with the unconscious patients, in order to sensitise health professionals to the importance of communication in the ICU, to contribute to improving the quality of care and help in establishing the prognosis.

1.1. Verbal Communication with the Comatose Patient

The ICU can create a great deal of tension and stress for patients and families (Elliot and Wright, 1999). Effective and appropriate communication is an important part of the healing process, not only for the patient, but also for the family.

The ICUs are busy places filled with all types of noise. Patients do not get much sleep, and sleep deprivation along with narcotic drugs can contribute to confusion in the ICUs (so called *ICU psychosis*) due to the critical care environment and illness itself. Many patients become confused just being removed from their homes into a different environment, or in the dark of the night, even when they are not particularly ill. As the patient continues to heal, medication dosage is typically decreased. Normal sleep patterns gradually return and the patient recovers from the confusion of ICU psychosis. The act of communication will aid the patient in the process of returning to normal orientation. Various types and levels of communication will be of benefit at different stages of the recovery process (Alasad and Almad, 2005).

The unconscious patient is under stress because he/she is likely to be misunderstood, and is unsure of what is happening. Furthermore, cognitive reappraisal is virtually impossible in the ICU environment. Verbal communication can reduce stress by facilitating the patient's use of adaptive coping mechanisms (Elliott and Wright, 1999).

Verbal communication is a way of orientating and providing meaningful sensory input to an unconscious patient. Familiar voices or simply using the patient's name and individualising care may provide better sensory input (Elliot and Wright, 1999; Holeckova et al., 2006; Perrin et al., 2006). Hence, verbal communication can help patients preserve self-identity and self-esteem and reduce social isolation (Elliot and Wright, 1999). These factors help to enhance feelings of wellbeing and optimism, a proposed mechanism for decreasing recovery time (Johnson et al., 1989; Doman et al., 1993; Baker and Melby, 1996; Walker et al., 1998; Elliot and Wright, 1999; Holeckova et al., 2006; Perrin et al., 2006).

Communication with a critically ill patient who might be sedated, ventilated, and therefore unable to speak and subject to distorted sensory input, is important (Baker and Melby, 1996; Verity, 1996): it allows access to and exchange of information and emotional support during stress (Elliot and Wright, 1999); it is a crucial means of preventing the feeling of isolation which could lead to the development of ICU psychosis (Verity, 1996).

1.2. Responses of Comatose Patients to Verbal Communication

Research to date has resulted in conflicting findings regarding the effects of stimulation on unconscious patients (Johnson, Omery and Nikas, 1989; Walker, Eakes and Siebelink, 1998). There is, however, sufficient evidence to support the idea that increased mentation and emotional arousal may affect the unconscious patient. Evidence reported in the literature (Johnson, Omery and Nikas, 1989; Jones et al., 1994; Walker, Eakes and Siebelink, 1998) also clearly shows a correlation between the auditory stimulation and increases in arterial blood pressure, pulse rate, respiratory rate, intracranial pressure, body movement and facial movement.

We also found evidence in the literature (Johnson, Omery and Nikas, 1989; Doman et al., 1993; Baker and Melby, 1996; Walker, Eakes and Siebelink, 1998; Elliot and Wright, 1999; Holeckova et al., 2006; Perrin et al., 2006) that unconscious patients have neurological and physiological alterations with auditory stimulation, and that most intensive care health professionals claim that verbal communication with unconscious patients is very important (Ashworth, 1987; Turnock, 1991; Baker and Melby, 1996; Elliot and Wright, 1999; Hoonakker et al., 2008), but there is still some ambiguity as to the unconscious patient's level of awareness.

Major factors influencing communication are the patient's level of consciousness, the amount of physical care being given and the presence of relatives (Baker and Melby, 1996). In the present study we found that the purpose of verbal communication by critical care health professionals focuses on professional aspects, and the purpose of patients' family communication focuses in personal aspects.

Evidence (Walker, Eakes and Siebelink, 1998) suggests that the majority of comatose patients have normal brainstem auditory evoked responses and indicates that they may hear. Indeed, LaPuma et al. (1990) have gone so far as to suggest that talking to comatose patients may have considerable therapeutic value.

1.3. Purpose of Verbal Communication by Health Professionals

Communication is the foundation of nursing and the art of caring holistically for patients', and much of health care is communication-centred (Verity, 1996). This becomes even more relevant for the unconscious patient who is particularly dependent on hearing for sensory stimulation. Health professionals give directions, offer reassurance, provide consolation, commiserate, interpret and receive information. Therefore, the most effectively and efficiently they communicate, the more accomplished they will become in fulfilling their role as health professionals.

There are a limited number of studies (Ashworth, 1987; Baker and Melby, 1996; Elliot and Wright, 1999) that investigated what health professionals actually say to their patients. However, studies to date suggest that intensive care health professionals are not engaging enough in verbal communication (Elliot and Wright, 1999), tend to concentrate more on the technical aspect of their work, and often fail to meet patient's psychological and social needs by insufficient and/or ineffective communication (Baker and Melby, 1996). Communication with unresponsive patients is limited, relating primarily to nursing interventions (Ashworth, 1987). There appear to be four main problem areas (Baker and Melby, 1996): basic difficulty in communicating with a patient who cannot respond; pressures of the working environment; limited knowledge about unconscious patients' needs; limited detailed knowledge of why or how to communicate with unconscious patients.

Increasing numbers of patients report their experiences, when they recover from coma. A common suggestion by these patients is the need for effective verbal communication (Elliot and Wright, 1999). Unconscious patients have a considerable need for information and support, and verbal communication is a way of orientating and providing meaningful sensory input to these patients.

Research undertaken by Ashworth (1987) showed that the communication aspect of care by intensive care nurses was less skilled than other aspects of their care. This theory was supported by Turnock (1991) who found that nurses did not provide adequate verbal and non-verbal communication. Also Elliott and Wright (1999) concluded that intensive care nurses may not be reflecting and understanding the importance of communication in their practice.

Baker and Melby (1996) concluded that at times verbal communication with unconscious patients was so minimal, that any potential benefit to the patients would have been negligible. Survival is a major consideration in intensive care (Ashworth, 1987), and communicating with the patient may become a low priority, whilst the health professional attends to the demands of highly technical equipment needed to support life and aid recovery (Verity, 1996).

Without adequate communication skills health professionals, can neither assess, plan, implement, nor evaluate care effectively. To achieve effective communication, especially in intensive care, where it is often vital to convey the essence of a message to the recipient in a language that they can understand, requires knowledge of how to convey a message, at the right level, without the use of jargon (Verity, 1996). This is a skill that needs to be taught theoretically and practically, so that barriers to communication may be minimised.

It is the duty of intensive care health professionals to ensure that they provide care and attention which cause the least possible discomfort and distress to patients (Baker and Melby, 1996). Ashworth (1987) suggested through friendly interaction, giving relevant information and encouraging the patient to communicate (if possible), health professionals could help to prevent such adverse effects.

It is the health professionals' responsibility to promote all avenues of communication in order to lessen the effects of the communication deficit, thereby promoting patient orientation which could reduce the risk of ICU psychosis. The necessity for reassurance, encouragement and reorientation is very important when the patient is deprived of some or all methods of communication. Therefore, formal support systems and continued education of health professionals about the benefits of verbal communication is deemed necessary.

1.4. Purpose of Verbal Communication by Patients' Family

Research about verbal communication by the patients' family has also resulted in conflicting findings. While families' visits to the ICU have been thought by some to precipitate detrimental changes in physiological variables such as heart rate and blood pressure (Walker, Eakes and Siebelink, 1998), others have observed no untoward effects associated with such stimulation. Still, opportunities for family interaction with comatose patients are usually very limited.

Familiar voices or simply using the patient's name and individualising care may provide better sensory input (Elliott and Wright, 1999). Jones et al. (1994) examined physiological measures (pulse rate, respiratory rate, body movement, and facial movement) using four different auditory stimuli (rock music, classical music, nature sounds, and family/friend voices). Their (Jones et al. 1994) findings suggested that taped voices of family and friends consistently resulted in greater increases in arousal than did other types of taped stimuli.

The study of Walker, Eakes and Siebelink (1998) revealed no negative effects associated with exposure to taped familial voices. However, no significant changes were evidenced in intracranial pressure (ICP), blood pressure, pulse, respiratory rate, mean arterial pressure, oxygen saturation level, or level of restlessness for any of the study participants. No experimental intervention had to be stopped because of an adverse reaction. The study (Walker, Eakes and Siebelink 1998) findings, therefore, add support to allowing family interactions with comatose head-injured patients during their hospitalisation in ICU settings.

2. Method

2.1. Aims

The primary aims of this study were: to understand the effects of auditory stimulation on comatose patients' physiological parameters and monitored curves; to explore the differences between the use of an auditory stimulus based on an unknown voice versus a familiar voice; to study the impact of this data on the prognosis of Coma.

2.2. Ethical Considerations

The Hospital's Ethics Committee gave full ethical approval and the study was registered with the Hospital's Research Office, thus fulfilling local research governance requirements. Meetings were held with ward staff prior to the study commencing. Ward staff contacted patients' relatives, so they would consent to be part of the study. All participants were given written information about the study and time to decide whether they wished that the patient take part, and they all gave written consent. Relatives were assured that there was no obligation to take part and that the care of the patients would not be affected should they decline. All data were confidential and kept securely in locked filing cabinets and password protected computers.

2.3. Participants

The study was carried out in a Medical/Surgical adult ICU (Serviço de Cuidados Intensivos 1 do Hospital de Santo António, Porto, Portugal).

Considering the temporal issues, the clinical characteristics of the population in study, the intention to reduce the variables that could affect the results and finally the data of the patients admitted to the ICU during the year of 2006, the sample consisted of10 comatose patients, which were selected using the following inclusion and exclusion criteria:

- Inclusion: Loss of consciousness; Absence of macroscopic response to auditory stimulation; Coma diagnosis; Total score of the Glasgow Coma Scale (Jennett, 2002) between 6 and 9; Aetiology medic, surgical or politrauma without brain injury; Hospitalised for more than 4 days; Identical care assistance; Identical environmental conditions.

- Exclusion: Hearing loss; A score higher than0in the “Hearing” sub-scale of the IARCUA (Simões, Jesus, Voegeli, Couto, Fernandes and Morgado 2011); History of excessive consumption of alcohol and/or narcotic substances; Epilepsy or degenerative diseases of the central nervous system; Diagnosed injury of the auditory pathways and the cerebral cortex; Significant changes of the auditory evoked responses; Induced Coma by infusion of anaesthetic medication.

2.4. Selection of the Social Group and Significant Person

As mentioned previously, the comatose patient was stimulated acoustically with a recorded message by an unknown person and the most meaningful person to him. Thus, we held a meeting with the significant people of the comatose patients to explain the purpose of the study and how they could collaborate. In this explanation we indicated that only one (significant) person could record the stimulus message. The following inclusion and exclusion criteria were defined:

- Inclusion: Belonging to the same social group as the comatose patient, i.e., having shared objectives, values, activities, characteristics and meanings, interrelating with each other during an extended period of time in a process of continuous communication and interaction, interdependence of individuals which is governed by a set rules or common standards, having group consciousness (Bolander, 1994; Prados, 2000; Stanhope and Lancaster, 2011); Accepting to participate in the study; Having a psychological profile to cooperate in interviews and recordings; Communicate verbally; Portuguese language speaker; Being able to read.
- Exclusion: No ability to communicate verbally.

Once selection of people had occurred, individual meetings were held with each significant person to apply sociometric tests. At this meeting we asked two questions. Firstly we asked the person to identify a list of people that would be meaningful to the comatose patient and who would constitute their social group. Secondly we asked them to elect the people who they considered most significant, so that a sociomatrix with only a reasonable number of participants could be built.

After the formation of the social group, we used sociometric techniques (Bastin, 1980; Kerlinger and Lee, 2000) to determine the member most significant for the comatose patient. The method consists of the description and evaluation of social situations and social processes by measuring the proportion of mutual acceptance and rejection observed among individuals in a group. This is determined through a survey of group members about who the comatose person would most or least like to have as company during particular activities (Kerlinger and Lee, 2000). In our case we just set out to assess the acceptability of the elements of the social group.

The aim was that individuals would place the previously identified people in descending order of importance to determine who was most important to the comatose patient. Each person identified as being a member of the social group of the comatose patient was informed of the objectives and relevance of the study, and what taking part would involve. They were asked for permission to access and use information provided by them and guaranteed confidentiality of all data. They then gave their informed consent to participate in the study. Then the question was made and their response recorded.

After collecting the data from each member of the social group, they were organised into a sociometric matrix or sociomatrix (Kerlinger and Lee, 2000) using the program *Sociogram* (Adit, 2006). We used the first 3 people prioritised by each participant to build our matrix. The numerical identifiers of each member of the social group were entered into the first column with row 1 being assigned to the comatose patient as seen in Figure 1. The numerical identifiers of the selected first 3 choices were then placed in the subsequent columns.

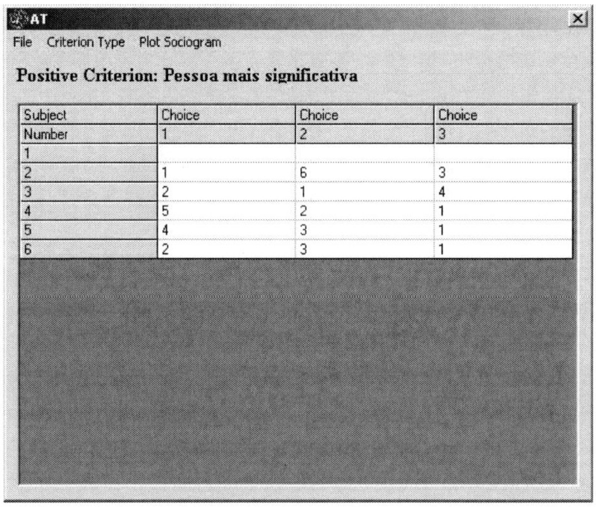

Positive Criterion: Pessoa mais significativa

Subject Number	Choice 1	Choice 2	Choice 3
1			
2	1	6	3
3	2	1	4
4	5	2	1
5	4	3	1
6	2	3	1

Figure 1. Sociomatrix of the social group of the comatose patient AT.

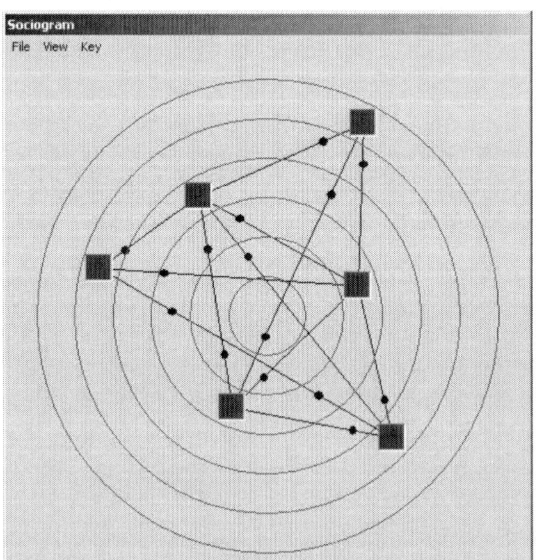

Figure 2. Group sociogram preferences of the comatose patient AT.

The *Sociogram* program then calculates a score corresponding to each choice, allocating maximum marks to the first choice and minimum marks for the third choice. These scores were then used to build sociograms which synthesize the data and present them visually (see Figure 2). Each sociogram then represents the sociometric status of each individual in the group by plotting the inter-relations between the people (Bastin, 1980).

The sociograms (see Figure 2) consist of concentric circles where the inner circle represents those closest to the comatose patient and the outer circle those with the lowest scores and therefore more distant. The relationships between the members of the group are represented by the black lines linking the person who chooses and the person chosen. A black dot is placed closest to the person who chooses. If both members mutually selected each other two black dots are used.

2.5. Stimulus Message

2.5.1. Construction of the Stimulus Message

In this study an initial stimulus message was used as a template to construct the stimulus message of the unknown and significant person. These messages were recorded and used later to stimulate the comatose patients.

The stimulus message was based on the results of a content analysis of the verbal communication of critical care nurses and patients' families, as reported in the literature (Simões, Jesus and Voegeli, 2013), and the cooperation of a group of experts (speech and language therapists, and psychologists).

The content analysis of papers about verbal communication with unconscious patients was performed by examining the relative frequencies with which thematic areas, categories and sub-categories occurred. The data were analysed manually using the framework approach (Ritchie and Spencer, 1994; Fischer, 2006; Anderson, 2007), combining pre-established themes with themes arising from the data to develop a coding framework. As with other qualitative research methods, analysis involved abstracting salient features from the immense detail of raw data. Re-reading, clustering and condensing the data lead to the emergence of a number of sub-categories and categories.

The thematic areas that were selected, attempt to justify the importance of the communication with the unconscious patients and also justify the content of the verbal communication of the intensive care health professionals and the patients' relatives.

The expert group proposed the construction of a message that would have an increasing degree of stimulation throughout its deployment. It was judged important that the stimulus, should include both pleasant contents (e.g., "Your family told me they really like you and that they wish you recover quickly") and orders (e.g., "Mr./Mrs. A, I'm here to help you, come on, open your eyes"). This order was thought to be more likely to provoke an exacerbated reaction in the stimulated patient.

The stimulus message consisted of three parts: Presentation and orientation; Information; Functional assessment and stimulation (see Figure 3).

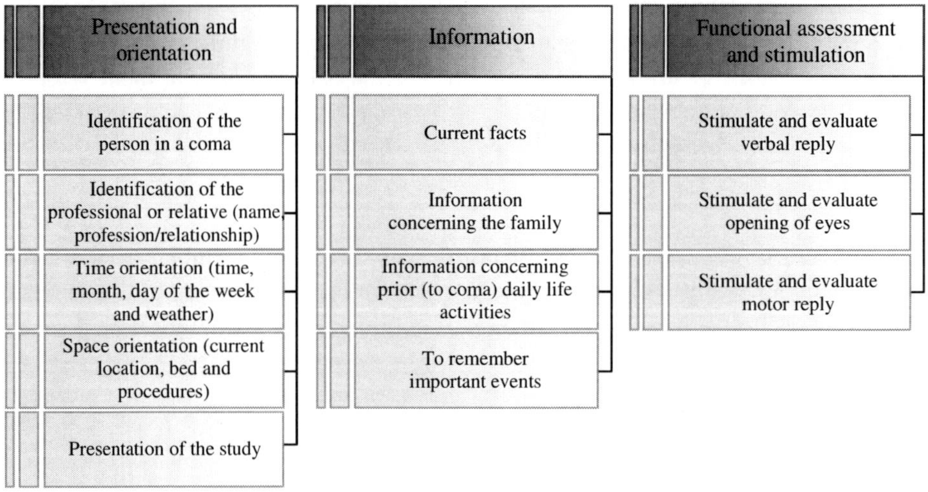

Figure 3. Composition of the stimulus message.

The first part included contents with the purpose of introducing the speaker and orientating the conversation in space and time. This part was composed of the acknowledging the person in coma, identifying the health professional or relative (name, profession or relationship), stating the date, day of the week and the weather, and explaining their current location. Finally the study was presented to the patient.

The second part consisted of providing the patient with information about: Current affairs; Family; Daily life activities prior to coma; Important events.

The last part of the stimulus message had the purpose of assessing the functional abilities and response to stimulation of the patients. This part was composed of instructions (e.g., open your eyes) and an evaluation of their responses, including verbal replies, opening of eyes and motor responses.

2.5.2. Recording and Editing the Stimulus Message

After identifying the significant people we explained, to each, the process of recording the stimulus message and how they could participate. Informed consent was required for them to be recorded whilst reading the stimulus message. Then a short interview was conducted to collect data about the experiences of the comatose patient and the names by which each significant person called him. These data enabled the adaptation of message to each individual patient. The final result was validated by the significant person to verify whether the information was entered properly.

Next, the recording of the message was performed in a room of the ICU that provided as much privacy and silence as possible.

Subsequently we proceeded to the choosing an unknown person. This selection was made by a convenience method meeting the following criteria: Accept to participate in the study; Portuguese speaking; Same gender as the significant person; Being unknown to the comatose patient; Not having had any contact with the comatose patient since he had been admitted to the ICU. The stimulus message of the unknown person was based on that of the significant person, and recorded in a room at the University of Aveiro which had identical conditions to the room where the recording of the significant person was conducted.

The recordings of the stimulus messages were performed with a directional microphone Shure PG 48, located 20 cm away from the mouth and in front of the person recorded. The acoustic signals were collected using an external sound card, Edirol-UA-25 to 16 bits and with a sampling frequency of 16 kHz, saved to a .wav file and edited with *Audacity 1.2.6* (Bland et al., 2006).

During the editing process, we avoided changing the content of the message and attempted to make the stimulus as natural as possible, with no repetition of parts of the written text. The editing process was based only on the insertion of pauses where necessary, clipping the messages when too long, removing mistakes and removing emotional reactions which were considered more extreme (e.g., crying).

2.6. Data Collection

The process of data collection began by explaining the study aims to the comatose patient and providing information about what would happen. The sample characteristic questionnaire, which collected demographic and clinical data, was then completed. The monitoring devices were placed correctly and we ensured that they were able to accurately record, and display on the multiparameter monitor all the required parameters for the study.

Before presenting the auditory stimuli to the comatose patient we ensured that they were haemodynamically stable and verified that they had not been subjected to any medical or nursing intervention in the previous 10 minutes. We also gained the agreement of the multidisciplinary team not to perform any clinical interventions during the stimulation, except in emergencies. In this case, we would repeat the stimulation later, once the patient stabilised.

A laptop computer was placed beside the comatose patient. This was used to: Produce an auditory alarm 30 seconds preceding the auditory stimulation; Generate the two stimuli; Collect the various parameters selected previously. The laptop was connected to the multiparameter monitor via an interface cable Datex-Ohmeda ADU-AS/3 Monitor Cable. Headphones (Roland RM-50) were placed on the comatose patient and connected to the internal sound card (SoundMAX Integrated Digital Audio) of the laptop computer. The stimulus message was played using the *Audacity 1.2.6* program (Bland et al., 2006). The researcher simultaneously listened to the playback of the stimulus message to the comatose patient, to make sure that the optimal playback was achieved. The playback volume used for all comatose patients was always thc samc.

We chose to divide the collection of data into two different time slots (stimulation with a familiar voice and stimulation with the unknown voice) each lasting a total of 45 minutes (see Figure 4).

The following data were monitored during auditory stimulation: Curves of ECG; Invasive arterial pressure; Pulse plethysmography; Pulse frequency; Blood pressures (systolic, diastolic and mean); Oxygen saturation level; Peripheral body temperature.

The time interval between stimulation with an unknown voice and a familiar voice was always at least 30 minutes. We considered this to be the minimum possible whilst allowing for the necessary patient care to be provided.

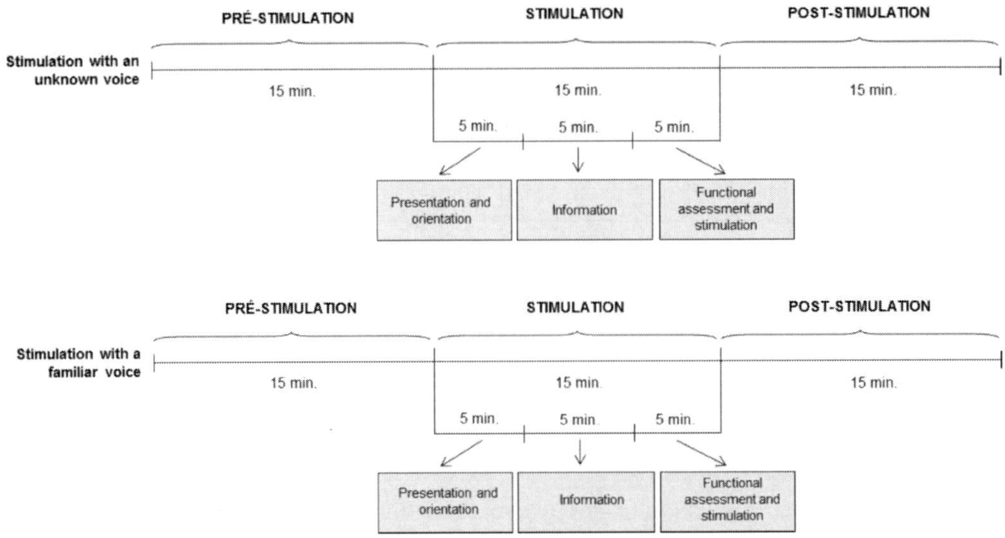

Figure 4. Auditory stimulation.

The stimulation and data collection was always conducted at the end of the afternoon shift. This was the quietest time of day and did not interfere with the patient's biorhythm (e.g., sleep) or relatives visiting times. Selecting a consistent time for stimulation also reduced variability in data which might have occurred due to fluctuations in the patients biorhythms.

The sampling frequency was 0.2 Hz for the values of the parameters and 300 Hz for the curves monitored. In the study we used a multiparameter monitor of the ICU (*Compact Critical Care Monitor* from *Datex-Ohmeda).* This was connected to the computer via a Datex-Ohmeda ADU-AS/3 Monitor Cable (Datex-Ohmeda, 2003).

The COM1 port for data export was selected and the program started. Data were always recorded in DRC format (.drc) and converted to ASCII (.asc) format.

2.7. Data Analysis

2.7.1. Analysis of the Monitored Parameters

The ASCII files of the monitored parameters were imported into the *Excel 2003* program to create individual databases for each comatose patient. This program enabled us to construct various plots for visual interpretation of the parameters monitored during the three periods of stimulation and the three phases of the stimulation message with an unknown voice and a familiar voice.

After this initial analysis, the data were imported into the *SPSS 17.0* program, constructing two databases: one for the periods of stimulation and other for the phases of stimulation. We calculated the minimum and maximum values, the mean and the standard deviation for each monitored parameter during the periods of stimulation with an unknown voice and a familiar voice and in each phase of the stimulus message (Johnson et al., 1989; Walker et al., 1998). This analysis aimed to investigate whether there were significant differences in these values in the different periods and phases mentioned.

We could not apply the commonly used statistical tests (parametric and nonparametric) to the complete data set because the samples are not random. That is, the observations made for each comatose patient are dependent throughout the periods and phases, and to measure the variability we need independent observations. So, we decided to build a new database in *SPSS* with all the minimum and maximum values, means and standard deviation of all the comatose patients, grouped by monitored parameter.

For the minimum, maximum and standard deviation we chose to apply the Friedman test (nonparametric), because the small sample size of our data did not allow us to perform normality tests (Kerlinger and Lee, 2000). For the values of the means we applied a simple analysis of variance (parametric test) because based on the Central Limit Theorem, we can say that the distribution approaches normality (Murteira et al., 2007).

These tests enabled us to determine whether there were any statistically significant differences between the three periods of stimulation and the three phases of the stimulus message. In this exploratory study, with such a small sample size, we used the statistical significance level $\alpha = 10\%$. We chose this value in order to increase the power of the tests which are strongly affected by the small sample size (Murteira et al., 2007).

Where differences were statistically significant, we performed multiple comparison tests: Wilcoxon test for the minimum, maximum and standard deviation; t tests for the mean values. In all cases we applied the Bonferroni correction, i.e., the significance level α was divided by 3.

After this process we also applied the Wilcoxon test to the minimum, maximum and standard deviation, and the t test to the means, now to identify significant differences between the stimulation with an unknown voice and a familiar voice during the periods of stimulation and post-stimulation and also for the three phases of the stimulus message.

We have chosen only to compare between the periods of stimulation and post-stimulation (see Figure 4), as the comatose patient had not been stimulated at baseline. The literature shows some evidence that comatose patients react more significantly to familiar voices while they are stimulated (Hoyt 1996; Walker et al., 1998; Alasad and Ahmad 2005). Furthermore, unilateral tests are recommended when it is expected to have results in one specific direction (Murteira et al., 2007). Therefore, unilateral tests were applied, as we initially expected that the comatose patients would react more relevantly to the familiar voice.

According to our knowledge, there are no studies which have applied stimulus messages composed by different phases. Therefore, for the phases of the stimulus message (see Figure 4), we used a bilateral analysis, because we did not know if the reaction would be further exacerbated by stimulation with a familiar or an unfamiliar voice.

Next, we conducted a separate analysis of each parameter monitored for each comatose patient to check if there were clinically relevant differences in each participant. This analysis was based on the methodology used by Johnson, Omery and Nikas (1989).

Variations in the parameter values were analysed separately in terms of their clinical relevance using the values shown in Table 1. These values were based on literature review (Bolander, 1994; Urden, Stacy and Lough, 2010), the professional experience of the researchers and validated by a group of experts in the field of Intensive Care Nursing.

Table 1. Reference values for the analysis of clinical relevance of the monitored parameters. A variation from previous values was considered clinically relevant when the values exceeded the range presented (for minimal and maximum values analysis)

Monitored parameters	Reference values
Pulse frequency	±10.0 p/m
Systolic blood pressure Diastolic blood pressure Mean blood pressure	±10.0 mmHg
Peripheral body temperature	±1.0 ºC
Oxygen saturation level	±5.0%

Mean and standard deviation analysis of clinical relevance was based on half the range shown.

2.7.2. Analysis of the Monitored Curves

The data of each monitored curve were exported to an ASCII file (.asc) and analysed individually. The duration of RR intervals were calculated and the differences in the three periods and phases of stimulation were analysed. Figure 5 shows the graphical representation of this interval, which is related to the pulse frequency of the comatose patient. This allowed us to verify if the pulse frequency increased or decreased with the auditory stimulation and to compare the analysis results of the ECG curve with the analysis of pulse frequency reported previously.

For the analysis of RR intervals we used a tool developed specifically for this purpose (*check_signals.m* program) using the program *Matlab R2007b* to calculate the intervals between R peaks of QRS complex.

With this tool, the heartbeats are detected by a simple thresholding method (threshold manually selected), applied to the band-pass filtered signals (3-15 Hz). It should be noted that depending on ECG morphology, either positive or negative-going peaks are used.

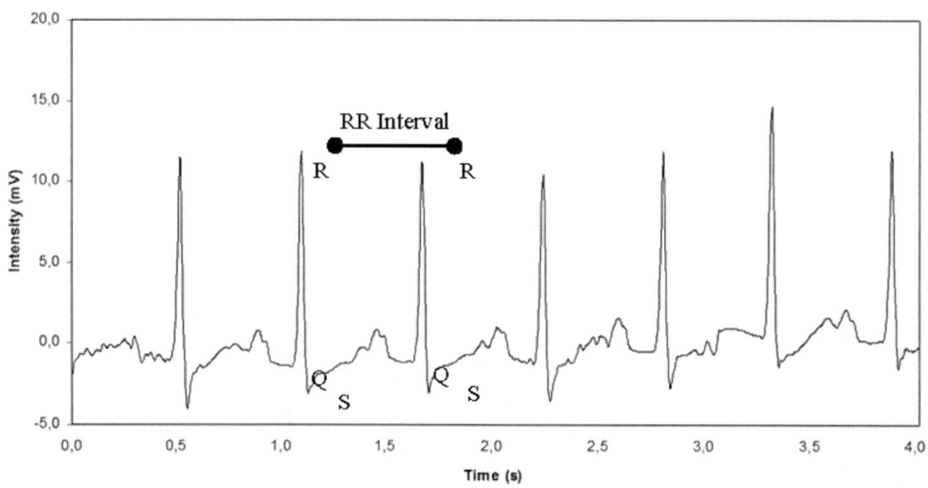

Figure 5. Graphical representation of the RR interval of the ECG curve of the participant AT during the stimulation with an unknown voice.

The isolated peaks (maximum two QRS complexes) in the RR interval time-series were visually inspected, and removed. They were first replaced by 'not a number' (NaN). The raw RR-interval time-series was then linearly interpolated at 4 Hz, and any missing heartbeats were replaced also by linear interpolation between adjacent RR intervals.

The output resulting from this review process was stored in the format .mat, accompanied by a file with the times in which several R peaks were detected. Thus, we had two files for each comatose patient (one for the unknown voice and another for the familiar voice), each with a different number of samples. These data were then imported to the *Excel 2003* program. We also constructed plots for a visual interpretation of the presence or absence of changes in RR intervals during the periods of stimulation with an unknown voice and a familiar voice.

After this initial analysis, the data were imported into *SPSS 17.0*. We then calculated the minimum and maximum values, means and standard deviations (Johnson et al., 1989) of the values of RR intervals during the periods of pre-stimulation, stimulation and post-stimulation with an unknown and a familiar voice. The same procedure was applied to the phases of the stimulus message.

The statistical analysis process followed the guidelines outlined above for the analysis of the parameters with the level of statistical significance set at a maximum of 10%.

Next, we conducted a separate analysis per comatose patient to check whether there were any clinically relevant differences in each participant.

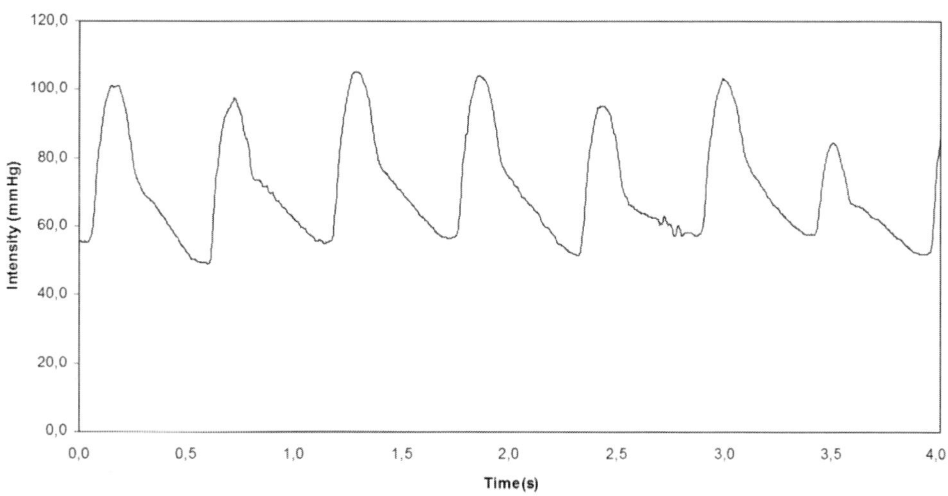

Figure 6. Graphical representation of the Arterial Pressure curve of the participant AT during the stimulation with an unknown voice.

In relation to the analysis of the curves of Blood Pressures and Pulse Plethysmography (represented in Figures 6 and 7, respectively) we conducted research into the potential of analysing these two curves (Allen, 2007; Rangayyan, 2002). This was performed in *Matlab R2007b*. As we had the aim of applying the same statistical tests, before mentioned on the analysis of the ECG wave, we chose to calculate the means and standard deviations in the different periods and phases of stimulation for each comatose patient and type of stimulation (unknown and familiar voice). In this case we did not calculate the minimum and maximum values, because a visual analysis of the graphical representations obtained in *Matlab R2007b*, identified artefacts in the signal, resulting in maximums and minimums that could lead to misinterpretation of results.

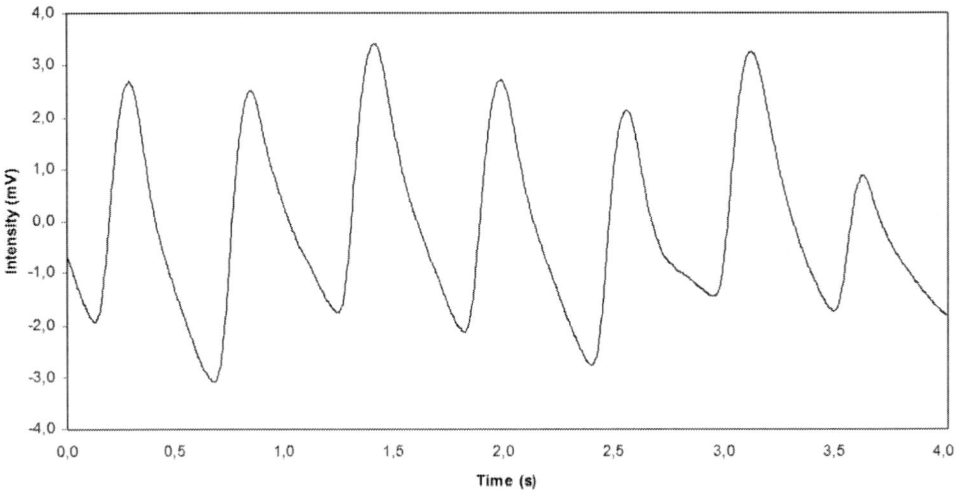

Figure 7. Graphical representation of the Pulse Plethysmography curve of the participant AT during the stimulation with an unknown voice.

After obtaining the means and standard deviations for the periods and phases of stimulation, two databases were built in *SPSS 17.0* with the aim of applying the Friedman test in the case of standard deviation values and a simple analysis of variance (parametric test) in the case of the mean values. The remaining data analysis followed the same guidelines described above in relation to the ECG curve.

Table 2. Reference values for the analysis of clinical relevance of the monitored curves. A variation from previous values was considered clinically relevant when the minimum and maximum values exceeded the range presented (for minimal and maximum values analysis)

Monitored curves	Reference values
RR intervals	±0.05 s
Arterial pressure	±5.0 mmHg
Pulse plethysmography	±0.05 mV

Mean and standard deviation analysis of clinical relevance was based on half the range shown

The clinical relevance of the variations in all values derived from the analysis of monitored curves was based in the values shown in Table 2. These values resulted from an extensive literature review (Allen, 2007, Castells et al., 2007, Clifford, Azuaje and McSharry, 2006; McSharry and Clifford, 2006; Oliveira, Maojo, Martin-Sanchez and Pereira, 2005; Rangayyan, 2002; Reisner, Clifford and Mark, 2006; Saritha et al. 2008; Simoliuniene et al. 2008; Urden et al., 2010), the professional experience of the researchers and were validated by a group of experts in Intensive Care Nursing.

3. Results

3.1. Demographic and Clinical Characteristics of the Sample

Four participants were eliminated after the start of the study because their clinical conditions worsened during the period of data collection, meaning they no longer met the study inclusion criteria. One participant was excluded from the study due to technical failures of the software used for the collection of monitored parameters and curves. Also one of the participants was excluded from the study because he was transferred out of the ICU during the data collection period. Only one potential participant was excluded because of changes in auditory evoked responses. Table 3 shows the characteristics of patients included in our study.

Table 3. Demographic and clinical characteristics of patients included in the study (*n*=10)

Characteristics	*n*	Mean ± SD	Range
Age (years)	10	56.0 ± 18.4	20-79
Gender			
Males	8		
Females	2		
Days in coma	10	8.8 ± 3.9	3-15
IARCUA total score	10	3.2 ± 0.6	2-4
GCS total score		6.8 ± 0.8	6-8
6	4		
7	4		
8	2		
9	0		
Aetiology			
Medical	6		
Emergency Surgery	3		
Transplant	1		

In our study the sample consisted of 10 patients in coma (8 male and 2 female). Their ages ranged between 20 and 79 years with a mean age of 56 years.

The number of days in coma ranged from 3 to 15 days, with an average of 8.8 days. This value is relatively low, given the urgency of collecting data due to the clinical instability of these patients. Regarding the aetiology six patients were hospitalised for medical reasons (e.g., Hepatic Coma and Sepsis), three for surgical emergencies and one patient following complications after a transplant.

The total IARCUA scores ranged from 2 to 4 and the total score of Glasgow Coma Scale from 6 to 8.

3.2. Statistical and Clinical Analysis of the Monitored Parameters

3.2.1 Statistical and Clinical Analysis in the Periods of Stimulation

The results of the Friedman and ANOVA tests for the parameters monitored during the three periods of stimulation with an unknown voice are presented in Table 4.

Assuming $\alpha = 0.10$ only the differences in minimum and maximum values of peripheral body temperature and the standard deviations of pulse frequency were statistically significant.

In these three cases we performed multicomparison tests as described previously. However, in all cases, we concluded that there were no statistically significant differences in the comparisons made, despite the Friedman test being significant when we clustered the three periods. This contradiction in results may be due to the high value of the significance level (increasing the probability of a Type I error in the first test) or to the small sample size (increasing the probability of a Type II error in the multiple comparisons test) or eventually to the existence of ties (repeated values).

For minimum values of the whole sample and from a clinical point of view, we found that the pulse frequency values remained very stable in the transition of the periods of data collection. This relative uniformity in the values of the monitored parameters was also found for the other parameters, with no relevant differences from a clinical point of view.

Table 4. Statistical analysis of the parameters monitored during the periods of pre-stimulation, stimulation and post-stimulation with an unknown voice (p-values of the various tests)

	Friedman test			ANOVA
Parameters	p-values of the minimum value	p-values of the maximum value	p-values of the standard deviation	p-values of the mean
Pulse frequency	0.227	0.382	**0.081**	0.187
Systolic blood pressure	0.897	0.601	1.000	0.864
Diastolic blood pressure	0.316	0.314	0.648	0.896
Mean blood pressure	0.897	0.222	0.437	0.914
Peripheral body temperature	**0.038**	**0.017**	1.000	0.172
Oxygen saturation level	0.142	0.899	0.396	0.974

Bold – Statistically significant (p-value < 0.10).

Analysis of the maximum values of the pulse frequency revealed a trend for these values to vary between periods, although this was not statistically significant ($p = 0.382$). So, from a clinical point of view, the maximum values of the whole sample increase between the pre-stimulation period and stimulation periods and decrease again when the stimulation ends.

We identified a clinically relevant change in the minimum and maximum values of the maximum systolic blood pressure in the transition between the periods of pre-stimulation and stimulation although this was not statistically significant. This increase could be associated with anxiety and stress reported by Alasad and Ahmad (2005) and Doman et al. (1993), or the excitatory effect of hearing someone speaking. When the stimulation ends, we noted a marked decrease in the minimum values. The mean values seem to decrease in the transition between the three periods of stimulation. The values tended to normalise when the auditory stimulation finished, returning to pre-stimulation baseline.

For the clinical analysis of the standard deviations resulting from the initial analysis of all the comatose patients of our sample, we can conclude that there were relevant differences in values for the pulse frequency between the period of pre-stimulation and the period of stimulation (minimum, maximum and mean values). After stimulation values decreased, returning to the values obtained before stimulation. This is in accordance with the Friedman test performed previously (we stated that there were significant differences between the three periods of stimulation). This increase in the values of standard deviation indicates the existence of a greater variability of values in the period of stimulation, which means that this parameter was more unstable, probably caused by the stimulation with the recording of the unknown voice.

The values of diastolic and mean blood pressure increased during the pre-stimulation and stimulation periods and decreased when the stimulation message ended, returning almost to the pre-stimulation values. These results are consistent with the variation of pulse frequency values, because these vital parameters are physiologically related.

The results of the Friedman test and ANOVA for the parameters monitored during the periods of pre-stimulation, stimulation and post-stimulation with a familiar voice are described in Table 5. As can be seen, statistically significant differences were identified for the maximum values and means of diastolic blood pressure, and the minimum and maximum values and means in peripheral body temperature.

As mentioned above, Walker et al. (1998) evaluated the effect of familiar voices in patients in coma caused by a Traumatic Brain Injury and found no statistically significant changes for the various vital signs. Data analysis revealed only variability in physiological measures and the pulse was a measure that had a greater variability in relation to other physiological parameters. The results of their (Walker et al., 1998) study agree in part with ours.

Proceeding as before, we applied multiple comparison tests in the cases where significant differences occurred. No statistically significant differences were identified between periods for the maximum values of diastolic blood pressure. The mean diastolic blood pressure was statistically significant when comparing the periods of pre-stimulation and stimulation (p-value = 0.070) and in the periods of pre-stimulation and post-stimulation (p-value = 0.065).

No relevant changes in peripheral body temperature were identified in any of the comparative analyses between the periods of stimulation for the minimum values. There were significant differences for the maximum values of the peripheral body temperature, between the periods of pre-stimulation and post-stimulation (p-value = 0.016) and between the periods of pre-stimulation and post-stimulation (p-value = 0.016).

Table 5. Statistical analysis of the parameters monitored during the periods of pre-stimulation, stimulation and post-stimulation with a familiar voice (p-values of the various tests)

	Friedman test			ANOVA
Parameters	p-values of the minimum value	p-values of the maximum value	p-values of the standard deviation	p-values of the mean
Pulse frequency	0.881	0.850	0.458	0.914
Systolic blood pressure	0.135	0.601	0.974	0.120
Diastolic blood pressure	0.763	**0.048**	0.808	**0.048**
Mean blood pressure	0.601	0.368	0.601	0.101
Peripheral body temperature	**0.001**	**0.002**	0.222	**0.023**
Oxygen saturation level	0.754	0.949	0.930	0.292

Bold – Statistically significant (p-value < 0.10).

For the mean values of peripheral body temperature there were significant differences when we compared the values of the period of pre-stimulation with the period of stimulation (p-value = 0.025) and when we compared the values of the periods of pre-stimulation and post-stimulation (p-value=0.032).

There was a clinically relevant increase in the minimum values of systolic, diastolic and mean blood pressure when stimulation began which continued to increase when the stimulus message ended. This increase was unexpected and had not been previously reported in the literature. We chose to designate this event “separation anxiety episode”, which has already been described for children (Bar-Haim, Dan, Eshel, and Sagi-Schwartz, 2007; Mofrad, Abdullah, Samah, Mansor, and Baba, 2009; Verissimo, Alves, Monteiro and Oliveira, 2003). Usually, such episodes are related to expressions of concern, anxiety and fear when a significant person leaves or is absent. Thus, we believe that there was a reaction linked to the termination of the stimulus message, and a generation of a sense of abandonment and absence. There was an increase in the values of standard deviation of the pulse frequency in the period of pre-stimulation, during the period of stimulation, followed by a decrease between the period of stimulation and the period of post-stimulation. These results indicate that the variability of values was higher in the period of stimulation.

Similarly, clinically relevant variability in the values of systolic, diastolic and mean blood pressure was found. The peripheral body temperature and oxygen saturation levels remained almost unchanged throughout the periods of stimulation.

Next we present in Table 6 the results of the analysis for comparing the unknown with the familiar voice.

As can be seen in the Table 6 there are significant difference in some of the parameters both during the stimulation and the post-stimulation periods. This indicates that the comatose patients reacted differently depending on the type of voice.

Table 6. Comparative statistical analysis of the minimum values, maximum values, standard deviation and mean of the parameters monitored during the periods of stimulation and post-stimulation with an unknown and familiar voice (p-values of the various tests)

	Stimulation period				Post-stimulation period			
	Wilcoxon test			t test	Wilcoxon test			t test
Parameters	p-values of the minimum value	p-values of the maximum value	p-values of the standard deviation	p-values of the mean	p-values of the minimum value	p-values of the maximum value	p-values of the standard deviation	p-values of the mean
Pulse frequency	**0.061**	0.150	0.396	**0.027**	**0.064**	**0.066**	0.357	**0.033**
Systolic blood pressure	0.357	0.423	0.246	0.348	0.285	0.216	**0.010**	0.446
Diastolic blood pressure	0.348	0.461	0.271	0.401	0.138	**0.040**	**0.002**	0.308
Mean blood pressure	0.275	0.461	0.210	0.384	0.278	**0.097**	**0.003**	0.472
Peripheral body temperature	**0.073**	**0.047**	0.250	**0.038**	**0.027**	**0.043**	0.750	**0.025**
Oxygen saturation level	0.215	**0.053**	0.234	0.127	0.148	0.219	0.438	0.197

Bold – Statistically significant (p-value < 0.10).

When we compared the responses with an unknown voice and those with familiar voices we found that, for all parameters, there were more significant differences post-stimulation than during the period of stimulation. The events that we designated as "separation anxiety episodes" occur more frequently when the voices of family end than for those of unfamiliar people.

A clinically relevant difference in standard deviation values between the stimulation with an unknown voice and a familiar voice for the transition between the period of stimulation and the period of post-stimulation was observed. There was a tendency for participants to normalise the values of the monitored parameters whilst stimulated with an unknown voice and therefore the majority of the participants showed a decrease of the standard deviation values. When stimulated with a familiar voice, we concluded that there was an increase in the number of participants showing higher standard deviation values after stimulation terminated. This provides additional evidence for the "separation anxiety episodes" described previously.

3.2.2. Statistical and Clinical Analysis in the Phases of the Stimulus Message

Table 7 contains the results of the tests performed to study the period of auditory stimulation with the stimulus message of the unknown voice. These tests compare the values during the transition of the three phases of presentation and orientation, information and functional assessment and stimulation.

We found that the Friedman test was significant only for the values of standard deviations of the oxygen saturation level. The ANOVA was significant for the mean of the pulse frequency.

As before, we performed multiple comparisons in both situations where differences were considered significant. The results of the Wilcoxon test for the different paired phases of standard deviation values of oxygen saturation level indicated that no statistically significant differences exist between them. For the mean of the pulse frequency we applied multiple t tests, which detected significant differences between the means of the phase of information and the phase of functional assessment (p-value = 0.066) and between the means of the phase of presentation and orientation and the phase of functional assessment and stimulation (p-value = 0.032). These changes may be related to the contents of the phase of functional assessment and stimulation being more provocative, or because the comatose patient knows that if unresponsive to verbal stimulation, a painful stimulus will be applied. Focusing on the clinical analysis, the minimum value of the pulse frequency did not change significantly. The same happened for the values of the other parameters that remain similar throughout the three phases of the stimulus message, which is consistent with the tests previously presented.

Table 7. Statistical analysis of the parameters monitored during the phases of presentation and orientation, information and functional assessment and stimulation with an unknown voice (p-values of the various tests)

	Friedman test			ANOVA
Parameters	p-values of the minimum value	p-values of the maximum value	p-values of the standard deviation	p-values of the mean
Pulse frequency	0.989	0.444	0.740	**0.020**
Systolic blood pressure	0.974	0.974	0.615	0.565
Diastolic blood pressure	0.710	0.830	0.330	0.628
Mean blood pressure	0.974	0.830	0.897	0.549
Peripheral body temperature	0.519	0.556	1.000	0.630
Oxygen saturation level	0.345	0.936	**0.050**	0.300

Bold – Statistically significant (p-value < 0.10).

From the clinical analysis of the maximum values of the pulse frequency, we concluded that no changes exist in this parameter between the phase of presentation and orientation and the phase of information. However when we observed the values of the phase of information and the phase of functional assessment and stimulation there was an increase in the values, which means that the participants reacted when they heard an unknown voice asking them to move. We also looked at the range of the maximum values of the systolic blood pressure. The minimum values of this range for the total sample, showed a relevant change between the phase of presentation and orientation and the phase of information. These values move back toward normal ranges in the functional assessment and stimulation phases.

We then clinically examined the minimum and maximum values, means and standard deviations of the means of the total sample. The global observation of the values, indicated that the differences were not significant (except in the minimum values of the mean of the systolic blood pressure that increased between the phase of presentation and orientation and the phase of information) in the various parameters evaluated or in the transition of the three phases of the stimulus message.

Analysing the variations in the values of standard deviations, we found that overall there were no clinically relevant changes in all parameters monitored when comparing the values of the phase of presentation and orientation and the phase of information. If we compare the values of the phase of information with the values of the phase of functional assessment and stimulation, we observed an increase in standard deviation values (and therefore variability) in pulse frequency, systolic blood pressure, diastolic blood pressure and mean blood pressure.

In the overall statistical analysis for the three phases of the familiar stimulus message, we found (see Table 8) that the Friedman test was significant for the maximum values of the diastolic blood pressure, minimum and maximum values of the peripheral body temperature and maximum values and standard deviation of the oxygen saturation level. ANOVA was significant for the mean values of peripheral body temperature.

Table 8. Statistical analysis of the parameters monitored during the phases of presentation and orientation, information and functional assessment and stimulation with a familiar voice (p-values of the various tests)

	Friedman test			ANOVA
Parameters	p-values of the minimum value	p-values of the maximum value	p-values of the standard deviation	p-values of the mean
Pulse frequency	0.516	0.856	0.156	0.430
Systolic blood pressure	0.710	0.437	0.135	0.331
Diastolic blood pressure	0.491	**0.092**	0.114	0.261
Mean blood pressure	0.601	0.135	0.176	0.271
Peripheral body temperature	**0.074**	**0.082**	1.000	**0.040**
Oxygen saturation level	0.938	**0.094**	**0.045**	0.894

Bold – Statistically significant (p-value < 0.10).

As before, we have carried out multiple comparisons. We found no statistically significant differences in most of the cases.

We only found statistically significant differences in the maximum values of oxygen saturation level between the phases of information and functional assessment and stimulation (p-value = 0.031) and in the mean values of peripheral body temperature between the phase of presentation and orientation and the phase of functional assessment and stimulation (p-value = 0.037).

Table 9. Comparative statistical analysis of the minimum values, maximum values, standard deviation and mean of the parameters monitored during the phases of presentation and orientation, information and functional assessment and stimulation with an unknown and familiar voice (p-values of the various tests)

	Presentation and orientation				Information				Functional assessment and stimulation			
	Wilcoxon test			t test	Wilcoxon test			t test	Wilcoxon test			t test
Parameters	p-values of the minimum value	p-values of the maximum value	p-values of the standard deviation	p-values of the mean	p-values of the minimum value	p-values of the maximum value	p-values of the standard deviation	p-values of the mean	p-values of the minimum value	p-values of the maximum value	p-values of the standard deviation	p-values of the mean
Pulse frequency	**0.063**	**0.004**	0.604	**0.047**	0.402	**0.021**	0.887	**0.079**	0.328	0.434	1.000	**0.070**
Systolic blood pressure	0.846	0.910	**0.078**	0.778	0.625	0.695	0.719	0.496	0.922	0.625	0.322	0.901
Diastolic blood pressure	0.574	0.264	**0.049**	0.692	0.846	0.922	0.826	0.873	0.643	0.922	0.668	0.649
Mean blood pressure	0.232	0.770	**0.047**	0.791	0.359	0.826	0.859	0.512	0.570	0.826	0.643	0.980
Peripheral body temperature	0.145	0.127	1.000	0.125	0.141	0.125	1.000	**0.076**	**0.090**	**0.094**	1.000	**0.051**
Oxygen saturation level	**0.063**	0.523	0.945	0.213	0.633	**0.066**	0.500	0.283	0.734	0.461	0.906	0.247

Bold – Statistically significant (p-value < 0.10).

We also analysed each of the variables to verify if any clinically relevant changes existed in the transition of the various phases of the stimulus message with a familiar voice. The standard deviations for the different pulse frequency values remained relatively stable in the transition between the phase of presentation and orientation and the phase of information, but there was a large increase in the values when we transited to the phase of functional assessment and stimulation. In the values of systolic, diastolic and mean blood pressure we observed a decrease in standard deviation values when we transit from the phase of presentation and orientation to the phase of information, which increased again when we transit to the phase of functional assessment and stimulation. The values for peripheral body temperature remained relatively unchanged during the three phases of stimulation and the values of oxygen saturation level followed the same line of changes described above for the pressures.

Table 9 contains the results of the tests performed in order to assess differences between the responses with the unknown and the familiar voices

We can observe that during the phase of presentation and orientation there were statistically significant differences in some of the variables in all the phases for the pulse frequency. The other parameters exhibited fewer significant differences with rather high p-values.

3.3. Statistical and Clinical Analysis of the Monitored Curves

3.3.1. Statistical and Clinical Analysis in the Periods of Stimulation

The curve analysis proved important because it allowed us to corroborate the results obtained previously for the parameters and allowed a more rigorous study of the influence of auditory stimulation, in the sense that we worked directly with physiological variables monitored but not processed by the multiparameter monitor. It was also important to analyse the curves, because we worked with a sampling frequency much higher than the sampling frequency of the parameters.

Table 10 shows that there were no statistically significant differences in any of the monitored curves with an unknown voice.

Table 10. Statistical analysis of the curves monitored during the periods of pre-stimulation, stimulation and post-stimulation with an unknown voice (p-values of the various tests)

	Friedman test			ANOVA
Curves	p-values of the minimum value	p-values of the maximum value	p-values of the standard deviation	p-values of the mean
RR intervals	0.726	0.114	0.938	0.157
Arterial pressure	-	-	0.222	0.868
Pulse plethysmography	-	-	0.897	0.304

We concluded that the most clinically relevant changes occurred in the means of the curve of arterial pressure. The standard deviations showed a progressive increase over the three periods of stimulation, which means that there was a greater variability in the values found. No relevant clinical changes were found for the curve of pulse plethysmography over the period of analysis.

Table 11 shows that there were no statistically significant differences in the values of the three periods of stimulation with the stimulus message of the familiar voice.

Table 11. Statistical analysis of the curves monitored during the periods of pre-stimulation, stimulation and post-stimulation with a familiar voice (p-values of the various tests)

	Friedman test			ANOVA
Curves	p-values of the minimum value	p-values of the maximum value	p-values of the standard deviation	p-values of the mean
RR intervals	0.994	0.205	0.852	0.973
Arterial pressure	-	-	0.601	0.105
Pulse plethysmography	-	-	0.947	0.685

The mean values of the curve of arterial pressure showed a clinically relevant increase when we started the stimulation, but also continued to increase when the auditory stimulation ended. For the standard deviation of the curve of pulse plethysmography we observed no changes for the transition between the pre-stimulation and stimulation period, however, when the stimulation ended there was a decrease in all values.

Table 12. Comparative statistical analysis of the minimum values, maximum values, standard deviation and mean of the curves monitored during the periods of stimulation and post-stimulation with an unknown and familiar voice (p-values of the various tests)

	Stimulation period				Post-stimulation period			
	Wilcoxon test			t test	Wilcoxon test			t test
Curves	p-values of the minimum value	p-values of the maximum value	p-values of the standard deviation	p-values of the mean	p-values of the minimum value	p-values of the maximum value	p-values of the standard deviation	p-values of the mean
RR intervals	0.107	0.201	**0.047**	**0.011**	**0.049**	0.332	**0.070**	**0.017**
Arterial pressure	-	-	0.385	0.379	-	-	0.385	0.443
Pulse plethysmo graphy	-	-	0.271	0.222	-	-	0.287	0.390

Bold – Statistically significant (p-value < 0.10).

Table 12 contains the results of the tests performed to compare the responses obtained with the unknown and the familiar voices. Unilateral test were selected because it was expected that the comatose patients would have a more exacerbated response during the stimulation with a familiar voice than during the stimulation with an unknown voice.

As can be seen, significant differences were only detected in the RR intervals, both in the stimulation and the post-stimulation periods.

From an overall clinical (relevance) analysis, we concluded that the reaction of participants was more marked on stimulation with a familiar voice than to an unknown voice on every curve and on the two transitions in question.

3.3.2. Statistical and Clinical Analysis in the Phases of the Stimulus Message

The results of the statistical tests used to study the monitored curves resulting from stimulation with and unknown voice are presented in Table 13.

Table 13. Statistical analysis of the curves monitored during the phases of presentation and orientation, information and functional assessment and stimulation with an unknown voice (p-values of the various tests)

	Friedman test			ANOVA
Curves	p-values of the minimum value	p-values of the maximum value	p-values of the standard deviation	p-values of the mean
RR intervals	0.519	0.586	0.737	**0.047**
Arterial pressure	-	-	0.974	0.588
Pulse plethysmography	-	-	0.285	**0.097**

Bold – Statistically significant (p-value < 0.10).

We found significant differences for the means of the RR intervals and the means of the curve of pulse plethysmography. Thus, in these cases we applied multiple comparison tests. We only found significant differences (p-value = 0.037) in the case of the RR intervals between the phase of presentation and orientation, and the phase of functional assessment and stimulation.

From a clinical point of view we found that the mean of the minimum values of the RR interval decreased considerably during the phase of functional assessment and stimulation. However, the values of standard deviation increased. This increase in the values of standard deviation means that the entire sample exhibits a greater variability in the phase of functional assessment and stimulation. The decrease in the other values means that, in this phase, there has been an increase in heart rate and consequently in the pulse frequency.

For the mean values of the curve of arterial pressure, the standard deviation value of the phase of functional assessment and stimulation was higher than the others, which reveals that, although the mean values are not relevantly different, there is a greater variability among the participants in the study in this phase. In the curve of pulse plethysmography, we concluded that higher values of standard deviation were recorded in the information phase. Relative to the stimulation with a familiar voice we can see in Table 14, that there were no statistically significant differences in any of the cases under analysis.

From a clinical perspective, we found that in the minimum values of RR intervals, the most significant alteration occurred during the phase of functional assessment and stimulation, registering a slight decrease in all the values. For the means of the RR interval the phase of functional assessment and stimulation registered the lowest values indicating that it is in this phase where the increased values of the pulse frequency occurred.

Table 15 shows the results of comparing the values obtained in the analysis of the curves monitored during the three phases of the stimulus message with an unknown voice and a familiar voice. In this analysis we chose to apply two-tailed tests because it was not possible to predict that if there was a reaction from the comatose patients at various phases of stimulation, this would be exacerbated in the case of stimulation with an unknown voice or a familiar voice.

Table 14. Statistical analysis of the curves monitored during the phases of presentation and orientation, information and functional assessment and stimulation with a familiar voice (p-values of the various tests)

	Friedman test			ANOVA
Curves	p-values of the minimum value	p-values of the maximum value	p-values of the standard deviation	p-values of the mean
RR intervals	0.993	0.774	0.177	0.428
Arterial pressure	-	-	0.763	0.236
Pulse plethysmography	-	-	0.682	0.766

Table 15. Comparative statistical analysis of the minimum values, maximum values, standard deviation and mean of the curves monitored during the phases of presentation and orientation, information and functional assessment and stimulation with an unknown and familiar voice (p-values of the various tests)

	Presentation and orientation				Information				Functional assessment and stimulation			
	Wilcoxon test			t test	Wilcoxon test			t test	Wilcoxon test			t test
Curves	p-values of the minimum value	p-values of the maximum value	p-values of the standard deviation	p-values of the mean	p-values of the minimum value	p-values of the maximum value	p-values of the standard deviation	p-values of the mean	p-values of the minimum value	p-values of the maximum value	p-values of the standard deviation	p-values of the mean
RR intervals	**0.016**	0.389	0.438	**0.014**	**0.027**	0.344	0.531	**0.036**	0.250	0.754	**0.094**	**0.056**
Arterial pressure	-	-	0.695	0.760	-	-	0.557	0.504	-	-	1.000	0.963
Pulse plethysmography	-	-	0.674	0.796	-	-	0.557	0.411	-	-	0.311	0.427

Bold – Statistically significant (p-value < 0.10).

As can be seen in Table 15, only the RR-intervals show significant differences. They occur in all phases but only in some of the variables.

Conclusion

It is considered good practice to communicate with comatose patients because they may be able to hear and some patients may benefit from verbal stimulation which may assist in reducing stress, helping patients preserve self-identity and self-esteem, and provide orientation, meaningful sensory input and reduce social isolation.

This study investigated the effects of recorded voice messages provided by an unknown person and by a significant person on key measures of physiological functioning in comatose patients in an ICU.

Results showed that the stimulation with an unknown and familiar voice generated an increase in heart rate values and in the values of arterial pressure. This was corroborated by the analysis of monitored curves. However, for a familiar voice the values of these parameters continued to increase post-stimulation, which revealed that participants developed episodes of separation anxiety. No changes in body temperature occurred during the stimulation.

When the phases of the stimulus message were analysed, during the stimulation with an unknown voice, the functional assessment and stimulation phase showed a larger variability in the values of heart rate, arterial pressure and oxygen saturation level. This was corroborated by the analysis of the curves. Stimulation with a familiar voice resulted in greater changes for the same parameters during the functional assessment and stimulation phase and participants also presented relevant change during the presentation and orientation phases.

Acknowledgments

The authors would like to thank AT, AF, AS, FF, JF, LB, MA, MC, NP, RG and their families, André Baldaia, Aníbal Marinho, Fernando Rua, José Pinho, Sara Demain and Rui Alcovia.

This work was partially funded by National Funds through FCT – Foundation for Science and Technology, in the context of the project PEst-OE/EEI/UI0127/2014.

References

Adit (2006). *Sociogram Software.* Dwyran: Adit.

Alasad, J. and Ahmad, M. (2005) Communication with critically ill patients. *Journal of Advanced Nursing* 50(4), 356-362.

Allen, J. (2007). Photopletysmography and its application in clinical physiological measurement. *Physiological Measurement 28*, R1-R39.

Anderson, R. (2007). *Thematic content analysis (TCA): descriptive presentation of qualitative data.* Palo Alto: Institute of Transpersonal Psychology.

Ashworth, P. (1987). The needs of the critically ill patient. *Intensive Care Nursing* 3(4), 182-190.

Baker, C. and Melby, V. (1996). An investigation into the attitudes and practices of intensive care nurses towards verbal communication with unconscious patients. *Journal of Clinical Nursing* 5, 185-192.

Bar-Haim, Y., Dan, O., Eshel, Y., and Sagi-Schwartz, A. (2007). Predicting children's anxiety from early attachment relationships. *Journal of Anxiety Disorders* 21, 1061-1068.

Bastin, G. (1980). Les Techniques Sociometriques. Paris: Seli Arsian.

Bland, W., Busam, V., Gunlogson, B., Mekkes, G., and Saunders, A. (2006). Audacity (Version 1.2.6). Source Forge.

Bolander, V. (1994). *Sorensen and Luckmann's Basic Nursing: A Psychophysiologic Approach* (3rd ed.). Philadelphia: W. B. Saunders.

Boly, M., Garrido, M., Gosseries, O., Bruno, M., Boveroux, P., Schnakers, C., Massimini, M., Litvak, V., Laureys, S. and Friston, K. (2011). Preserved feedforward but impaired top-down processes in the vegetative state. *Science* 332(5), 858-862.

Castells, F., Laguna, P., Sornmo, L., Bollmann, A., and Roig, J. M. (2007). Principal component analysis in ECG signal processing. *EURASIP Journal on Advances in Signal Processing* 74580, 1-21.

Clifford, G., Azuaje, F., and McSharry, P. (2006). *Advanced Methods and Tools for ECG Data Analysis*. Norwood: Artech House.

Datex-Ohmeda. (2003). *Datex-Ohmeda S/5 Collect - User's Reference manual*. Helsinki: Datex-Ohmeda.

Doman, G., Wilkinson, R., Dimancescu, M. and Pelligra, R. (1993). The effect of intensive multi-sensory stimulation on coma arousal and recovery. *Neuropsychological Rehabilitation* 3(2), 203-212.

Elliot, R. and Wright, L. (1999). Verbal communication: what do critical care nurses say to their unconscious or sedated patients? *Journal of Advanced Nursing* 29(6), 1412-1420.

Fischer, C. (2006). *Qualitative research methods for psychologists: introduction through empirical studies*. New York: Academic Press.

Holeckova, I., Fischer, C., Giard, M., Delpuech, C. and Morlet, D. (2006). Brain responses to a subject's own name uttered by a familiar voice. *Brain Research* 1082, 142-152.

Hoonakker, P., Carayon, P., Douglas, S., Schultz, K., Walker, J. and Wetterneck, T. (2008). Communication in Intensive Care Units and the Relation with Quality of Care and Patient Safety from a Nurse Perspective. In L. Sznelwar, F. Mascia and U. Montedo (Eds.), *Human Factors in Organizational Design and Management – IX* (pp. 715-721). Santa Monica: IEA Press.

Hoyt, J. (1996). A gentle approach: interacting with a person who is semi-conscious or presumed in coma. *Issues in Law and Medicine* 12(1), 71-78.

Jennett, B. (2002). The Glasgow Coma Scale: History and current practice. *Trauma* 4, 91-103.

Johnson, S., Omery, A. and Nikas, D. (1989). Effects of conversation on intracranial pressure in comatose patients. *Neurologic Aspects of Critical Care* 18(1), 56-63.

Jones, R., Hux, K., Morton-Anderson, A. and Knepper, L. (1994). Auditory stimulation effect on a comatose survivor of traumatic brain injury. *Archives of Physical Medicine Rehabilitation* 75, 164-171.

Kerlinger, F. and Lee, H. (2000). *Foundations of behavioral research* (4th ed.). Belmont: Wadsworth - Thomson Learning.

LaPuma, J., Schiedermayer, D., Galias, A. and Siegler, M. (1990). Talking to comatose patients. *Neurotrauma Medical Report* 4, 1-3.

McSharry, P. and Clifford, G. (2006). Models for ECG and RR Interval Processes. In C. Gari D, F. Azuajeand P. E. McSharry (Eds.), *Advanced Methods and Tools for ECG Data Analysis* (pp. 101-133). Norwood: Artech House.

Mofrad, S., Abdullah, R., Samah, B., Mansor, M. and Baba, M. (2009). Maternal psychological distress and separation anxiety disorders in children. *European Journal of Social Sciences* 8(3), 386-394.

Murteira, B., Ribeiro, C., Silva, J. and Pimenta, C. (2007). *Introdução à Estatística* (2nd ed.). Lisboa: McGraw-Hill.

Oliveira, J., Maojo, V., Martin-Sanchez, F. and Pereira, A. (2005). *Biological and Medical Data Analysis*. Berlin: Springer.

Perrin, F., Schnakers, C., Schabus, M., Degueldre, C., Goldman, S., Brédart, S., Faymonville, M., Lamy, M., Moonen, G., Luxen, A., Maquet, P. and Laureys, S. (2006). Brain Response to one's own name in vegetative state, minimally conscious state, and locked-in syndrome. *Archives of Neurology* 63, 562-569.

Prados, J. (2000). *Sociologia de Los Grupos Escolares: Sociometría y Dinámica de Grupos*. Almería: Universidad de Almería.

Puggina, A. (2006) *O uso da música e de estímulos vocais em pacientes em estado de coma: relação entre estímulo auditivo, sinais vitais, expressão facial e escalas de Glasgow e Ramsay*. M.Sc. Thesis, Universidade de São Paulo, São Paulo, Brasil.

Rangayyan, R. (2002). Biomedical Signal Analysis - A case-Study Approach. New York: Wiley-Interscience.

Reisner, A., Clifford, G., and Mark, R. (2006). The Physiological Basis of the Electrocardiogram. In G. Clifford, F. Azuage and P. Mcsharry (Eds.), *Advanced Methods and Tools for ECG Data Analysis* (pp. 1-25). Norwood: Artech House.

Ritchie, J., and Spencer, L. (1994). Qualitative data analysis for applied policy research. In A. Brymanand and R. Burgess (Eds.), *Analysing Qualitative Data* (pp. 173-194). London: Routledge.

Saritha, C., Sukanya, V., and Murthy, Y. N. (2008). ECG Signal Analysis Using Wavelet Transforms. *Bulg. J. Phys. 35*, 68-77.

Simões, J., Jesus, L. and Voegeli, D. (2013). Verbal communication with unconscious patients. *Acta Paulista Enfermagem* 26(5), 506-513.

Simões. J., Jesus, L., Voegeli, D., Couto, P., Fernandes, J. and Morgado, M. (2011). Assessment of comatose patients: a Portuguese instrument based on the Coma Recovery Scale – Revised and using nursing standard terminology. *Journal of Advanced Nursing* 67(5), 1129-1141.

Simoliuniene, R., Krisciukaitis, A., Macas, A., Baksyte, G., Saferis, V., and Zaliunas, R. (2008). Principal Component Analysis Based Method for Detection and Evaluation of ECG T-Wave Alternans. *Computers in Cardiology* 35, 757-760.

Stanhope, M., and Lancaster, J. (2011). *Public Health Nursing: Population-Centered Health Care in the Community (8th ed.)*. St. Louis: Mosby.

Turnock, C. (1991). Communicating with patients in ICU. *Nursing Standard* 9(5), 38-40.

Urden, L., Stacy, K., and Lough, M. (2010). *Critical Care Nursing: Diagnosis and Management (6th ed)*. St. Louis: Mosby.

Veríssimo, M., Alves, S., Monteiro, L., and Oliveira, C. (2003). Ansiedade de separação materna e adaptação psicossocial ao pré-escolar. *Psicologia, Saúde e Doenças* 4(2), 221-229.

Verity, S. (1996). Communicating with sedated ventilated patients in intensive care: focusing on the use of touch. *Intensive and Critical Care Nursing* 12, 354-358.

Walker, J., Eakes G. and Siebelink E. (1998). The effects of familial voice interventions on comatose head-injured patients. *Journal of Trauma Nursing* 5 (2), 41-45.

In: Horizons in Neuroscience Research. Volume 15
Editors: Andres Costa and Eugenio Villalba
ISBN: 978-1-63321-523-8

Chapter 6

The Splitting of the Brain: A Reorientation towards Fiber Tracts Damage in Amnesia

Angelica Staniloiu[1,2,3*] and Hans J. Markowitsch[1]
[1] Physiological Psychology, University of Bielefeld, Bielefeld, Germany
[2] Centre for Addiction and Mental Health, Toronto, Canada
[3] University of Toronto, Toronto, Canada

Abstract

Mnemonic processing involves large scale networks, where several fiber bundles play significant roles, as they enable adequate and smooth flow of information between and within networks. The damage of these bundles might lead to various degrees of memory impairments, including amnesia. Cases of amnesia after the damage of white matter tracts have been reported for a long time, especially after frank alterations of white matter pathways due to strokes. The recent wider employment of structural brain imaging methodologies, such as diffusion tensor imaging or magnetization transfer ratio measurements has however improved the ability to estimate the integrity of the brain white matter and resulted in an increase in findings of fiber bundles' damage in association with memory impairments, including amnesia. This led to a rejuvenated interest in the mechanism of disconnection underlying amnesia in the scientific literature. Several medical, neurological and substance use- related conditions might be accompanied by both white matter damage and memory impairments. Furthermore, recent research data suggest that subtle micro- structural white matter tracts' changes may account for memory impairments in several psychiatric conditions, including dissociative amnesia. In the present chapter, after a brief introduction of the historical conceptualization of amnesia as a disconnection syndrome, an overview is provided of the fiber bundles that are important for mnemonic processing and causes and conditions that might alter their integrity and lead to amnesic conditions are discussed.

* Address for correspondence: Angelica Staniloiu; University of Bielefeld; Physiological Psychology; POB 100131; 33501 Bielefeld; Germany.

Introduction

The white matter of the central nervous system makes up almost half of the human brain (Paus, 2010) and undergoes considerable changes over the life span (Lebel, Leemans, Phillips, & Beaulieu, 2008; Paus, 2010; Paus, Keshavan, & Gjedd, 2008; Malloy, Correia, Stebbins, & Laidlow, 2007). Changes in white matter are modulated by genes, sex hormones, environmental factors and epigenetic mechanisms (L. S. Allen & Gorski, 1991; L. S. Allen, Richey, Chai, & Gorski, 1991). Incidentally, the heritability of white matter volume was found to generally increase with age (Paus et al., 2008), suggesting a cumulative effect of experience on brain structure or a gene-environment correlations effect. Aside from the numerous locally interconnecting fibers, the central nervous system contains several dozens of fiber tracts which, similar to big cables, run over long distances (Schmahmann, Smith, Eichler, & Filley, 2008). In humans, knowledge about the long distance fibers had initially come from findings from gross dissection. The advent of new techniques, such as diffusion tensor imaging has enabled the characterization of long distance fibers in the living human brain.

Disruption of such major fiber tracts may have devastating effects on brain function and consequently may lead to major cognitive, affective and behavioral disturbances. Several decades ago, Geschwind (1965) termed the consequences of such fiber damage 'disconnection syndrome'. With this expression he intended to point out that the damage of fiber bundles may, in comparison to the damage of single grey matter structures, have more devastating effects on cognition; a disruption of fiber bundles typically impacts on the (coordinated) function of at least two structures, namely the structure from which or in the proximity of which the fiber bundle originates and that in which it terminates, respectively. Neuroimaging and pathological studies suggest that some fiber tracts continue their development over a period of decades, while others mature early in life (Lebel et al., 2006). The majority achieves maturation by the time human beings enter their third decade of life. Data from imaging studies show that white matter deterioration due to ageing follows complex patterns. According to "the last in first out" hypothesis (N. Davis, Dennis et al., 2009), age related white matter changes in the frontal and temporal association regions had been expected to be more pronounced in comparison to the changes in the posterior brain regions. However, according to one diffusion tensor imaging (DTI) study, white matter deficits related to ageing increased from the posterior to anterior areas within specific fiber bundles crossing the frontal and parietal cortices, but not the temporal cortex (N. Davis, Dennis et al., 2009).

The largest fiber system in primates is the corpus callosum which interconnects the two hemispheres. Given that both hemispheres (at least in the human brain) are engaged in different intellectual functions, the corpus callosum is responsible for integrating activities leading to synchronized conscious behavior. In patients with pharmaceutically-intractable forms of epilepsy, one method of treatment has been the surgical dissection of parts of the corpus callosum. This surgical technique had led to partly bizarre manners, as if two human beings struggle with each other (Bogen, 1979; Gazzaniga, 1970).

Fiber –Behavior Interactions with a Focus on Conscious Memory

In the healthy brain, fibers are building highly interacting networks for attention (Mesulam, 1981), memory, language (Markowitsch, 1988a; Mesulam, 1990), logical reasoning (Reverberi et al., 2012), creativity (Jung et al., 2010) and other cognitive functions (Power et al., 2011). In this review we will focus on relations between processing within conscious memory systems (semantic and episodic–autobiographical) and the white matter (Markowitsch & Staniloiu, 2012). We will start by providing a short overview of the current understanding of memory and memory disorders; thereafter we will present fiber systems that are relevant for memory processing and comment on conditions which affect the white matter of the brain.

Memory – Systems and Divisions

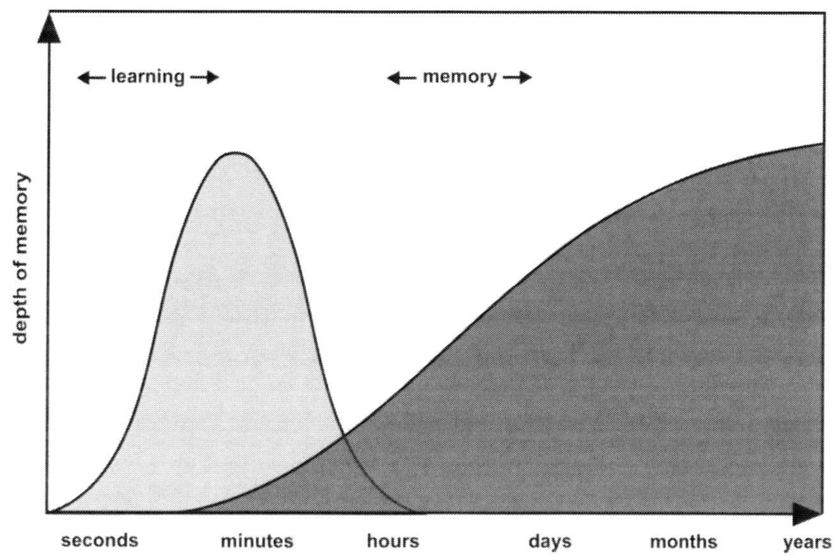

Figure 1. The most uniformly accepted division of memory into short-term and long-term memory. Note, however, that short-term memory is considered to last a few minutes at most (digit span, about 5 bits).

Memory and Time

The most common distinction in memory research is between short-term and long-term memory (Fig. 1). Short-term memory is defined as the online-holding of information such as several digits; it has been limited to a few bits (4-7) and a time range of seconds to minutes (G. A. Miller, 1956; Cowan, 2000). The introduction of the term 'working memory' by Alan Baddeley (Baddeley & Hitch, 1974; Baddeley, 2000) further extended the time-related dichotomy of memory. As its name suggests, working memory refers to working with memory – that is with the online-holding not only of new information, but also with the retrieval of portions of old, already stored information. Any information that is not lost (due to unsuccessful encoding) or exceeds the limited capacity of short-term and working memory is considered to have been assigned to long-term memory stores.

When considering memory as being embedded in time (Markowitsch, 2005), the sequence of psychological processes of the memory has to be mentioned. Information is perceived, encoded, consolidated, stored, retrieved and re-encoded after retrieval.

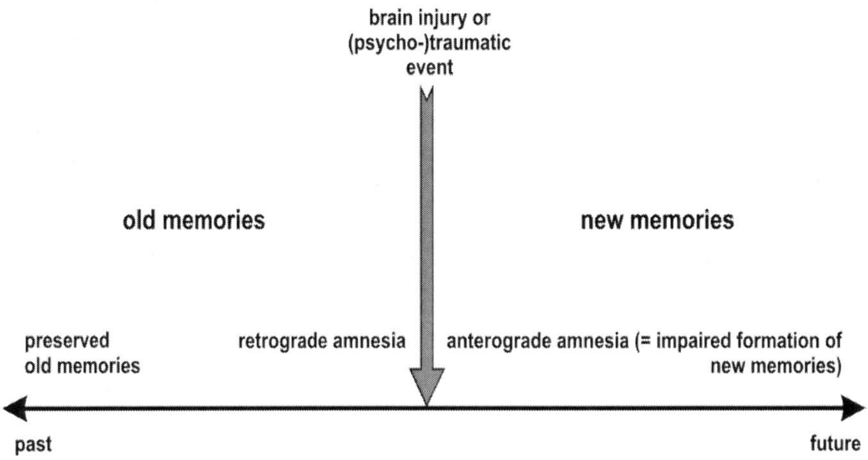

Figure 2. Relations between anterograde and retrograde amnesia. The arrow symbol represents the time point of a brain infarct or of a major psychic traumatic event, leading to either anterograde or retrograde amnesia or to both. Note that for retrograde amnesia the frequently observed gradient – termed Ribot's law (Ribot, 1882), is indicated by stating that usually very old memories are preserved in retrograde amnesia, while those close to the point of the event are impaired.

An additional time-related categorization of memory with clinical applicability is the distinction between old and new memories, and anterograde and retrograde memories, respectively. The compromised ability to access information that happened before the memory-impairing incident (brain injury or psychogical trauma) corresponds to retrograde memory impairment, while anterograde memory impairment refers to the compromised capacity to long-term acquire new information after the incident (Fig. 2). Anterograde amnesia can occur with minimal or no retrograde amnesia, and vice versa, retrograde amnesia can occur with minimal or no anterograde amnesia (Markowitsch & Staniloiu, 2012). Furthermore, there may be changes over time such as the anterograde amnesia following a neurological insult may be present initially, but may then resolve or become subtle with the passage of time (Barbizet, 1970; Levine et al., 1998).

Memory and Content

More than 100 years ago, scientists observed and studied that not all forms or kinds of memory are equally affected after brain damage (Markowitsch, 1992a; Schneider, 1912, 1928; Semon, 1904). Systematic research focused on memory systems or subdivisions began only about 40 years ago. Especially Endel Tulving (1972, 1983) promoted the idea that memory is not a unity. He in particular advanced and emphasized the distinction between episodic memory (requiring a superior level of consciousness named "autonoetic consciousness") and semantic memory (requiring noetic consciousness). Markowitsch (2003a) and Markowitsch & Staniloiu (2011a) have commented on these different forms of memory and their accompanied levels of consciousness in detail. In comparison to noetic consciousness that is found in animals as well and is used to gain awareness of objects, situations and states of the world, not present to senses, "autonoetic" consciousness, is considered uniquely human. Autonoetic consciousness features a superior ability to reflect upon oneself and to distinguish oneself from the social and biological environment. It entails a "sense of self in time and the ability to relive subjective experiences from the encoding context by mentally travelling back in time" (Lemogne et al., 2006, p.260). Mental time traveling through subjective time from present to both past and future is according to Piolino, Desgranges and Eustache (2009) one of the last features of episodic-autobiographical memory that becomes fully functional, but the first one to be affected by memory changes due to ageing or amnesic disorders.

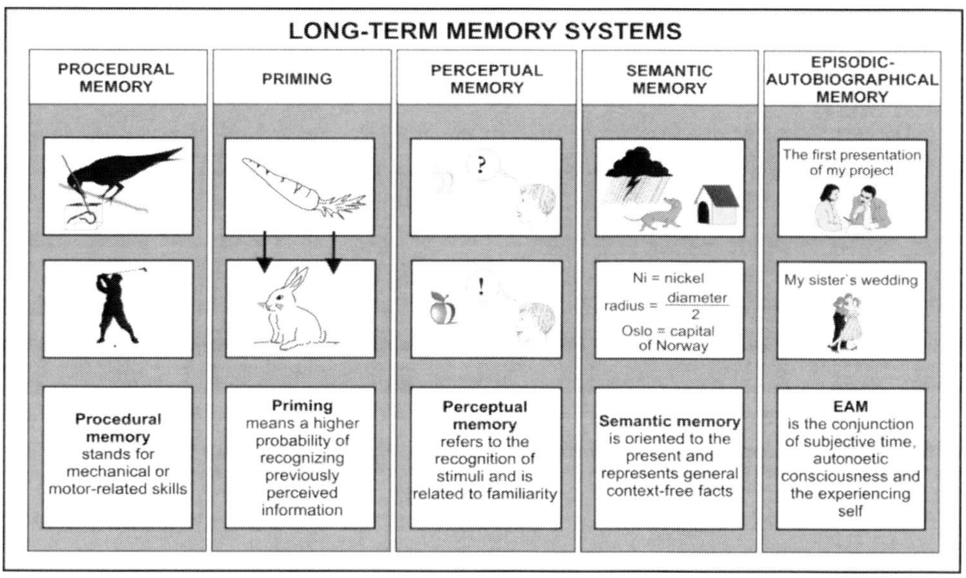

Figure 3. The five long-term memory systems and their assumed brain bases. Procedural memory is largely motor-based, but includes also sensory and cognitive skills ("routines"). Priming refers to a higher likeliness of re-identifying previously perceived stimuli. Perceptual memory allows distinguishing an object, item, or person on the basis of distinct features. Declarative memory is context-free and refers to general facts. It is termed semantic memory or the knowledge systems as well. The episodic-autobiographical memory system is context-specific with respect to time and place. It allows mental time travel. Examples are events such as the last vacation or the dinner of the previous night. Tulving (2005) defined it as the conjunction of autonoetic consciousness, subjective time, and the experiencing self.

With respect to the contribution of animal research to the classification of memory systems, Mishkin and Petri (1984) distinguished between a 'habit' and a 'memory' system; the habit system referred to procedures and routines, the memory system to the acquisition of facts and relations between objects. Later, Tulving and other researchers (e.g., Squire, 1982) expanded the classification of memory systems by adding other memory systems, in particular those acting on an automatic, implicit and subconscious ("anoetic"; without conscious reflection) level – procedural memory and the priming system. Procedural memory refers to complex motor acts like skiing, biking, driving a car, playing a musical instrument. Priming is defined by a higher probability of choosing a stimulus that was perceived earlier in the same or a similar manner. An example of priming is the presently frequently observed sequence of advertisement in TV news, where in the break within a movie spots for different products are followed by a short repetition of the previously advertized product, implying that the initial spot only reached a subconscious level of realization in the brain (a "prime"), while the repetition of the spot in a similar (e.g., abbreviated) form brought it to a conscious level with the consequence of increasing the likelihood that the product will be purchased. Tulving and Markowitsch later added a further, fifth memory system to the four systems described above, namely the 'perceptual memory system'; this was then published first in the paper of Markowitsch (2003b). This system acts consciously ("noetically"), but on a presemantic level and relies on familiarity judgments. An example is the identification of an apple, no matter what color it has or whether it is already half eaten or not, and the ability to easily distinguish the apple from a pear or peach (without having to use language).

The conceptualizations of the various memory systems have undergone several revisions over the years; this applies in particular to the episodic memory system, which we nowadays term episodic-autobiographical memory system (cf. Fig. 3 and Markowitsch & Staniloiu, 2011a, 2012) and define it as the conjunction of subjective time, autonoetic consciousness and the experiencing self (Tulving, 2005). Though the term autobiographical is often used interchangeably with the term episodic, we would like to emphasize that some elements of autobiographical memory do not have an episodic quality. A distinction should therefore be made between autobiographical-episodic memory and autobiographical-semantic memory. The latter refers to knowledge of name, date of birth or self-traits and may be preserved or updated in spite of blocked access to episodic-autobiographical memory (= memory for personal events).

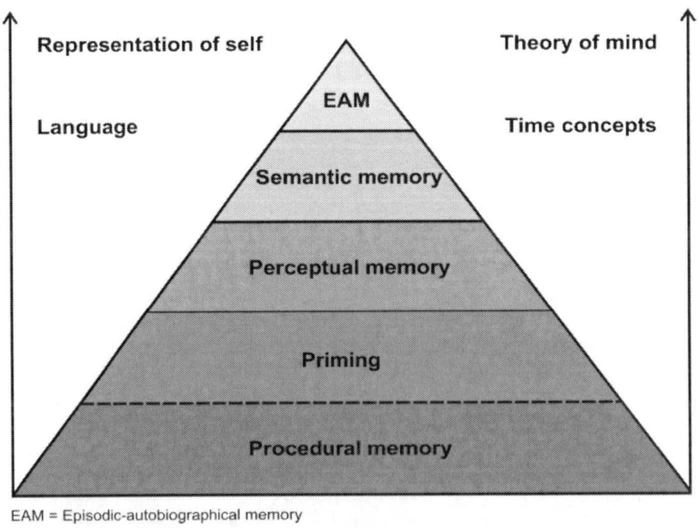

Figure 4. Diagram depicting Tulving's (2002, 2005) assumption of a hierarchical organization of long-term memory systems and demonstrating the developmental aspects in the formation of memory systems.

While Figure 3 sketches the memory systems, Figure 4 illustrates the idea that the memory systems are hierarchically organized according to their assumed ontogenetical and phylogenetical development and that the emergence of episodic-autobiographical memory occurs in the context of the attainment of a new stage of self awareness , on one hand and provides the foundation for further self development, on the other hand.

Episodic-autobiographical memory is the last ontogenetic acquisition of human beings (Nelson & Fivush, 2004; Tulving, 2005; Piolino, Hisland et al., 2007) and is also considered to have developed latest in phylogeny, being – in the form as defined above – uniquely human (Tulving , 2005; Clayton & Dickinson, 1998; Brand & Markowitsch, 2008). It is assumed, that the emergence of episodic-autobiographical memory is a gradual process that occurs in concert with the development of other cognitive abilities such as executive functions, personal semantic knowledge, emotional knowledge, Theory of Mind functions, language and working memory (Picard, Reffuveille, Eustache, & Piolino, 2009; Nelson & Fivush, 2004; Wang, 2008). Although language is not an essential condition for the establishment of episodic-autobiographical memory, it supports and enriches its development (Tulving, 2005). The earliest childhood episodic-autobiographical memories typically date to around age 3.5 years, but individual, cultural and gender differences exist (Harpaz-Rotem & Hirst, 2005; Wang, 2001; N. Davis, Gross, & Hayne, 2008). Developmental changes of episodic-autobiographical memory extend however beyond childhood into early adolescent years (Picard et al., 2009) and may, on the brain level, reflect the extensive developmental structural and functional reorganization in childhood and adolescence of different components of the brain network supporting episodic-autobiographical memory (Shing, Werkle-Bergner, Brehmer, Müller, Li, & Lindenberger, 2010).

Memory Disorders

Memory disorders occur in connection with numerous etiologies (Table 1). They can be found after focal brain damage, such as in diencephalic amnesia (Markowitsch, 1988b, 1992b; Markowitsch & Staniloiu, 2012), after widespread brain damage such in dementia or traumatic brain injury (Markowitsch, Kessler, Schramm, & Frölich, 2000), or after psychological stress, such as in dissociative or psychogenic amnesias. The latter variants of amnesia typically happen without evidence of significant brain damage as measured by standard static neuroimaging techniques (Markowitsch, 1996; Staniloiu & Markowitsch, 2010a). As mentioned above, it has to be emphasized that processing of memory (as of many other cognitive functions) is never a matter of a single brain structure; instead we assume that memory processing occurs in a complex, *Gestalt*-like manner (Markowitsch, 1984, Staniloiu & Markowitsch, 2010b), reflecting that both specialization and integration are features of the human brain (Paus, 2010).

Information enters the brain via the sensory organs and is then further processed depending on the kind of information (cf. Fig. 3). Subconsciously processed information is directly led to unimodal neocortical structures (priming), or engages the basal ganglia, premotor and other motor-related areas (procedural learning). Consciously processed information recruits more widespread networks which are still largely neocortical for perceptual learning, but include limbic regions for the other two memory systems – the knowledge system/semantic memory and episodic-autobiographical memory. For these two memory systems incoming information first enters limbic structures where its biological and social relevance is extracted, the information is compared with already existing related memories and later bound to and integrated with these. This process is also named synchronization (Axmacher, Mormann, Fernandez, Elger, & Fell, 2006). Further consolidation occurs during sleep immediately after encoding (Stickgold & Walker, 2005), but may extend to years (Haist, Bowden Gore, & Mao, 2001).

Table 1. Etiologies of Amnesic Disorders

Head injuries (closed or penetrating)
Strokes (ischemic or hemorrhagic)
Degenerative brain diseases (e.g., Alzheimer's disease)
Intracranial tumors
Limbic encephalitis
Aneurysmal bleedings, aneurysm surgeries
Surgical procedures on brain
Medial temporal lobe epilepsies
Neurosyphilis
Drug abuse
Anticholinergics
Neurotoxins
Vitamin deficiencies (e.g., Vitamin-B_1 deficiency)
Transient global amnesias (TGA)
Transient epileptic amnesias (TEA)
Chronic alcohol abuse, Korsakoff's syndrome
Electroconvulsive therapy (ECT)
Psychiatric illnesses (e.g., schizophrenia)

The cerebral cortex is considered to be the principal storage place of memories. For episodic-autobiographical memories, limbic brain structures may be involved in storage in addition. Retrieving facts and events requires an engagement of at least three closely interacting networks, namely activating brain stem structures comprising portions of the reticular activating system, the main information of the respective fact or event containing neocortical network, and – especially for the events or episodes – a limbic network providing the emotional flavor (Anderson, Wais, & Gabrieli, 2006). For the retrieval of episodic-autobiographical events, a regional combination of right hemispheric inferior lateral prefrontal and anterior temporal structures, which are interconnected via the uncinate fascicle, is viewed as being central (Markowitsch, Calabrese et al., 1993; Calabrese et al., 1996; Fink et al., 1996; Kroll, Markowitsch, Knight, & Cramon, 1997).

For the present topic especially two limbic circuits are relevant, as both of them work in the processing of newly incoming information, evaluate it and transfer it to cortical storage networks. These are the Papez-circuit (Papez, 1937; Markowitsch, 1999a, b) and the basolateral limbic circuit (Livingston & Escobar, 1971) (Fig. 5). Bilateral damage to one of the structures or fiber tracts within these circuits usually leads to anterograde amnesia, especially in the episodic-autobiographical domain (Markowitsch & Pritzel, 1985; Markowitsch, 1998; Brand & Markowitsch, 2008; Markowitsch & Staniloiu, 2012).

Fiber Systems Relevant for Conscious Memory Processing

Schmahmann et al. (2008) stated that every area of the cerebral cortex is linked with other cortical and subcortical areas by pathways which can be grouped into five bundles, namely

1) association fibers running ipsilaterally between cortical areas (short, neighboring, and long fibers)
2) striatal fibers coursing to the basal ganglia (Muratoff bundle, external capsule)
3) commissural fibers
4) thalamic fibers, named thalamic peduncles
5) pontinc fibcrs travclling from dicnccphalon to spinal cord.

Using this gross subdivision we will comment on the importance of fibers for memory processing.

Association Fibers with Yet Unclear Contribution to Conscious Mnemonic Processing

The superior longitudinal bundle consists of three subcomponents (Schmahmann et al., 2008). One subdivision principally interconnects parietal and supplementary and premotor cortex and is involved in higher motor functions. Another one interconnects parieto-occipital with dorsolateral and dorsomedial prefrontal areas and is engaged in visuo-attentive functions. And the third one interconnects parietal and frontal opercula with somatosensory cortex and is engaged in somatomotor and language functions. There is no clear evidence that the superior longitudinal bundle is implicated in episodic-autobiographical memory processing; however, one study suggested that its integrity might correlate in older adults with better performance on tasks that preponderantly tapped on semantic memory (Kennedy & Rafaz, 2009).

Some authors consider the arcuate fascicle as being part of the anterior limb of the superior longitudinal fasciculus (Nieuwenhuys, Voogt, & van Huizen, 2008), while others describe it as a separate entity. The arcuate fascicle mainly has to do audition and interconnects temporal lobe/Sylvian fissure with the dorsolateral prefrontal cortex. The (left) arcuate fascicle was traditionally believed to connect temporal (Wernicke) and inferior frontal (Broca) areas of speech and its damage had been in the past believed to cause the so named conduction aphasia. This view is currently debated, however. There is at present no robust evidence to support the arcuate fascicle's involvement in conscious mnemonic processes. One DTI imaging study performed in patients with multiple sclerosis suggested correlations between performance on a semantic memory task and the integrity of the parietal portion of the left superior longitudinal fasciculus, the inferior longitudinal fasciculus and arcuate fasciculus (Dineen et al., 2009).

Located between claustrum and insula and claustrum and orbital frontal cortex, respectively the extreme capsule is assumed to play a role in language. The extent to which damage to the capsula extrema may lead to memory impairments is unclear. The claustrum is considered to be some kind of integrative hub, as many projections from cortical areas converge on it (Markowitsch, Irle, Bang-Olsen, & Flindt-Egebak, 1984; Guldin, Markowitsch, Lampe, & Irle, 1986). Interestingly, some authors proposed that the nuclei of the basolateral amygdaloid group are an extension of the claustrum (Nieuwenhuys et al., 2008). Furthermore, a part of the claustrum was found to be embedded in the uncinate fascicle by Kier and co-workers (2004). The uncinate fascicle, which also has an insular part, has been implicated in emotional and mnemonic processing (see below). Both the claustrum and insula have been attributed functions in consciousness and memory. Activations of the claustrum have recently been reported during performance of episodic-like memory tasks or in conjunction with category fluency tasks (Botzung, Rubin, Miles, Cabeza, & LaBar, 2010). The insular cortex has also been activated during verbal memory tasks and portions of the orbito-frontal cortex have been assumed to play roles in episodic-autobiographical memory (Brand & Markowitsch, 2006).

Another fascicle engaged in language processing is the middle longitudinal fascicle. This fascicle runs between the border region of inferior parietal and superior temporal areas and paralimbic, parahippocampal and cingulate/prefrontal areas. As these regions have to do with various functions, including memory, damage to the middle longitudinal fasciculus may lead to verbal memory disturbances.

Visual agnosia and prosopagnosia are associated with the inferior longitudinal fascicle which is part of the ventral visual pathway interconnecting mainly occipito-temporal and in part also parieto–occipital regions. The fronto-occipital fasciculus on the other hand is part of the dorsal visual pathway linking parieto-occipital regions with the premotor and prefrontal cortices. Kier and co-workers (2004) described the inferior fronto-occipital fasciculus as being an important part of the anterior temporal lobe, temporal stem (that also includes the uncinate fascicle, Meyer's loop of the optic radiation, the anterior commissure, inferior thalamic fibers, and the inferior fronto-occipital fasciculus; Ebeling & Cramon, 1992) and the external and extreme capsules. Similarly to the uncinate fascicle, the inferior fronto-occipital fasciculus is an important pathway of seizure spread.

The degree to which all of the above named long cortical association bundles are involved in memory processing is currently under investigation. A single case study of a patient with functional amnesia found with modern imaging methods evidence for alteration of fibers of the superior longitudinal and the fronto-occipital fascicles (Tramoni et al., 2009). The methods employed to investigate these fibers were the magnetization transfer ratio and magnetic resonance spectroscopic imaging; these methods are sensitive to microstructural and metabolic brain damage. Functional amnesia is defined as amnesia without evidence of brain damage on conventional structural imaging, and of an unsure cause (Markowitsch & Staniloiu, 2012). The authors concluded that their advanced imaging methods "showed the presence of significant metabolic and structural changes deep within the white matter of the right prefrontal lobe, a structure which is known to play a key role in the retrieval of autobiographical episodes" (p. 622). They assumed that one of the explanations for the patient's amnesia has to be sought in a "hypo-retrieval" mechanism, which they tentatively attributed to damage of the two long association tracts mentioned above.

Another study, done by Philippi, Mehta, Grabowski, Adolphs and Rudrauf (2009), suggested that the lesions of the inferior fronto-occipital fascicle (which has a segment that passes close to the amygdala) might indirectly contribute to episodic-autobiographical memory (without affecting the performance on tasks preponderantly tapping on semantic memory) by possibly interfering with the emotional tagging of personal events (Markowitsch & Staniloiu, 2011b). And a quarter decade before them, Bauer (1982) had described a patient without damage to the amygdala itself, but with damage to the inferior longitudinal fascicles, terminating in the basolateral limbic circuit (Fig. 5) and therefore in a regional array, supporting episodic-autobiographical memory formation. The patient suffered from an inability to create or to associate an emotional colorization to visual stimuli, leading him, for instance, to unsubscribe to the magazine *Playboy*.

In a large group of patients with multiple sclerosis correlations were found between the integrity of various fasciculi, including the inferior longitudinal fasciculus and fronto-occipital fasciculus, and performance on verbal memory tasks (Sepulcre et al., 2008). In a DTI study performed in older adults, the inferior longitudinal fascicle as well as other tracts linking more posterior regions, such as the left cingulum and the splenium of corpus callosum highly correlated with performance on visual memory tests (N. Davis et al., 2009).

Association Fibers with (Likely) Functions in Memory

Schmahmann et al. (2008) mention two other long cortical association bundles, namely the cingulum and the uncinate fascicle. The cingulum is described by Nieuwenhuys et al. (2008) as a bundle of short and long association fibers surrounding the corpus callosum. The short fibers interconnect subregions of the cingulate gyrus, the longer ones fan out from its posterior part into the medial temporal lobe (subicular cortex as part of the hippocampal region [cf. Fig. 5]). As part of the Papez circuit, the cingulum should have an important role in memory processing, although this hypothesis is according to our knowledge still insufficiently empirically explored. Patients with cingular damage have problems which range from attenuations in attention and concentration to mutism (Ochsner et al., 2001). Also attention deficit/hyperactivity disorders (Bush et al., 1999), psychiatric disorders (Carter, Mintun, Nichols, & Cohen, 1997; Vuilleumier, 2005), alexithymia (Borsci et al., 2009) and psychopathic traits (Boes, Tranel, Anderson, & Nopoulos, 2008; De Brito et al., 2009) are associated with a malfunctioning cingulate cortex. Most studies concerning functions of the cingulate gyrus are based on functional imaging results. Here we (Labudda et al., 2008) and others (Awh & Gehring, 1999; Badgaiyan & Posner, 1998; Carter, Macdonald et al., 2000; Catafau et al., 1998; Cohen, Heller, & Ranganath, 2005; Elliott & Dolan, 1998; Fellows & Farah, 2005) found an involvement in decision making, evaluation, and response selection. Another main involvement of the cingulate gyrus is in pain (Albanese, Duerden, Rainville, & Duncan, 2007; Cheng et al., 2007; Vogt, 2005). Conflict monitoring (Botvinick, Cohen, & Carter, 2004) and the modulation of emotions and affect have also been reported as cingulate functions (Britton et al., 2006; Cooney, Joorman, Atlas, Eugene, & Gotlib, 2007; K. D. Davis et al., 2005; Lutcke & Frahm, 2008). For the anterior cingulate gyrus a memory-related function was found with respect to working memory (Lenartowicz and McIntosh, 2005) and in retrieving old autobiographical memories (Fink et al., 1996; Svoboda, McKinnon, & Levine, 2006). Also in a patient with damage to the right uncinate fascicle and severe episodic-autobiographical amnesia a reduced bold response was observed in functional magnetic resonance imaging in the posterior cingulate cortex (Levine, Svoboda, Turner, Mandic, & Mackey, 2009). The retrosplenial/ posterior cingulate cortices were proposed to be part of the core network involved in autobiographical memory processing (Svoboda et al., 2006). Most of the above-mentioned reports however focus on the grey and not the white matter within the cingulate cortex. Evidence for correlations between the integrity of various portions of the cingulum and performance on standardized memory tasks comes from imaging studies of patients with temporal lobe epilepsy (McDonald et al., 2008), multiple sclerosis (Dineen et al., 2009; Sepulcre et al., 2008) or healthy elderly (N. Davis et al., 2009). In patients with Alzheimer's dementia the cingulum bundle is altered in the early stages of the illness. Findings from the longitudinal follow up of patients with amnestic mild cognitive impairment suggest that both changes in uncinate fascicle and the cingulum bundle might underlie the progression to the Alzheimer's disease (Fouquet et al., 2009).

There is another main intra-cortical pathway, the uncinate fascicle, which interconnects anterior temporal and orbitofrontal and medial prefrontal regions. The uncinate fascicle can be evidenced at 15 weeks of gestation, but seems to mature later and more slowly than other connections, continuing its development beyond 30 years. The uncinate fascicle is a hook-like shaped fiber tract with a fan like arrangement of fibers. It links the anterior three temporal convolutions and the amygdale with gyrus rectus, orbito-frontal cortex and subcallosal area. Its ventromedial part connects the amygdala, uncus and the tip of the third convolution with the gyrus rectus and the subcallsosal area (Ebeling & von Cramon, 1992). In the fronto-orbital cortices (especially near the frontal pole) the fibers of the uncinate fascicle and those of the inferior occipito-frontal fasciculus are greatly intermingled. The uncinate fascicle also shows an intermingling with the anterior commissure. The ventral part of the uncinate fascicle is implicated in the retrieval (and possibly also in the formation) of episodic-autobiographical memories. Changes of the uncinate fascicle have been reported in several conditions, such as chronic alcohol misuse, psychopathy, schizophrenia, unipolar depression, bipolar disorder, generalized social anxiety disorder, in children with a history of neglect, and in patients with multiple sclerosis, epilepsy, mild cognitive impairment, or Alzheimer's disease. Incidentally, in all these conditions impairments of processing in long term conscious memory systems have been described as well (Staniloiu & Markowitsch, 2010b). In the past, damage of the uncinate fascicle had also resulted during temporal lobe surgery that damaged the temporal stem. The amnesic syndrome following surgical removal of medial temporal lobes was suggested to at least partly be accounted for by damage to fibers of the temporal stem (in particular, the uncinate fascicle) (Horel, 1978; Kier et al., 2004).

We and others have stressed in some older publications that the regional combination of inferolateral prefrontal and temporopolar cortex is intimately involved in the retrieval of episodic-autobiographical events and – possibly to a lesser degree – also in semantic knowledge. Especially damage to the right hemispheric combination of these regions seems to disrupt the retrieval of episodic-autobiographical memory, resulting in full-blown retrograde amnesia (Kapur, Ellison, Smith, McLellan, & Burrowa, 1992; Markowitsch et al., 1993; Kroll et al., 1997; Markowitsch & Staniloiu, 2012). Levine et al. in 1998 published a case with selective disruption of the right uncinate fascicle and almost complete retrograde amnesia for episodic-autobiographical episodes, stressing thereby the importance of the right temporo-frontal cortex for episodic-autobiographical memory retrieval. Confirmation for this view comes also from a review of LaBar and Cabeza (2006) who wrote on page 59: "Studies of retrograde amnesia support Markowitsch's proposal that retrieval of remote personal memories involves interactions between the inferior PFC [prefrontal cortex] and its connections with the anteromedial temporal lobe that course through the uncinate fasciculus."

Though evidence for a relation between damage of these regions and episodic-autobiographical amnesia continues to be found (e.g., Markowitsch & Ewald, 1997; Diehl et al., 2008), it was questioned whether the tissue damage itself was "responsible" for the amnesia, or whether there might be psychological/psychogenic components to it (Kapur, 1999, 2000; Kopelman, 2000). In any way it seems that even if the patients had dissociative or psychogenic amnesia (Staniloiu, Markowitsch, & Brand, 2010), it might be possible to infer that temporo-frontal brain connectivity alterations lead to or are associated with psychogenic/dissociative amnesia.

Of interest is a study investigating relations between white matter and declarative memory in children and adolescents (Mabbott, Rovet, Noseworthy, Smith, & Rockel, 2009). They found that auditory-verbal memory proficiency was related to white matter integrity of the left uncinate fasciculus (as well as parietal and cerebellar white matter).

Striatal Fibers

The Muratoff bundle (subcallosal fasciculus of Muratoff) and the fibers constituting the external capsule have apparently not been found to be related to memory processing. It should however be mentioned that the inner fibers of the uncinate fascicle pass through the capsula externa and according to Ebeling and Cramon (1992) the inferior occipito-frontal fasciculus is located both in the extreme and external capsules.

Commissural Fibers

There are several important commissural fibers which interconnect cortical regions. First, the corpus callosum – the largest fiber bundle interconnecting portions of the cerebral hemispheres – has to be mentioned. Second, the anterior commissure links according to Schmahmann et al. (2008) the caudal part of the orbital frontal cortices, the temporal poles, portions of the superior and inferior temporal cortices and the parahippocampal cortices. As mentioned before, some fibers of the anterior commissure intermingle with those of the uncinate fascicle. Third, there are so called hippocampal commissures.

Though it seems logical that damage to the commissural fibers might contribute to memory impairments, the objective evidence has remained insufficient. Damage to the callosal fibers that interconnect the Wernicke language area and visual regions might lead to a disconnection syndrome, which may affect word-based semantic and episodic-autobiographical memory ("associative agnosia") (Kavcic & Doty, 2002; Zaidel & Sperry, 1974). Concept formation has been reported to be impaired in callosotomized patients, which may have an indirect effect on memory (Cronin-Golomb, 1986). Papanicolaou, Hasan, Boake, Eluvathingal, & Kramer (2007) reported a case of a male patient that suffered profound amnesia after head injury and in whom DTI provided support for damage of several tracts, including the corpus callosum. The profound amnesia was however etiologically linked to damage of the fornix, in the light of an overwhelming amount of data supporting such an etiological connection (see below).

Suggestions that amnesic syndromes accompanied by changes of the corpus callosum might preponderantly stem from damage to other structures than corpus callosum were made long ago. Zingerle in 1900 described a "50-year-old Catholic married construction worker from the Steiermark" (p. 367) who died 15 days after hospital admission and had shown a major deterioration of memory. He was disoriented in time, unable to make simple calculations spontaneously and behaved rather motion- and emotionless. Post mortem examination of his brain revealed a tumor which was confined to the corpus callosum; it "at no place invaded the white matter, ... and is largest at the level of the third ventricle" (p. 369). Zingerle assumed that the tumor had originated on the right thalamic surface and then infiltrated the corpus callosum. Subsequently, the major memory deterioration might have reflected an impairment of the thalamic nuclei involved in mnemonic processing.

In line with the proposed functional specificity of various portions of the corpus callosum (e.g., Peters et al., 2002), the DTI study performed in older adults by N. Davis and colleagues (2009) found that in this population the intactness of the splenium of the corpus callosum highly correlated with performance on visual memory tests, while the integrity of the genu of the corpus callosum correlated with measures of executive functions.

The hippocampal commissures usually might only be indirectly affected, namely when there is – e.g., based on encephalitis – bilateral damage to the hippocampal formation. This then would have a major impact on new memory formation and one could speculate that this has an affect on both verbal and nonverbal memory and on the interaction between the two forms.

Thalamic Fibers (Peduncles)

The thalamus is considered to be the gate to the cortex (Dercum, 1925). As thalamic fibers reach all cortical areas (and receive projections from them), and as (bilateral) damage to thalamic nuclei such as the mediodorsal (Cramon, Hebel, & Schuri, 1985; Markowitsch et al., 1993; Hodges & McCarthy, 1993; Isaac et al., 1998; L. A. Miller et al., 2001; Perren, Clarke, & Bogousslavsky, 2005) and the anterior nuclear group (Aggleton & Brown, 1999; Caulo et al., 2005; Harding, Halliday, Caine, & Kril, 2000) lead to severe and lasting amnesia, probably by disrupting memory processing circuits (Papez circuit, basolateral limbic circuit; Fig. 5), it is likely that the thalamic peduncles are involved in memory. This was stated already in 1988b by Markowitsch. Especially the pedunculus thalami anterior, which interconnects the mediodorsal thalamus with the prefrontal cortex, might be of interest here. Again, however, selective damage of the thalamic peduncles and memory disturbances associated with such damage was apparently not reported. It should, however, be noted that in thalamic amnesia the fibers surrounding thalamic nuclei (lamina medullaris thalami) and the cortical and subcortical projection fibers (e.g., mammillothalamic tract) frequently contribute to the memory symptomatology (Markowitsch, 1988b; Cramon & Markowitsch, 1992).

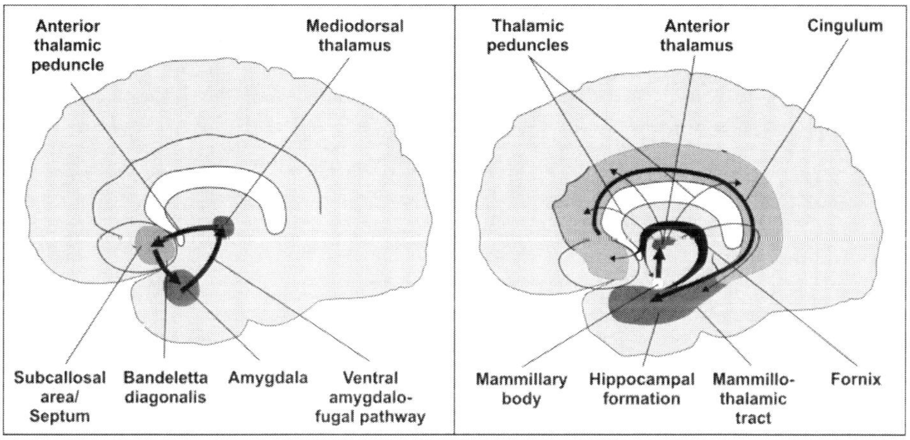

Figure 5. The basolateral limbic circuit (left) and the medial or Papez circuit (right). The medial circuit is probably associated with cognitive acts of memory processing and the basolateral circuit with the affective evaluation of information. Both circuits interact. The Papez circuit interconnects the hippocampal formation via the (postcommissural) fornix to the mammillary bodies, which are connected via the mammillothalamic tract to the anterior thalamus. The anterior thalamus has cortical projection targets that reach the cingulate gyrus; likewise, the subicular part of the hippocampal formation and the cingulum fibres project back from the cingulate gyrus into the hippocampal formation. (The precommissural fornix also provides a bidirectional connection between the hippocampal formation and the basal forebrain.) The basolateral limbic circuit links the amygdala, mediodorsal thalamic nucleus, and area subcallosa together by distinct fibre projections, namely the ventral amygdalofugal pathway, the inferior thalamic peduncle, and the bandeletta diagonalis.

Of special interest was a case reported by Tanji, Suzuki, Fujii, Higano, and Yamadori in 2003. The patient had bilateral small subcortical infarcts in two regions and was described as being densely amnesic with respect to both verbal and non-verbal materials. Recognition memory, on the other hand, was less affected. Symptoms persisted at a follow-up eight months after his initial investigation. One lesion was found in the left mediodorsal nucleus of the thalamus and the other in the anterior limb of the right internal capsule and the right globus pallidus. Mamillothalamic tracts and anterior thalamic nuclei were clearly spared. The authors concluded that the patient's amnesia might have been caused by a disruption of both left and right thalamofrontal circuits.

The Capsula Interna

The capsula interna contains a number of thalamic fibers. Of interest with respect to memory is especially the anterior limb of the internal capsule as this contains among others fibers running between mediodorsal and anterior thalamus and prefrontal and anterior cingulate cortex. Infarcts in the area of the genu of internal capsule frequently result in severe memory disturbances. Markowitsch, Cramon, Hofmann, Sick, and Kinzler (1990) found after a left-hemispheric capsular infarct severe verbal memory deficits in an adolescent patient. At the same time Bogousslavky and Regli (1990) wrote about the "Capsular genu syndrome" and Caplan et al. (1990) also described cases with infarcts in this region, leading to memory disturbances. Other cases were described by Schnider, Gutbrod, Hess, & Schroth (1996), and – more indirectly – by Cramon, Markowitsch and Schuri (1993). The patient of Cramon et al.

(1993) in addition to nuclear damage of the septum had left-sided damage of the anterior limb of the internal capsule and the inferior thalamic peduncle. Also a a minute hyperintense spot in the antero-basal limb of the right-sided internal capsule was observed. The patient had major emotion-related memory disturbances.

The anterior limb of the internal capsule can consequently be viewed as a major bottleneck structure with respect to memory processing.

Other Cortico-Subcortical Fiber Bundles Related to Memory Processing

The most clear-cut evidence for the necessity of a fiber connection for (anterograde) memory stems from the fornix (cf. Fig. 5). The fornix is usually regarded as being part of the Papez circuit, projecting from the subicular cortex of the hippocampal formation into the mammillary bodies at the bottom of the diencephalon (Markowitsch, 1999a; Irle & Markowitsch, 1982; Stephan, 1975). However, there is an additional, bidirectionally projecting part between subiculum and basal forebrain whose anterior portion is named the precommissural fornix. The fornix is consequently the only fiber bundle that interconnects all three major regions (medial temporal lobe, medial diencephalon, basal forebrain) that are important for memory consolidation and transfer into long-term memory (Markowitsch & Staniloiu, 2012).

Fornix damage had been associated with memory impairment since long (Markowitsch, 1992a). Similarly to unilateral hippocampal damage, also unilateral fornix damage can lead to amnesia (Gaffan, Gaffan, & Hodges, 1991) and was interpreted as a hippocampal disconnection syndrome (Tucker, Roeltgen, Tully, Hartmann, & Boxell, 1988). Usually fornix damage is surgically induced – for example to remove a cyst in the third ventricle (Sweet, Talland, & Ervin, 1959; Berti, Arienta, & Papagno, 1990; McMackin, Cockburn, Anslow, & Gaffan, 1995; Calabrese, Markowitsch, Harders, Scholz, & Gehlen, 1995; Hodges & Carpenter, 1991; Tucker et al., 1988; Aggleton et al., 2000; Garcia-Bengochea & Friedman, 1987). The severity and persistance of the amnesia has usually triggered the writing of case reports with fornix damage (Brion, Pragier, Guerin, & Teitgen, 1969; Gaffan et al., 1991; Chatterjee et al., 1993; Cramon & Schuri, 1992; Morris, Bowers, Chatterjee, & Heilman, 1992; Park, Hahn, Kim, Na, & Huh, 2000; Poreh et al., 2006; Yasuno et al., 1999). In a study of 38 patients with colloid cyst removal significant negative correlations were found between the fornix volume and recall in some standardized memory tests (Tsivilis et al., 2008).

Even when conventional imaging failed to reveal fornix changes, the application of diffusion tensor imaging revealed fornix alterations, explaining the dense amnesia observed (Papanicolaou et al., 2007). Similarly, when using diffusion-weighted imaging, Rudebeck et al. (2009) found in neurologically healthy participants that a higher microstructural integrity in the fornix was associated with significantly better recollection memory. No correlation was obtained for simple familiarity memory or non-mnemonic tasks.

The widely acknowledged role of the fornix in conscious mnemonic processing has prompted studies investigating therapeutic interventions (e.g., deep brain stimulation) targeting this fiber bundle in different populations with severe memory impairments (Laxton et al., 2010).

Conditions Affecting the Integrity of the Brain's White Matter

Conditions which might impair white matter functioning are numerous. Most frequently – with respect to number of affected patients – ischemic and hemorrhagic strokes and traumatic brain injury affect the white matter (e.g., Bendlin et al., 2008; Nakayama et al., 2006). Traumatic brain injury can result in diffuse (multifocal) axonal injury, which might lead to disturbances in anterograde and retrograde memory and other cognitive domains (Schmahmann et al., 2008; Kopelman, 2008; Fradera & Kopelman, 2009). Most consistently, on the other hand, multiple sclerosis involves white matter damage (Fu et al., 1998) and may be accompanied by memory disturbances (Markowitsch, Calabrese, Haupts, & Gehlen, 1996) and other cognitive and affective symptoms (Pujol, Bello, Deus, Marti-Vilalta, & Capdevila, 1997). Conditions of hypoxia and anoxia may affect the white matter and memory performance (J. S. Allen, Tranel, Bruss, & Damasio, 2006). Epilepsy is another form of brain disease which may affect white matter (Concha, Beaulieu, Collins, & Gross, 2009). Dementic conditions usually affect white matter in addition to grey matter (Borroni et al., 2007; Burns et al., 2005; Davies et al., 2005), but also in normal aging white matter changes are frequently reported. Using diffusion tensor tractography, N. Davis et al. (2009) detected evidence for myelodegenerative effects of aging. Posterior tract affections were correlated with visual memory performance and anterior tract changes with changes in executive functions.

In psychiatric disorders changes in white matter, such as in schizophrenia (Ashtari et al., 2007), bipolar disorders (Barnea-Goraly, Chang, Karchemskiy, Howe, & Reiss, 2009) or major depression (Frodl et al., 2002) are quite common. Similarly, drug abuse (e.g., ketamine) and alcohol abuse have frequently been reported to alter neuronal connectivity (Bjork, Grant, & Hommer, 2003; Pfefferbaum et al., 2000). Chronic use of ketamine can lead to memory problems and was found to be associated with changes in white matter within bilateral prefrontal cortex cortices and left-temporoparietal regions (Liao et al., 2010).

Patients with autism have been found to show white matter alterations as well (Herbert et al., 2003). And also other medical conditions such as diabetes, hypertension, or electrocution (e.g., Bryan, Andrews, Hurley, & Taber, 2009) can affect the white matter. The wide range of medical conditions which may affect white matter sometimes globally and sometimes focally is also illustrated in a review paper by Schmahmann and colleagues (2008).

Conclusion

Empirical findings support the idea that amnesic conditions and memory impairments cannot only occur in relation to selective damage to key grey matter structures involved in mnemonic processes, but also in relationship to macro- or even micro-structural damage of fiber tracts connecting these structures. Functional imaging methods have established the existence of large and complexly interwoven networks within the human brain (Vogel, Power, Petersen, & Schlaggar, 2010) such as default mode networks (Buckner& Carroll, 2007; Hagmann et al., 2008; Raichle et al., 2001; Raichle & Snyder, 2007). The default mode network is described to contain several brain areas (including medial prefrontal cortex, anterior and posterior cingulate, precuneus and retrosplenial cortex) that display high baseline metabolic activity during rest compared to a large range of tasks. It is ascribed functions in mind-wandering, introspection, prospection as well as episodic-autobiographical memory processing (Buckner, 2008). Abnormal default mode network activity was, for example, reported in patients with amnestic mild cognitive impairment, the early stage of Alzheimer's disease or some forms of amnesia.

Disruptions of the functional connectivity might be underlain by abnormalities of the white matter, which is essential for a proper flow of information in brain networks. Structural imaging methods, such as diffusion tensor imaging and magnetization-transfer-ratio improved the ability to estimate the integrity of the white matter, leading to a rejuvenated interest in the contribution of the mechanism of disconnection ("disconnection syndrome") to various forms of amnesia. At present there are several promising neuroimaging tools available for estimating the integrity of fibers, including diffusion tensor imaging (DTI). In order to overcome the limitations of these techniques, research is now focusing on finding ways to better distinguish between true fiber bundles abnormalities and biases intrinsic to various image analysis methodologies and, within true matter abnormalities, between axonal damage, myelin defects or fiber orientation variance (Kubicki, 2010).

Despite limitations of the current neuroimaging armamentarium, we expect that the larger use and the refinements of the neuroimaging techniques employed for the visualization of white matter will provide new insights into the underpinnings of various amnesic conditions, in particular of those with a presumed mixed (psychogenic and neurological) etiology, such as the ones arising after mild brain concussions, primary blast injuries or mild electrocuting accidents (Reinhold & Markowitsch, 2009; Tramoni et al., 2009; Staniloiu et al., 2010; Hoge et al., 2008; Silver et al., 2009; Kopelman, 2008; Squire & Bayley, 2007).

References

Aggleton, J. P., & Brown, M. W. (1999). Episodic memory, amnesia, and the hippocampal-anterior thalamic axis. *Behavioral and Brain Sciences, 22*, 425-489.

Aggleton, J. P., McMackin, D., Carpenter, K., Hornak, J., Kapur, N., Wiles, C. M. et al. (2000). Differential cognitive effects of colloid cysts in the third ventricle that spare or compromise the fornix. *Brain, 123*, 800-815.

Albanese, M.-C., Duerden, E. G., Rainville, P., & Duncan, G. H. (2007). Memory traces of pain in human cortex. *Journal of Neuroscience, 27(17)*, 4612-4620.

Allen, J. S., Tranel, D., Bruss, J., & Damasio, H. (2006). Correlations between regional brain volumes and memory performance in anoxia. *Journal of Clinical and Experimental Neuropsychology, 28(4)*, 457-476.

Allen, L. S., & Gorski, R. A. (1991). Sexual dimorphism of the anterior commissure and massa intermedia of the human brain. *Journal of Comparative Neurology, 312*, 97-104.

Allen, L. S., Richey, M. F., Chai, Y. M., & Gorski, R. A. (1991). Sex differences in the corpus callosum of the living human being. *Journal of Neuroscience, 11*, 933-942.

Anderson, A. K., Wais, P. E., & Gabrieli, J. D. (2006). Emotion enhances remembrance of neutral events past. *Proceedings of the National Academy of Sciences of the USA, 103*, 1599-1604.

Ashtari, M., Cottone, J., Ardekani, B. A., Cervellione, K., Szeszko, P. R., Wu, J. et al. (2007). Disruption of white matter integrity in the inferior longitudinal fasciculus in adolescents with schizophrenia as revealed by fiber tractography. *Archives of General Psychiatry, 64(11)*, 1270-1280.

Awh, E., & Gehring, W. J. (1999). The anterior cingulate cortex lends a hand in response selection. *Nature Neuroscience, 2*, 853-854.

Axmacher, N., Mormann, F., Fernandez, G., Elger, C. E., & Fell, J. (2006). Memory formation by neuronal synchronization. *Brain Research Reviews, 52*, 170-182.

Baddeley, A. (2000). The episodic buffer: a new component of working memory?. *Trends in Cognitive Sciences, 4*, 417-423.

Baddeley, A. D., & Hitch, G. J. (1974). Working memory. In G. A. Bower (Eds.), *Recent advances in learning and motivation* (pp. 47-89). New York: Academic Press.

Badgaiyan, R. D., & Posner, M. I. (1998). Mapping the cingulate cortex in response selection and monitoring. *Neuroimage, 7*, 255-260.

Barbizet, J. (1970). *Human memory and its pathology.* San Francisco, CA: W. H. Freeman & Company.

Barnea-Goraly, N., Chang, K. D., Karchemskiy, A., Howe, M. E., & Reiss, A. L. (2009). Limbic and corpus callosum aberrations in adolescents with bipolar disorder: a tract-based spatial statistics analysis. *Biological Psychiatry, 66(3)*, 238-244.

Bauer, R. M. (1982). Visual hypoemotionality as a symptom of visual-limbic disconnection in man. *Archives of Neurology, 39*, 702-708.

Bendlin, B. B., Ries, M. L., Lazar, M., Alexander, A. L., Dempsey, R. J., Rowley, H. A. et al. (2008). Longitudinal changes in patients with traumatic brain injury assessed with diffusion-tensor and volumetric imaging. *Neuroimage, 42(2)*, 503-514.

Berti, A., Arienta, C., & Papagno, C. (1990). A case of amnesia after excision of the septum pellucidum. *Journal of Neurology, Neurosurgery, and Psychiatry, 53*, 922-924.

Bjork, J. M., Grant, S. J., & Hommer, D. W. (2003). Cross-sectional volumetric analysis of brain atrophy in alcohol dependence: Effects of drinking history and comorbid substance use disorder. *American Journal of Psychiatry,160(11)*, 2038-2045.

Boes, A. D., Tranel, D., Anderson, S. W., & Nopoulos, P. (2008). Right anterior cingulate: a neuroanatomical correlate of aggression and defiance in boys. *Behavioral Neuroscience, 122(3)*, 677-684.

Bogen, J. E. (1979). The callosal syndrome. In K. M. Heilman & E. Valenstein (Eds.), *Clinical neuropsychology* (pp. 308-359). New York: Oxford University Press.

Bogousslavsky, J., & Regli, F. (1990). Capsular genu syndrome. *Neurology, 40*, 1499-1502.

Borroni, B., Brambati, S. M., Agosti, C., Gipponi, S., Bellelli, G., Gasparotti, R. et al. (2007). Evidence of white matter changes on diffusion tensor imaging in frontotemporal dementia. *Archives of Neurology, 64(2)*, 246-251.

Borsci, G., Boccardi, M., Rossi, R., Rossi, G., Perez, J., Bonetti, M. et al. (2009). Alexithymia in healthy women: a brain morphology study. *Journal of Affective Disorders, 114(1-3)*, 208-215.

Botvinick, M. M., Cohen, J. D., Carter, C. S. (2004). Conflict monitoring and anterior cingulate cortex: an update. *Trends in Cognitive Sciences, 8(12)*, 539-546.

Botzung, A., Rubin, D. C., Miles, A., Cabeza, R., & LaBar, K. S. (2010). Mental hoop diaries: Emotional memories of a college basketball game in rival fans. *Journal of Neuroscience, 30,* 2130-2137.

Brand, M., & Markowitsch, H. J. (2006). Memory processes and the orbitofrontal cortex. In D. Zald & S. Rauch (Eds.), *The orbitofrontal cortex* (pp. 285-306). Oxford: Oxford University Press.

Brand, M., & Markowitsch, H. J. (2008). The role of the prefrontal cortex in episodic memory. In E. Dere, A. Easton & J.P. Huston (Eds.), *Handbook of behavioral neuroscience: episodic memory research* (Vol. 18, pp. 317 - 342). Amsterdam: Elsevier.

Brion, S., Pragier, G., Guerin, R., & Teitgen, M. (1969). Syndrome de Korsakoff par ramollisement bilatéral du fornix. *Revue Neurologique, 120*, 255-262.

Britton, J. C., Phan, K. L., Taylor, S. F., Welsh, R. C., Berridge, K. C., & Liberzon, I. (2006). Neural correlates of social and nonsocial emotions: An fMRI study. *Neuroimage, 31(1)*, 397-409.

Bryan, B. C., Andrews, C. J., Hurley, R. A., & Taber, K. H. (2009). Electrical injury, part i: Mechanisms. *Journal of Neuropsychiatry and Clinical Neurosciences, 21(3)*, 241-244.

Buckner R. L., Andrews-Hanna, J. R., & Schacter, D. L. (2008). The brain's default network. Anatomy, function, and relevance to disease. *Annals of the New York Academy of Sciences, 1124*, 1–38.

Buckner, R. L., & Carroll, D. C. (2007). Self-projection and the brain. *Trends in Cognitive Science, 11(2)*, 49-57.

Burns, J. M., Church, J. A., Johnson, D. K., Xiong, C., Marcus, D., Fotenos, A. F. et al. (2005). White matter lesions are prevalent but differentially related with cognition in aging and early Alzheimer disease. *Archives of Neurology, 62(12)*, 1870-1876.

Bush, G., Frazier, J. A., Rauch, S. L., Seidman, L. J., Whalen, P. J., Jenike, M. A. et al. (1999). Anterior cingulate cortex dysfunction in attention-deficit/hyperactivity disorder revealed by fMRI and the counting stroop. *Biological Psychiatry, 45*, 1542-1552.

Calabrese, P., Markowitsch, H. J., Durwen, H. F., Widlitzek, B., Haupts, M., Holinka, B., & Gehlen, W. (1996). Right temporofrontal cortex as critical locus for the ecphory of old episodic memories. *Journal of Neurology, Neurosurgery, and Psychiatry, 61*, 304-310.

Calabrese, P., Markowitsch, H. J., Harders, A. G., Scholz, A., & Gehlen, W. (1995). Fornix damage and memory: A case report. *Cortex, 31*, 555-564.

Caplan, L., Schmahmann, J. D., Kase, C. S., Feldmann, E., Baquis, G., Greenberg, J. P., Gorelick, P. B., Helgason, C., & Hier, D. B. (1990). Caudate infarcts. *Archives of Neurology, 47*,133-143

Carter, C. S., Macdonald, A. M., Botvinick, M., Ross, L. L., Stenger, V. A., Noll, D. et al. (2000). Parsing executive processes: Strategic vs evaluative functions of the anterior cingulate cortex. *Proceedings of the National Academy of Sciences of the USA, 97*, 1944-1948.

Carter, C. S., Mintun, M., Nichols, T., & Cohen,, J. D. (1997). Anterior cingulate gyrus dysfunction and selective attention deficits in schizophrenia: [15O]H2O PET study during single-trial Stroop task performance. *American Journal of Psychiatry, 154*, 1670-1675.

Catafau, A. M., Parellada, E., Lomena, F., Bernardo, M., Setoain, J., Catarineu, S. et al. (1998). Role of the cingulate gyrus during the Wisconsin Card Sorting Test: a single photon emission computed tomography study in normal volunteers. *Psychiatry Research - Neuroimaging, 83*, 67-74.

Caulo, M., Van Hecke, J., Toma, L., Ferretti, A., Tartaro, A., Colosimo, C. et al. (2005). Functional MRI study of diencephalic amnesia in Wernicke-Korsakoff syndrome. *Brain, 128(7)*, 1584-1594.

Chatterjee, A., Morris, M. K., Bowers, D., Williamson, D. J., Doty, L., & Heilman, K. M. (1993). Cholinergic treatment of an amnestic man with a basal forebrain lesion: theoretical implications. *Journal of Neurology, Neurosurgery, and Psychiatry, 56*, 1282-1289.

Cheng, Y., Lin, C.-P., Liu, H.-L., Hsu, Y.-Y., Lim, K.-E., Hung, D. et al. (2007). Expertise modulates the perception of pain in others. *Current Biology, 17(19)*, 1708-1713.

Clayton, N. S., & Dickinson, A. (1998). Episodic-like memory during cache recovery in scrub-jays. *Nature, 395*, 272-274.

Cohen, M. X., Heller, A. S., & Ranganath, C. (2005). Functional connectivity with anterior cingulate and orbitofrontal cortices during decision-making. *Cognitive Brain Research, 23(1)*, 61-70.

Concha, L., Beaulieu, C., Collins, D. L., & Gross, D.W. (2009). White-matter diffusion abnormalities in temporal-lobe epilepsy with and without mesial temporal sclerosis. *Journal of Neurology, Neurosurgery, and Psychiatry, 80(3)*, 312-319.

Cooney, R. E., Joormann, J., Atlas, L.Y., Eugene, F., & Gotlib, I. H. (2007). Remembering the good times: neural correlates of affect regulation. *Neuroreport, 18(17)*, 1771-1774.

Cowan, N. (2000). The magical number 4 in short-term memory: A reconsideration of mental storage capacity. *Behavioral and Brain Sciences, 24*, 87-185.

Cramon, D. Y. von, Hebel, N., & Schuri U. (1985). A contribution to the anatomical basis of thalamic amnesia. *Brain, 108*, 993-1008.

Cramon, D. Y. von, & Markowitsch, H. J. (1992). The problem of "localizing" memory in focal cerebro-vascular lesions. In L. R. Squire & N. Butters (Eds.), *Neuropsychology of memory* (2nd. ed.) (pp. 95-105). New York: Guilford Press.

Cramon, D. Y. von, Markowitsch, H. J., & Schuri, U. (1993). The possible contribution of the septal region to memory. *Neuropsychologia, 31*, 1159-1180.

Cramon, D.Y. von, & Schuri, U. (1992). The septo-hippocampal pathways and their relevance to human memory: A case report. *Cortex, 28*, 411-422.

Cronin-Golomb, A. (1986). Comprehension of abstract concepts in right and left hemispheres of complete commissurotomy subjects. *Neuropsychologia, 24*, 881-887.

Davies, R. R., Hodges, J. R., Kril, J. J., Patterson, K., Halliday, G. M., & Xuereb, J. H. (2005). The pathological basis of semantic dementia. *Brain, 128(9)*, 1984-1995.

Davis, K. D., Taylor, K. S., Hutchison, W. D., Dostrovsky, J. O., McAndrews, M. P., Richter, E. O. et al. (2005). Human anterior cingulate cortex neurons encode cognitive and emotional demands. *Journal of Neuroscience, 25(37)*, 8402-8406.

Davis, N., Dennis, N. A., Buchler, N. G., White, L. E., Madden, D. J., & Cabeza, R. (2009). Assessing the effects of age on long white matter tracts using diffusion tensor tractography. *NeuroImage, 46(2)*, 530-541.

Davis, N., Gross, J., & Hayne, H. (2008). Defining the boundary of childhood amnesia. *Memory, 16(5)*, 465-474.

De Brito, S. A., Mechelli, A., Wilke, M., Laurens, K. R., Jones, A. P., Barker, G. J. et al. (2009). Size matters: increased grey matter in boys with conduct problems and callous-unemotional traits. *Brain, 132(Pt 4)*, 843-852.

Dercum, F. X. (1925). The thalamus in the psysiology and pathology of the mind. *A.M.A. Archives of Neurology and Psychiatry, 14*, 289-302.

Diehl, B., Busch, R. M., Duncan, J. S., Piao, Z., Tkach, J., & Lüders, H. O. (2008). Abnormalities in diffusion tensor imaging of the uncinate fasciculus relate to reduced memory in temporal lobe epilepsy. *Epilepsia, 49*, 1409-1418.

Dineen, R. A., Vilisaar, J., Hlinka, J., Bradshaw, C. M., Morgan, P. S., Constantinescu, S. C., & Auer, D. P. (2009). Disconnection as a mechanism for cognitive dysfunction in multiple sclerosis. *Brain, 132*, 239-249.

Ebeling, U., & Cramon, D. von (1992). Topography of the uncinate fascicle and adjacent temporal fiber tracts. *Acta Neurochrirugica (Vienna), 115*, 143-148.

Elliott, R., & Dolan, R. J. (1998). Activation of different anterior cingulate foci in association with hypothesis testing and response selection. *NeuroImage, 8*, 17-29.

Fellows, L. K., & Farah, M. J. (2005). Is anterior cingulate cortex necessary for cognitive control? *Brain, 128*, 788-797.

Fink, G. R., Markowitsch, H. J., Reinkemeier, M., Bruckbauer, T., Kessler, J., & Heiss, W. D. (1996). Cerebral representation of one's own past: neural networks involved in autobiographical memory. *Journal of Neuroscience, 16*, 4275-4282.

Frodl, T., Meisenzahl, E. M., Zetzsche, T., Born, C., Groll, C., Jäger, M., & Leinsinger, G. (2002). Hippocampal changes in patients with a first episode of major depression. *American Journal of Psychiatry, 159*, 1112-1118.

Fouquet, M., Desgranges, B., Landeau, B., Duchesnay, E., Mezenge, F., de la Sayette, V., Viader, F., Baron, J.C., Eustache, F., & Chetelat, G. (2009). Longitudinal brain metabolic changes from amnestic mild cognitive impairment to Alzheimer's disease. *Brain, 132*, 2058-2067.

Fradera, A., & Kopelman, M. D. (2009). Memory disorders. In L. R. Squire (Ed.), *Encyclopedia of neuroscience* (Vol. 5) (pp. 751-760). Oxford: Academic Press.

Fu, L., Matthews, P. M., De Stefano, N., Worsley, K. J., Narayanan, S., Francis, G. S. et al. (1998). Imaging axonal damage of normal-appearing white matter in multiple sclerosis. *Brain, 121*, 103-113.

Gaffan, E. A., Gaffan, D., & Hodges, J. R. (1991). Amnesia following damage to the left fornix and to other sites. A comparative study. *Brain, 114*, 1297-1313.

Garcia-Bengochea, F., & Friedmann, W. A. (1987). Persistent memory loss following section of the anterior fornix in humans: A historical review. *Surgical Neurology, 27*, 361-364.

Gazzaniga, M. S. (1970). *The bisected brain*. New York: Appelton Century Crofts.

Geschwind, N. (1965). Disconnection syndromes in animals and man. *Brain, 88*, 237-294, 585-644.

Guldin, W. O., Markowitsch, H. J., Lampe, R., & Irle, E. (1986). Cortical projections originating from the cat's insular area and remarks on claustro-cortical connections. *Journal of Comparative Neurology, 243*, 468-487.

Hagmann, P., Cammoun, L., Gigandet, X., Meuli, R., Honey, C. J., Wedeen, V. J. et al. (2008). Mapping the structural core of human cerebral cortex. *Public Library of Science Biology, 6*(7), e159.

Haist, F., Bowden Gore, J., & Mao, H. (2001). Consolidation of human memory over decades revealed by functional magnetic resonance imaging. *Nature Neuroscience, 4*, 1139-1145.

Harding, A., Halliday, G., Caine, D., & Kril, J. (2000). Degeneration of anterior thalamic nuclei differentiates alcoholics with amnesia. *Brain, 123*, 141-154.

Harpaz-Rotem, I., & Hirst, W. (2005). The earliest memories in individuals raised in either traditional and reformed kibbuz or outside the kibbuz. *Memory, 13*, 51-62.

Herbert, M. R., Ziegler, D. A., Deutsch, C. K., O'Brien, L. M., Lange, N., Bakardjiev, A. et al. (2003). Dissociation of cerebral cortex, subcortical and cerebral white matter volumes in autistic boys. *Brain, 126*, 1182-1192.

Hodges, J. R., & Carpenter, K. (1991). Anterograde amnesia with fornix damage following removal of IIIrd ventricle colloid cyst. *Journal of Neurology, Neurosurgery, and Psychiatry, 54*, 633-638.

Hodges, J. R., & McCarthy, R. A. (1993). Autobiographical amnesia resulting from bilateral paramedian thalamic infarction. *Brain, 116*, 921-940.

Hoge, C. W., McGurk, D., Thomas, J. L., Cox, A. L., Engel, C. C., & Castro, C. A. (2008). Mild traumatic brain injury in U.S. soldiers returning from Iraq. *New England Journal of Medicine, 358*, 453-462.

Horel, J. A. (1978). The neuroanatomy of amnesia. A critique of the hippocampal memory hypothesis. *Brain, 101*, 403-445.

Irle, E., & Markowitsch, H. J. (1982). Connections of the hippocampal formation, mamillary bodies, anterior thalamus and cingulate cortex. A retrograde study using horseradish peroxidase in the cat. *Experimental Brain Research, 47*, 79-94.

Isaac, C. L., Holdstock, J. S., Cezayirli, E., Roberts, J. N., Holmes, C. J., & Mayes, A. R. (1998). Amnesia in a patient with lesions limited to the dorsomedial thalamic nucleus. *Neurocase, 4*, 497-508.

Jung, R. E., Grazooplene, R., Caprihan, A., Chavez, R. S., & Haier, R. J. (2010). White matter integrity, creativity, and psychopathology: Disentangling constructs with diffusion tensor imaging. *PLoS ONE, 5(3)* e9818.

Kapur, N. (1999). Syndromes of retrograde amnesia: A conceptual and empirical synthesis. *Psychological Bulletin, 125*, 800-825.

Kapur, N. (2000). Focal retrograde amnesia and the attribution of causality: An exceptionally benign commentary. *Cognitive Neuropsychology, 17*, 623-637.

Kapur, N., Ellison, D., Smith, M. P., McLellan, D. L., & Burrowa, E. H. (1992). Focal retrograde amnesia following bilateral temporal lobe pathology. *Brain, 115*, 73-85.

Kavcic, V., & Doty R. W. (2002). Efficiency of the forebrain commissures: memory for stimuli seen by the other hemisphere. *Acta Neurobiologiae Experimentalis, 62*, 235-242.

Kennedy, K. M., & Raz, N. (2009). Aging white matter and cognition: Differential effects of regional variations in diffusion properties on memory, executive functions, and speed. *Neuropsychologia, 47*, 916-927.

Kier, E. L., Staib, L. H., Davis, L. M., & Bronen. R. A. (2004). MR Imaging of the temporal stem: Anatomic dissection tractography of the uncinate fasciculus, inferior occipitofrontal fasciculus, and meyer's loop of the optic radiation. *American Journal of Neuroradiology, 25*, 677-691.

Kopelman, M. D. (2000). Comments on focal retrograde amnesia and the attribution of causality: An exceptionally benign commentary by Narinder Kapur. *Cognitive Neuropsychology, 17*, 639-640.

Kopelman, M. D. (2008). Retrograde memory loss. In G. Goldenberg & B. L. Miller (Eds.), *Handbook of clinical neurology, Vol. 88: Neuropsychology and behavioral neurology* (pp. 185–202). Amsterdam: Elsevier.

Kroll, N., Markowitsch, H. J., Knight, R., & Cramon, D. Y. von (1997). Retrieval of old memories - the temporo-frontal hypothesis. *Brain, 120*, 1377-1399.

Kubicki, M. (2010). Neurocognition and white matter imaging: Can the relationship be reliably quantified? *American Journal of Psychiatry, 167(4)*, 373-375.

LaBar, K. S., & Cabeza, R. (2006). Cognitive neuroscience of emotional memory. *Nature Reviews Neuroscience, 7*, 54-64.

Labudda, K., Woermann, F. G., Mertens, M., Pohlmann-Eden, B., Markowitsch, H. J., & Brand, M. (2008). Neural correlates of decision making with explicit information about probabilities and incentives in elderly healthy subjects. *Experimental Brain Research, 187*, 641-650.

Laxton, A. W., Tang-Wai, D. F., McAndrews, M. P., Zumsteg, D., Wennberg, R., Keren, R., Wherrett, J., Naglie, G. et al. (2010). A phase I trial of deep brain stimulation of memory circuits in Alzheimer's disease. *Annals of Neurology, 68*, 521-534.

Lebel, C. W., Leemans, L., Phillips, L., & Beaulieu, C. (2008). Microstructural maturation of the human brain from childhood to adulthood. *NeuroImage, 40*, 1044-1055.

Lemogne, C., Piolino, P., Friszer, S., Claret, A., Girault, N., Jouvent, R. et al. (2006). Episodic autobiographical memory in depression: Specificity, autonoetic consciousness, and self perspective. *Consciousness and Cognition, 15*, 258-268.

Lenartowicz, A., & McIntosh, A. R. (2005). The role of anterior cingulate cortex in working memory is shaped by functional connectivity. *Journal of Cognitive Neuroscience, 17(7)*, 1026-1042.

Levine, B., Black, S. E., Cabeza, R., Sinden, M., McIntosh, A. R., Toth, J. P., Tulving, E., & Stuss, D. T. (1998). Episodic memory and the self in a case of isolated retrograde amnesia. *Brain, 121*, 1951-1973.

Levine, B., Svoboda, E., Turner, G. R., Mandic, M., & Mackey, A. (2009). Behavioral and functional neuroanatomical correlates of anterograde autobiographical memory in isolated retrograde amnesic patient M.L. *Neuropsychologia, 47*, 2188-2196.

Liao, Y., Tang, J., Ma, M., Wu, Z., Yang, M., Wang, X., Liu, T., Chen, X., Fletcher, P. C., & Hao, W. (2011). Frontal white matter abnormalities following chronic ketamine use: a diffusion tensor imaging study. *Brain, 69(1)*, 42-48.

Livingston, K. E., & Escobar, A. (1971). Anatomical bias of the limbic system concept: a proposed reorientation. *Archives of Neurology, 24*, 17-21.

Lutcke, H., & Frahm, J. (2008). Lateralized anterior cingulate function during error processing and conflict monitoring as revealed by high-resolution fMRI. *Cerebral Cortex, 18(3)*, 508-515.

Mabbott, D. J., Rovet, J., Noseworthy, M. D., Smith, M. L., & Rockel, C. (2009). The relations between white matter and declarative memory in older children and adolescents. *Brain Research, 1294*, 80-90.

Malloy, P., Correia, S., Stebbins, G., & Laidlaw, D. H. (2007). *Clinical Neuropsychologist, 21*, 73-109.

Markowitsch, H. J. (1984). Can amnesia be caused by damage of a single brain structure? *Cortex, 20*, 27-45.

Markowitsch, H. J. (1988a). Individual differences in memory performance and the brain. In H. J. Markowitsch (Ed.), *Information processing by the brain* (pp. 125-148). Toronto: H. Huber; 1988.

Markowitsch, H. J. (1988n). Diencephalic amnesia: a reorientation towards tracts? *Brain Research Reviews, 13*, 351-370.
Markowitsch, H. J. (1992a). *Intellectual functions and the brain. An historical perspective.* Toronto: Hogrefe & Huber Publs.
Markowitsch, H. J. (1992b). Diencephalic amnesia. In D. Barcia Salorio (Ed.), *Trastornos de la memoria* (pp. 269-336). Barcelona: Editorial MCR.
Markowitsch, H. J. (1996). Organic and psychogenic retrograde amnesia: two sides of the same coin? *Neurocase, 2*, 357-371.
Markowitsch, H. J. (1998). Cognitive neuroscience of memory (Editorial). *Neurocase, 4*, 429-435.
Markowitsch, H. J. (1999a). Gestalt view of the limbic system and the Papez circuit – another approach on unity and diversity of brain structures and functions. *Behavioral and Brain Sciences, 22*, 459-460.
Markowitsch, H. J. (1999b). Limbic system. In R. Wilson & F. Keil (Eds.), *The MIT encyclopedia of the cognitive sciences* (pp. 472-475). Cambridge, MA: MIT Press.
Markowitsch, H. J. (2003a). Autonoëtic consciousness. In A. S. David & T. Kircher (Eds.), *The self in neuroscience and psychiatry* (pp. 189-186). Cambridge: Cambridge University Press.
Markowitsch, H. J. (2003b). Psychogenic amnesia. *NeuroImage, 20*, 132-138.
Markowitsch, H. J. (2005). Time, memory, and consciousness. A view from the brain. In R. Buccheri, A. C. Elitzur & M. Saniga (Eds.), *Endophysics, time, quantum, and the subjective* (pp. 131-147). Singapur: World Scientific Publ.
Markowitsch, H. J., Calabrese, P., Haupts, M., & Gehlen, W. (1996). Memory disturbances in patients with multiple sclerosis. In M. Prosiegel, R. Vermote & P. Ketelaer (Eds.), *Neuropsychology and psychology* (pp. 53-72). Genova: Associazione Italiana Sclerosi Multipla.
Markowitsch, H. J., Calabrese, P., Liess, J., Haupts, M., Durwen, H. F., & Gehlen, W. (1993). Retrograde amnesia after traumatic injury of the temporo-frontal cortex. *Journal of Neurology, Neurosurgery, and Psychiatry, 56*, 988-992.
Markowitsch, H. J., Cramon, D. Y. von, Hofmann, E., Sick, C.-D., & Kinzler, P. (1990). Verbal memory deterioration after unilateral infarct of the internal capsule in an adolescent. *Cortex, 26*, 597-609.
Markowitsch, H. J., & Ewald, K. (1997). Right-hemispheric fronto-temporal injury leading to severe autobiographical retrograde and moderate anterograde episodic amnesia. *Neurology, Psychiatry and Brain Research, 5*, 71-78.
Markowitsch, H. J., Irle, E., Bang-Olsen, R., & Flindt-Egebak, P. (1984). Claustral efferents to the cat's limbic cortex studied with retrograde and anterograde tracing techniques. *Neuroscience, 12*, 409-425.
Markowitsch, H. J., Kessler, J., Schramm, U., & Frölich, L. (2000). Severe degenerative cortical and cerebellar atrophy and progressive dementia in a young adult. *Neurocase, 6*, 357-364.
Markowitsch, H. J., & Pritzel, M. (1985). The neuropathology of amnesia. *Progress in Neurobiology, 25*, 189-287.
Markowitsch, H. J., & Staniloiu, A. (2011a). Memory, autonoetic consciousness, and the self. *Consciousness and Cognition, 20*, 16-39.

Markowitsch, H. J., & Staniloiu, A. (2011b). Amygdala in action: Relaying biological and social significance to autobiographic memory. *Neuropsychologia, 49*, 718-733.

Markowitsch, H. J., & Staniloiu, A. (2012). Amnesic disorders. *Lancet*, in press.

McDonald, C. R., Ahmadi, M. E., Hagler, D. J., Iragui, V. J., Gharapetian, L., Dale, A. M., & Halgren, E. (2008). Diffusion tensor imaging correlates of memory and language impairments in temporal lobe epilepsy. *Neurology, 71*, 1869-1876.

McMackin, D., Cockburn, J., Anslow, P., & Gaffan, D. (1995). Correlation of fornix damage with memory impairment in six cases of colloid cyst removal. *Acta Neurochirurgica, 135*, 12-18.

Mesulam, M.-M. (1981). A cortical network for directed attention and unilateral neglect. *Annals of Neurology, 10*, 309-325.

Mesulam, M.-M. (1990). Large-scale neurocognitive networks and distributed processing for attention, language, and memory. *Annals of Neurology, 28*, 597-613.

Miller, G. A. (1956). The magical number seven plus minus two. Some limits on our capacity for processing information. *Psychological Review, 63*, 244-257.

Miller, L. A., Caine, D., Harding, A., Thompson, E. J., Large, M., & Watson, J. D. G. (2001). Right medial thalamic lesion causes isolated retrograde amnesia. *Neuropsychologia, 39*, 1037-1046.

Mishkin, M., & Petri, H. L. (1981). Memories and habits: Some implications for the analysis of learning and retention. In L. R. Squire & N. Butters (Eds.), *Neuropsychology of memory* (pp. 287-296). New York: Guilford Press.

Morris, M. K., Bowers, D., Chatterjee, A., & Heilman, K. M. (1992). Amnesia following a discrete basal forebrain lesion. *Brain, 115*, 1827-1847.

Nakayama, N., Okumura, A., Shinoda, J., Yasokawa, Y. T., Miwa, K., Yoshimura, S. I., & Iwama, T. (2006). Evidence for white matter disruption in traumatic brain injury without macroscopic lesion. *Journal of Neurology, Neurosurgery, and Psychiatry, 77*, 850-855.

Nelson, K., & Fivush, R. (2004). The emergence of autobiographical memory: a social cultural developmental theory. *Psychological Review, 111*, 486-511.

Nieuwenhuys, R., Voogt, J., & van Huizen, C. (2008). *The human central nervous system. A synopsis and atlas* (4th ed.). Berlin: Springer.

Ochsner, K. N., Kosslyn, S. M., Cosgrove, G. R., Cassem, E. H., Price, B. H., Nierenberg, A. A. et al. (2001). Deficits in visual cognition and attention following bilateral anterior cingulotomy. *Neuropsychologia, 39*, 219-230.

Papanicolaou, A. C., Hasan, K. M., Boake, C., Eluvathingal, T. J., & Kramer, L. (2007). Disruption of limbic pathways in a case of profound amnesia. *Neurocase, 13(4)*, 226-268.

Papez, J. W. (1937). A proposed mechanism of emotion. *A.M.A. Archives of Neurology and Psychiatry, 79*, 217-224.

Park, S. A., Hahn, J. H., Kim, J. I., Na, D. L., & Huh, K. (2000). Memory deficits after bilateral anterior fornix infarction. *Neurology, 54*, 1379-1382.

Paus, T. (2010). Growth of white matter in the adolescent brain: Myelin or axon? *Brain and Cognition, 72*, 26-35.

Paus, T., Keshavan, M., & Giedd, J. N. (2008). Why do many psychiatric disorders emerge during adolescence? *Nature Reviews Neuroscience, 9*, 947-957.

Perren, F., Clarke, S., & Bogousslavsky, J. (2005). The syndrome of combined polar and paramedian thalamic infarction. *Archives of Neurology, 62(8)*, 1212-1216.

Peters, M., Oeltze, S., Seminowicz, D., Steinmetz, H., Koeneke, S., & Jäncke, L. (2002). Division of the corpus callosum into subregions. *Brain and Cognition, 50*, 62-72.

Pfefferbaum, A., Sullivan, E. V., Hedehus, M., Adalsteinsson, E., Lim, K. O., & Moseley, M. (2000). In vivo detection and functional correlates of white matter microstructural disruption in chronic alcoholism. *Alcoholism: Clinical and Experimental Research, 24*, 1214-1221.

Philippi, C. L., Mehta, S., Grabowski, T., Adolphs, R., & Rudrauf, D. (2009). Damage to association fiber tracts impairs recognition of the facial expression of emotions. *Journal of Neuroscience, 29*, 15089-15099.

Picard, L., Reffuveille, I., Eustache, F., & Piolino, P. (2009). Development of autonoetic autobiographical memory in school-age children: Genuine age effect or development of basic cognitive abilities? *Consciousness and Cognition, 18*, 864-876.

Piolino, P., Desgranges, B., & Eustache, F. (2009). Episodic autobiographical memories over the course of time: cognitive, neuropsychological and neuroimaging findings. *Neuropsychologia, 47*, 2314-2329.

Piolino, P., Hisland, M., Ruffeveille, I., Matuszewski, V., Jambaque, I., & Eustache, F. (2007). Do school-age children remember or know the personal past? *Consciousness and Cognition, 16*, 84-101.

Poreh, A., Winocur, G., Moscovitch, M., Backon, M., Goshen, E., Ram, Z. et al. (2006). Anterograde and retrograde amnesia in a person with bilateral fornix lesions following removal of a colloid cyst. *Neuropsychologia, 44(12)*, 2241-2248.

Power, J. D., Cohen A. L., Nelson, S. M., Wig, G. S., Barnes, K. A., Church, J. A., Vogel, A. C., Laumann, T. O., Miezin, F. M., Schlaggar, B. L., & Petersen, S. E. (2011). Functional network organization of the human brain. *Neuron, 72*, 665-678.

Pujol, J., Bello, J., Deus, J., Marti-Vilalta, J., & Capdevila, A. (1997). Lesions in the left arcuate fasciculus region and depressive symptoms in multiple sclerosis. *Neurology, 49*(4), 1105-1110.

Raichle, M. E., MacLeod, A. M., Snyder, A. Z., Powers, W. J., Gusnard, D. A., & Shulman, G. L. (2001). A default mode of brain function. *Proceedings of the National Academy of Sciences of the USA, 98*, 676-682.

Raichle, M. E., & Snyder, A. Z. (2007). A default mode of brain function: A brief history of an evolving idea. *Neuroimage, 37*(4), 1083-1090.

Reinhold, N., & Markowitsch, H. J. (2009). Retrograde episodic memory and emotion: a perspective from patients with dissociative amnesia. *Neuropsychologia, 47*, 2197-2206.

Reverberi, C., Bonati, L. L., Frackowiak, R. S. J., Paulesu, E., Cherubini, P., & Macaluso, E. (2012). Large scale brain activations predict reasoning profiles. *NeuroImage, 59*, 1752-1764

Ribot, T. (1882). *Diseases of memory*. New York: D. Appleton and Co.

Rudebeck, S. R., Scholz, J., Millington, R., Rohenkohl, G., Johansen-Berg, H., Lee, A. C. H. (2009). Fornix microstructure correlates with recollection but not familiarity memory. *Journal of Neuroscience, 29(47)*, 14487-14992.

Schmahmann, J. D., Smith, E. E., Eichler, F. S., & Filley, C. M. (2008). Cerebral white matter: neuroanatomy, clinical neurology, and neurobehavioral correlates. *Annals of the New York Academy of Sciences, 1142*, 266–309.

Schneider, K. (1912). Über einige klinisch-psychologische Untersuchungsmethoden und ihre Ergebnisse. Zugleich ein Beitrag zur Psychopathologie der Korsakowschen Psychose.

[On some clinical-pathological methods of investigation and their results. Simultaneously a contribution to the psychopathology of Korsakoff's psychosis]. *Zeitschrift für die gesamte Neurologie und Psychiatrie/ Originalien, 8*, 553-615.

Schneider, K. (1928). Die Störungen des Gedächtnisses [Disturbances of memory]. In O. Bumke (Ed.). *Handbuch der Geisteskrankheiten (Vol. 1)* (pp. 508-529). Berlin: Springer.

Schnider, A., Gutbrod, K., Hess, C. W., & Schroth, G. (1996). Memory without context: amnesia with confabulations after infarction of the right capsular genu. *Journal of Neurology, Neurosurgery, and Psychiatry, 61*, 186-193.

Semon, R. (1904). *Die Mneme als erhaltendes Prinzip im Wechsel des organischen Geschehens.* Leipzig: Wilhelm Engelmann.

Sepulcre, J., Masdeu, J. C., Sastre-Garriga, J., Goni, J., Velez-de-Mendizabal, N., Duque, B., Pastor, M. A., Bejarano, B., &Villoslada, P. (2008). Mapping the brain pathways of declarative verbal memory: Evidence from white matter lesions in the living human brain. *NeuroImage, 42*, 1237-1243.

Shing, Y. L., Werkle-Bergner, M., Brehmer, Y., Müller, V., Li, S.-C., & Lindenberger, U. (2010). Episodic memory across the lifespan: The contributions of associative and strategic components. *Neuroscience and Biobehavioral Reviews, 34*, 1080-1091.

Silver, J. M., McAllister, T. W., & Arciniegas, D. B. (2009). Depression and cognitive complaints following mild traumatic brain injury. *American Journal of Psychiatry, 166*, 653–61.

Squire, L. R. (1982). The neuropsychology of human amnesia. *Annual Review of Neuroscience, 5*, 242-273.

Squire, L. R,, & Bayley, P. J. (2007). The neuroscience of remote memory. *Current Opinion of Neurobiology, 17*, 185-196.

Staniloiu, A., & Markowitsch, H.J. (2010a). Understanding psychogenic amnesia and psychiatric disorders as causes of dementia. *Journal of General Medicine, 22*, 41-49.

Staniloiu, A., & Markowitsch, H. J. (2010b). Looking at comorbidity through the glasses of neuroscientific memory research: a brain network perspective. *Behavioral and Brain Sciences, 33*, 171-172.

Staniloiu, A., Markowitsch, H. J., & Brand, M. (2010). Psychogenic amnesia – A malady of the constricted self. *Consciousness and Cognition, 19*, 778-801.

Stephan, H. (1975). Allocortex. Handbuch der mikroskopischen Anatomie des Menschen [Handbook of microsocopic anatomy of man]. (Vol. 4, Part 9). Berlin: Springer.

Stickgold, R., & Walker, M. P. (2005). Memory consolidation and reconsolidation: what is the role of sleep? *Trends in Neuroscience, 28*, 408-415.

Svoboda, E., McKinnon, M. C., & Levine, B. (2006). The functional neuroanatomy of autobiographical memory: a meta-analysis. *Neuropsychologia, 44(12)*, 2189-2208.

Sweet, W. H., Talland, G. A., & Ervin, F. R. (1959). Loss of recent memory following section of fornix. *Transactions of the American Neurological Association, 84*, 76-82.

Tanji, K., Suzuki, K., Fujii, T., Higano, S., & Yamadori, A. (2003). A case of frontal network amnesia. *Journal of Neurology, Neurosurgery, and Psychiatry, 74*, 106-109.

Tramoni, E., Aubert-Khalfa, S., Guye, M., Ranjeva, J. P., Felician, O., & Ceccaldi, M. (2009). Hypo-retrieval and hyper-suppression mechanisms in functional amnesia. *Neuropsychologia, 47*, 611-624.

Tsivilis, D., Vann S. D., Denby, C., Roberts, N., Mayes, A. R., Montaldi, D., & Aggleton, J. P. (2008). A disproportionate role for the fornix and mammillary bodies in recall versus recognition memory. *Nature Neuroscience, 11*, 834-842.

Tucker, D. M., Roeltgen, D. P., Tully, R., Hartmann, J., & Boxell, C. (1988). Memory dysfunction following unilateral transection of the fornix: A hippocampal disconnection syndrome. *Cortex, 24*, 465-472.

Tulving, E. (1972). Episodic and semantic memory. In E. Tulving & W. Donaldson (Eds.), *Organization of memory* (pp. 381-403). New York: Academic Press.

Tulving, E. (1983). *Elements of episodic memory*. Oxford: Clarendon Press.

Tulving, E. (2002). Episodic memory: from mind to brain. *Annual Reviews of Psychology, 53*, 1-25.

Tulving, E. (2005). Episodic memory and autonoesis: Uniquely human? In H. S. Terrace & J. Metcalfe (Eds.), *The missing link in cognition: Self-knowing consciousness in man and animals* (pp. 3-56). New York: Oxford University Press.

Vogel, A. C., Power, J. D., Petersen, S. E., & Schlaggar, B. L. (2010). Development of the brain's functional network architecture. *Neuropsychological Reviews, 20*, 362-375.

Vogt, B. A. (2005). Pain and emotion interactions in subregions of the cingulate gyrus. *Nature Reviews Neuroscience, 6(7)*, 533-544.

Vuilleumier, P. (2005). Hysterical conversion and brain function. *Progress in Brain Research, 150*, 309-329.

Wang, Q. (2001). Culture effects on adults' earliest childhood recollection and self-description: Implications for the relation between memory and the self. *Journal of Personality and Social Psychology, 81*, 220-233.

Wang, Q. (2008). Emotion memory and autobiographical memory across the preschool years: a cross-cultural longitudinal investigation. *Cognition, 108*, 117-135.

Yasuno, F., Hirata, M., Takimoto, H., Taniguchi, M., Nakagawa, Y., Ikejiri, Y. et al. (1999). Retrograde temporal order amnesia resulting from damage to the fornix. *Journal of Neurology, Neurosurgery, and Psychiatry, 67*, 102-105.

Zaidel, D., & Sperry, R.W. (1974). Memory impairment after commissurotomy in man. *Brain, 97*, 263-272.

Zingerle, H. (1900). Zur Symptomatik der Geschwülste des Balkens [On the symptoms of the tumors of the corpus callosum]. *Jahrbücher für Psychiatrie und Neurologie, 19*, 367-379.

In: Horizons in Neuroscience Research. Volume 15
Editors: Andres Costa and Eugenio Villalba

ISBN: 978-1-63321-523-8

Chapter 7

The Role of Mineralocorticoid Receptors in Major Depressive Disorder

***Natalia Petros*[2] *and L. A. Carvalho*[1,*]**
[1] Department of Epidemiology and Public Health,
University College London, London, UK
[2] Department of Life Sciences, University of Roehampton, London, UK

Abstract

Corticosteroids regulate the Hypothalamic-Pituitary-Adrenal (HPA) axis via two types of receptors: the glucocorticoid receptor (GR) and the mineralocorticoid receptor (MR). The dysfunction of both receptors and a disturbed balance in the MR/GR ratio is thought to be implicated in stress related illness such as Major Depressive Disorder (MDD). Until recently there has been a lack of investigation into MR function in MDD. This review presents findings that reveal the vital role of MR in MDD. MR function is overactive in MDD, possibly compensating for a disturbed GR function, also dysfunction of MR is associated with treatment resistant depression and could also be an indicator of poor outcome following treatment. MR gene variants MRI180V and 215G/C could increase vulnerability to MDD. Future therapies for MDD should perhaps target the MR/GR ratio, MR function and expression and the interactions MR have with serotonin's 5HT1A receptors and with corticotrophin releasing hormone, in order to accelerate and improve treatment response in MDD. The exact mechanisms that underlie the role of MR in MDD are still enigmatic, thus future *in vitro* and *in vivo* investigations are required.

Keywords: Major Depressive Disorder, The Hypothalamic-Pituitary-Adrenal (HPA) axis, Glucocorticoid Receptor, Mineralocorticoid Receptor, Corticotrophin Releasing Hormone, Treatment-Resistant Depression

[*] Corresponding author: Dr. Livia A Carvalho, PhD MPharm BSc; UCL Research Department of Epidemiology and Public Health; 1-19 Torrington Place, London WC1E 7HB, Rm356a; T: +44 (0)20 7679 5973 F: +44 (0)20 7813 0242; e-mail: livia.a.carvalho@ucl.ac.uk.

Introduction

The Hypothalamic-Pituitary-Adrenal (HPA) axis has been well documented as playing a vital role in the pathophysiology of major depressive disorder (MDD). [1-4] This internal stress response system is regulated by the secretion of corticotrophin releasing hormone (CRH) and arginine vasopressin (AVP) in the periventricular nucleus of the hypothalamus (See Figure 1). In healthy individuals the HPA axis possesses a functional negative feedback mechanism; corticosteroids bind to their receptors in the hypothalamus and pituitary and suppress the release of CRH and AVP (See Figure 1). [5] Evidence has revealed that in MDD, patients present with an abnormal HPA axis; exhibiting prolonged activation of this neuroendocrine system with hypercortisolemia. [3] Additionally, MDD patients have demonstrated increased adrenocorticotropic hormone (ACTH) [6] and CRH levels, [7] depletion of corticosteroid receptors [8] and inflammation of the pituitary and adrenal glands. [9] These findings have left researchers to postulate that perhaps normalising the HPA axis could attenuate the pathophysiology of MDD. [3, 10-12]

There are two main receptors in which corticosteroids have been found to bind to; Type 1 – mineralocorticoid receptors (MR), found mainly on the hippocampal neurons and Type 2 – glucocorticosteroid receptors (GR) (see Table 1), found all over the central nervous system (CNS). [13] Differences between species in terms of MR and GR hippocampal expression exist; in humans; both are abundantly co-expressed in the hippocampus, [14] whereas in rodents; MR are not co-expressed with GR on all neurons. [15]

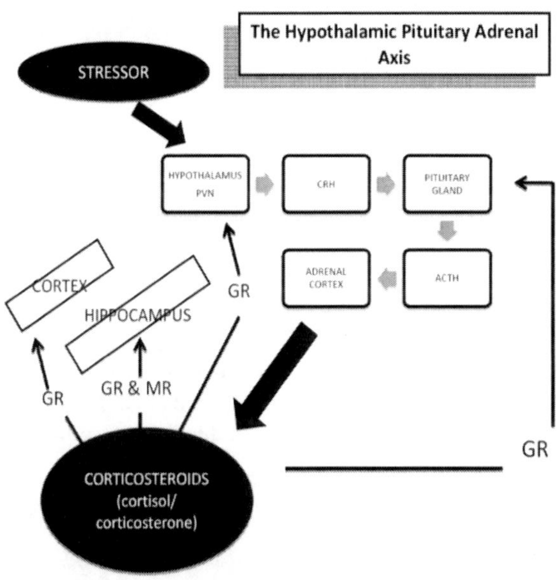

CRF = Corticotropin Releasing Hormone
ACTH = Adrenocorticotropic Hormone
McEwen et al (1968).

Figure 1. The Hypothalamic Pituitary Adrenal Axis: Stress targets the hypothalamus, which releases corticotrophin releasing hormone and arginine vasopressin. The consequence of the secretion of these two neuropeptides is the release of adrenocorticotropic hormone in the anterior pituitary, which travels to the adrenal cortex and stimulates the production of corticosteroids/glucocorticoids (cortisol in humans, corticosterone in rodents). GR and MR signalling are performed as a negative feedback mechanism to attenuate the activity of the axis once the stressor has been removed.

Table 1. The Main Functional Differences between MR and GR in the CNS

RECEPTOR TYPE	Capacity for endogenous corticosteroids	Affinity for endogenous corticosteroids	Occupation of corticosteroids	Expression within the CNS	Function
MINERALOCORTICOID RECEPTOR	Low	High (but not for dexamethasone) High affinity for aldosterone	Constant throughout low basal and high levels	Within specific areas, mainly hippocampus (co-localisation of MR and GR in majority of human hippocampal neurons)	Maintain homeostasis – involved in the sensitivity of the stress response system
GLUCOCORTICOID RECEPTOR	High	Low (Higher affinity for dexamethasone than MR)	Mainly during high level phases	Ubiquitous expression - rich on hypothalamic CRH neurons and pituitary corticotropes	Activated in response to stress Restoration of homeostasis and termination of stress response

De Kloet & Reul (1987) Joels & Dekloet (1994) [102, 122].

Traditionally, the MR were viewed as receptors that maintain the basal levels of the HPA axis - stress response system, whereas GR were regarded as the regulators of homeostasis and negative feedback. However, research has revealed that MR and GR work in partnership at peak levels and during stress. [16] More interestingly, MR are capable of facilitating negative feedback by themselves in response to challenge in humans [17] and animals. [18] Furthermore, impeding the MR function increases physiological and stress induced HPA activity. [19]

This review will focus on MR with particular focus on the dysregulation of the HPA axis negative feedback system in depressed patients, the involvement of MR and GR on this dysregulation; and the effect of antidepressants on correcting such disturbances. Furthermore this review will briefly discuss the role of MR gene variants, the interaction between MR and CRH and also cover investigations into the underlying mechanisms of MR dysfunction in MDD.

1. The Hypothalamic Pituitary Adrenal Axis in Major Depressive Disorder

1.1. Poor Negative Feedback in Major Depressive Disorder

One of the most widespread findings in psychiatry is the hyperactivity of the HPA axis of patients with MDD. [4, 20-22] The negative feedback system of the HPA axis in depressed individuals is thought to be impaired at the level of the GR, [23] in addition to the hypersecretion of CRH in the hypothalamus. [22] These hindrances are both suggested causes of the hyperactivity of the axis, which would normally be inhibited by endogenous corticosteroids and GR. [24] A combination of two neuroendocrine functions tests have been used to measure the negative feedback mechanism of the HPA axis in MDD. These two tests are: the Dexamethasone Suppression Test (synthetic corticosteroid which binds to GR) [25] and the CRH Stimulation Test (injection of CRH to assess ACTH response). [26] The DST has revealed that individuals with MDD are unable to supress cortisol and ACTH following dexamethasone treatment. [6, 27, 28] However, the CRH test was introduced due to the reduced sensitivity of the DST; [29] as it was deemed more responsive to dysfunctional feedback at the level of the pituitary and to increased CRH and vasopressin levels subsequent to impaired feedback in the brain. [20, 30] Together the tests support the impaired feedback inhibition found in MDD, particularly when combined. [16, 23, 27, 31-34] This impairment of the negative feedback loop is a major facet in the pathophysiology of depression. [3, 22, 23, 35-37] Furthermore, consistent findings reveal that MDD patients that continuously non-suppress on the DST/CRH test are more likely to relapse, particularly with longer time-course of HPA axis re-establishment, which produces a modest outcome following treatment. [38-42] It is possible that underlying mechanisms acting on the HPA axis of some MDD patients produces a persistent activation of the system that could be immune to any type of treatment. The reason why this dysfunction occurs is not well known. Reduction of MR and GR expression of these receptors [8] and their disturbed function [43] could be a potential cause of the weak feedback response and hyperactivity of the HPA axis observed in some patients with MDD.

More recently, mechanisms of action that are in play in the regulation of the HPA axis in major depression have started to be clarified. Kats et al showed decreased GR sensitivity due to decreased GR co-chaperones associated with depressive symptoms in pregnant women [123]. Menke et al recently published studies showing the involvement of NFKappaB. NFKappaB was decreased in the blood of patients with major depression [124] after ex-vivo stimulation. Polymorphisms in the gene enconding the GR regulating co-chaperone FKBP5 has also been shown to alter receptor sensitivity. The presence of the risk allele for FKBP5 is associated with GR resistance in patients with major depression. [124]

1.2. The Effects of Anti-Depressants on the Hypothalamic Pituitary Adrenal Axis

According to Holsboer and Barden antidepressants exert their influence on neurotransmitters and also upon the neuroendocrine system. [30] There is evidence to show that the HPA axis returns to a fully functional level following effective treatment with antidepressants in some patients with MDD. [2, 44-46] Experimental studies have revealed that antidepressants directly and indirectly act on the GR and MR, by modulating their expression, number and function. [8, 47-51] As an example Barden [52] revealed that amitriptyline increased GR expression and GR binding and reduced CRH expression in the hypothalamus suggestive of a possible mechanism for the reinstatement of the HPA axis to a functional state. This indicates that the activity of the negative feedback loop of the HPA axis is increased and driving towards normalisation of the HPA axis. [30, 40, 53] These findings are further supported by clinical evidence revealing that successful anti-depressant treatment in MDD patients attenuates cortisol, [28, 54, 55] ACTH [56] and CRH levels. [57, 58] Some inconsistencies have also been found; for example Pintor et al. [59] and Van Den Eede et al. [60] reported no difference in HPA axis status according to clinical presentation. Pintor et al. revealed that most of the MDD patients they studied continued to have a dysregulated HPA axis despite clinical remission and this could be explained by the existence of varied subtypes of depression, such as melancholic depression. Overall, however, research supports the influential effects of antidepressants on the HPA axis in MDD. Furthermore, restoration of GR function could possibly predict effective response to treatment and long term outcome. [61] However this only seems the case with some antidepressants and the same effect is not established in all patients, thus leaving researchers curious to investigate further into the pathophysiological mechanisms of the HPA axis in MDD and newer ways of treating the illness. As an example research has demonstrated the therapeutic effects of GR agonists and antagonists in patients resistant to traditional treatment and those with psychotic or atypical depression. [62-64]

1.3. *In Vitro* Studies

In vitro studies have been conducted in order to understand the clinical findings by investigating the mechanisms of action of antidepressants on corticosteroid receptor gene expression and function. However, findings have not been as clear-cut as expected. Carvalho et al. utilised peripheral human whole-blood cells to investigate the in vitro effects of antidepressants on GR. [61] Interestingly they found that clomipramine reduced GR function in healthy controls, whilst having no effect on patients with MDD. Here, GR function was measured by corticosteroid inhibition of lipopolysaccharide-induced interleukin 6 levels. The findings of this study emphasized that blood cells of patients exhibiting treatment resistant depression also lacked any anti-depressive outcome in vitro. This emphasises the pathophysiological differences that may be occurring in treatment resistant MDD, where antidepressants are having no effect on GR.

Animal in vitro studies have demonstrated that antidepressants can have a down-regulatory effect on GR function, which may be due to the differences in experimental conditions. For example, cell pre-treatment with anti-depressants prior to incubation with corticosteroids reduces GR function. While co-incubation of anti-depressants and cortisol increases GR function, on the other hand however replacing cortisol with corticosterone reduces GR function. [69, 70] The authors here measured GR function in several ways i.e., by GR expression, GR translocation to the nucleus and by cortisol intracellular concentration. Carvalho and Pariante proposed that increase in GR function is facilitated by a rise in intracellular corticosteroid due to the block of p-glycoprotein, which drives corticosteroids out from the cell. [50] Whereas reduced GR function is possibly an effect of antidepressant induced down-regulated GR mRNA. Indeed, they highlight that antidepressants impose a direct inhibition on membrane steroid transporters that facilitate the ejection of corticosteroids from cells and this effect is free from neurotransmitter influence. Furthermore, antidepressants inhibit the effect of p-glycoprotein in expelling cortisol from cells, thus increasing the intracellular levels. This followed later by an increase in GR function, leaving the ultimate outcome of antidepressant effect reliant on the balance between what is going inside the cell and receptor expression. Such findings have been supported by studies using both tricyclic and newer antidepressants. [71, 72] Therefore, it is possible that GR dysfunction is only one part in the pathophysiology of MDD involving the HPA axis and potentially individuals with treatment resistant MDD could also possess a deficit of p-glycoprotein. However, further investigation on this is required. Unfortunately there is a lack of evidence measuring anti-depressant effects on MR in vitro; one study by Lai et al. found no effect on the hippocampus. [73]

2. Approaches to Understanding Mineralocorticoid Receptor Function in Major Depressive Disorder

Research suggests the importance of an undisturbed MR/GR ratio on the regulation of the stress response. As an example Semont et al. demonstrated that inhibiting the synthesis of serotonin (5HT) in rats significantly increases hippocampal MR mRNA and decreases the mRNA of GR; thus producing an imbalance of the MR:GR ratio. [74] Furthermore, MR signalling at the commencement of the stress response occurs prior to any GR actions that work towards homeostasis. MR and GR work in concert, therefore disturbing the balance between the mechanisms at the activation stage and at the termination stage of the stress response causes dysregulation of the HPA axis. [13, 19] Indeed, the maintained balance between these two receptors is crucial to a responsive HPA axis, functional negative feedback mechanism and consequently an effective stress reaction. [5] Therefore, any disturbance to this could render the host more vulnerable to stress-related illness; alter behavioural adaptation and affect mental performance. [13, 75, 76]

2.1. Clinical Studies Demonstrating Selective Impairment of Mineralocorticoid Receptor in Subtypes of Depression

Dysregulation of the HPA axis negative feedback could be due to a hyperactivity of the MR. Young and colleagues conducted one of the first clinical investigations into the function of MR in MDD utilising the administration of MR antagonist spironolactone, which raises corticosteroid levels. [77] Young et al. found that in individuals with MDD, MR maintained higher levels of activity when exposed to increased levels of cortisol compared to controls. Moreover they found this over-activation to be associated with marked ACTH and cortisol secretion. While this study did not measure GR function, it did suggest that there is a shift in the functional levels of both receptors in MDD, which could indicate a reduction in the MR/GR ratio. Furthermore, they proposed the notion that the role of MR in MDD is significant; specifically via over-activation MR are perhaps attempting to maintain homeostasis of the HPA axis and reduce cortisol levels. With this in mind, it is possible that MR are attempting to compensate for the disturbance of GR function in MDD, however, these exact mechanisms are not completely understood.

Juruena et al. encouraged the use of a test similar to the DST/CRH but more sensitive for predicting outcome in patients with treatment resistant depression - for the purpose of measuring MR function. [78] Using another synthetic corticosteroid; the prednisolone suppression test (PST), binds to both GR and MR. In an earlier study, comparing depressed patients with controls they investigated the efficiency of both the DST and PST. Findings revealed that depressed patients had less suppression when compared to controls for the DST, but interestingly, for the PST, both patients and controls exhibited normal suppression. [79] Moreover, they found no difference in suppression of prednisolone between admission to hospital and at discharge in the depressed sample. The authors argued that this discovery suggests selective impairment of GR function in MDD - implied by the DST, whereas MR function seems to remain relatively preserved - as suggested by the PST. They proposed that the MR effects on negative feedback of the HPA axis could be unharmed in depressed individuals. Furthermore Juruena and colleagues supported the idea that this reasonably 'normal' response to prednisolone by the depressed patients could be a compensatory effect of MR due to weakened GR function.

More recent clinical investigation of the MR and its role in MDD has revealed that despite an overall 'normal' response to the PST in patients with depression, those that demonstrate a lack of response to multimodal inpatient treatment also exhibit an impaired response on the PST. [80, 81] Findings signify that MR functional impairment appears to be a characteristic held only by those with treatment resistant and severe MDD. This finding is consistent with the view that treatment resistant depression is also associated with disturbed GR function and failure of HPA axis re-establishment. These authors also hypothesized that MR function is a fixed unchanging component of the HPA axis of responders to treatment and also that MR efficiency is essential to the therapeutic effect of antidepressants. Support for this hypothesis stems from studies that utilise MR antagonists and agonists to manipulate the function of MR. [77, 82, 83]

2.2. Polymorphisms of the Mineralocorticoid Receptor in Major Depressive Disorder

Gene variants of MR could play a role in the variability of HPA axis competency and thus risk to stress and illness. According to DeRijk et al., humans carrying the MRI180V variant display an augmented cortisol response to psychosocial stressors compared to non-carriers and this could possibly render these individuals more vulnerable to MDD. [84] In contrast, the C-allele 215G/C MR gene is associated with baseline levels of cortisol. [85] Interestingly, Kuningas and colleagues found that elderly individuals carrying the C-allele gene variant displayed lower cortisol levels than non-carriers, supporting the evidence that proposes the MR inhibitory influence over the HPA axis. [86] However, individuals carrying the G-allele gene variant of 215G/C MR have higher levels of cortisol. [85] Although van Leeuwen et al. found that this was sex-specific; with males experiencing weakened suppression of dexamethasone compared to females. [87]

In relation to cortisol awakening response (CAR), the influence of MR gene variants remains unclear. van Leeuwen et al. also proposed the connection to be sex specific and dependent on dexamethasone administration. [87] Additionally in MDD CAR appears to be raised in remitted and unwell MDD patients and those at risk of the disorder. [88] Klok et al. discovered that in patients of both sex with MDD the MR -2G/C variant (G or C allele) can modify the CAR, however this effect was dependant on the individual use of selective serotonin reuptake inhibitors (SSRIs). [89] The authors highlighted that variability in CAR based on MR gene variants could characterise both a risk and protective factor for the course of MDD. Furthermore, the interaction between MR and SSRIs on CAR was dependent on MR genotype, with carriers of G-allele displaying higher CAR compared to other genotype carriers; this could have implications for development of pharmacological treatment for MDD.

Overall studies investigating MR polymorphisms hold several implications for the treatment of depression. Since MR function is associated with timing of treatment response [81, 82] and MR gene variants appear to be involved in HPA axis regulation during basal and stress-induced conditions. [84, 86, 87] However, further investigation on sex and age differences, MR polymorphisms and cortisol levels in relation to MDD is clearly needed.

3. Mechanisms That Underlie Mineralocorticoid Receptor Function in Major Depressive Disorder

3.1. Removal and over Expression of the Mineralocorticoid Receptor Gene

Many of the molecular and cellular mechanisms that underlie the exact role of MR in MDD continue to remain unidentified; therefore approaches such as removal or overexpression of the MR gene have been adopted in order to decipher this phenomenon and support clinical findings. When comparing wild-type mice with MR knock-out mice, Berger et al. reported that loss of the MR gene impairs behavioural plasticity. [90] This supports the notion of a required MR/GR balance and the significant position of MR in the pathophysiology and symptomology of MDD; emphasising the necessity of a present and functioning MR. Most investigations do not selectively target the deletion of the hippocampal MR. In view of the influence that hippocampal MR have on neuroendocrine control and behaviour, further investigation is required where MR gene removal is limited to the hippocampus. This is particularly significant since in the hippocampal formation of non-human primates, MR expression prevails over GR expression. [91] Rozeboom et al. demonstrated that MR over-expression in transgenic mice down-regulates the expression of GR and yet intriguingly they found an increase of 5HT1A receptors. [92] This is particularly relevant for the treatment of MDD since 5HT1A receptors have been shown to be depleted in depressed patients. [93] Therefore possible MR-mediated up-regulation of 5HT1A receptors could be a future treatment target for MDD, which is convincing as both receptor types are found within the hippocampus. However a study to validate this distinctly is required and Rozeboom and colleagues point out that this was an unexpected finding. Moreover, experimental studies have revealed that MR-mediated actions in the hippocampus have a down-regulatory effect on 5HT1A receptors [94-97] and lessen the release of 5HT. [98] Furthermore, 5HT1A receptors have been found to be up-regulated following treatment with tricyclic antidepressants. [99, 100] Welner and colleagues postulate that up-regulation of 5HT1A receptors could essentially assist the efficacy of antidepressants by augmenting 5HT transmission.

3.2. Post Mortem Studies

Collectively, much research suggests MR over-activation and GR mRNA down-regulation are associated with depression. [19, 101, 102] Variability in MR expression may result from genetic and/or environmental influences, [20] old age [103] and can also ensue from stress during lifetime [104] or stress prior to death. [105] Lopez et al. conducted a two-component post mortem study on humans to compare the results with animals. [106] They demonstrated that chronic stress causes high levels of corticosterone in rats. This in turn down-regulates mRNA and binding of 5HT1A receptors and produces a disturbed MR/GR ratio. They also postulated that these mechanisms could be responsible for instigating and upholding depressive state in MDD. Furthermore, they found reduced expression of hippocampal 5HT1A, MR and GR and altered MR/GR balance in suicide victims with a history of MDD, similar to that found in the rodents (See Table 2). However, there are some inconsistencies in the findings; where MR levels in areas other than the hippocampus have exhibited unpregulation [107] or normal levels despite reduced GR expression [108, 109] in MDD patients. This is suggestive of the dynamic nature of MR regulation, transforming between different areas of the brain (See Table 2). Furthermore, recently, Klok et al. found no significant difference in MR/GR ratio in the brains of MDD patients compared to controls (See Table 2). [110] Overall, however findings generally point towards a reduction of MR in the hippocampus and an increase in the hypothalamus, with a reduced MR/GR ratio, which essentially corresponds with the hyperactivity of the HPA axis found in depressed individuals.

Table 2. MR and GR Expression in Major Depressive Disorder (MDD) patients

AUTHOR	MR (mRNA)	GR (mRNA)	MR/GR RATIO
Lopez et al. [106]	↑ Areas abundant with 5HT1A - Granular layer of Dentate Gyrus → CA3 of Hippocampus ↓ CA1 & CA2 of Hippocampus	→ Granule cells of the Dentate Gyrus and CA3 of Hippocampus ↓ In all other CA subfields	↓
Xing et al. [108]	X Prefrontal Cortex of Brodmann's Area 9	Not measured	
Klok et al. [110]	↓ Hippocampus, Inferior Front Gyrus & Cingulate Gyrus	↓ Inferior Front Gyrus & Cingulate Gyrus	X
Wang et al. [107]	↑ Periventricular tissue of Hypothalamus accompanied by amplified expression of CRH		

Up arrow indicates high levels of MR, down arrow indicates low levels of MR or MR/GR ratio, right arrow indicates moderate levels of MR and X indicates no significant changes/difference compared to controls.

3.3. Aberrant Interaction between Corticotrophin Releasing Hormone and the Mineralocorticoid Receptor

Hippocampal MR exert an inhibitory influence over HPA axis actions, therefore any disruption to this effect can result in an increase in basal waking CRH and vasopressin mRNA expression and HPA axis activity. [111, 112] A study by Gesing et al. discovered an upsurge in MR density in rodents following an acute psychological stressor. [113] Interestingly, they proposed that this augmentation was facilitated by CRH and consequently an increase in MR density was accompanied by further inhibition of the HPA axis. These findings show that MR are dynamic and actively involved in the resilience of the brain when under stress but also signifies a reciprocal interaction between CRH and MR.

The recognition of this interaction following a stressful event is important since studies on the post-mortem brains of previously depressed suicide victims exhibit increased levels of CRH in their cerebrospinal fluid. [7] However, Gesing and colleagues later expressed that this interaction between CRH and MR has the potential to become aberrant. [104] They suggest that chronic stress causes high levels of CRH in the brain; due to the CRH receptors and post-receptors becoming desensitised. Therefore, a possible consequence is poor regulation of MR levels (usually facilitated by CRH) and a subsequent decline in HPA axis regulation. However, it remains unclear whether in MDD an initial deficiency lies within MR causing advanced secretion of CRH or whether CRH hypersecretion can instigate MR dysfunction leading to hyperactivity of the HPA axis. Furthermore, to what extent the interaction between

CRH and MR is abnormal and the degree of HPA axis dysfunction is dependent upon various host characteristics, such as genetic influences and early life experiences. [5, 20, 37, 114, 115]

4. Manipulation of Mineralocorticoid Receptor Function and Expression with Antidepressants

4.1. Increasing MR Function

Although the weight that the HPA axis regulation holds in antidepressant efficacy still remains controversial, as previously emphasised the balance between MR and GR expression and function is critical in MDD. Support for this hypothesis stems from investigations into the dual actions of antidcprcssants on rcstoration of monoamines but also re establishment of a normalised HPA axis; analogous to an increase in MR (and GR) expression and/or function.

An unpublished double-blind pilot study by Hundt et al. (un-published findings cited in Holsboer) [116] was the first to provide any evidence for the clinical effects of a drug that acts on MR function; demonstrating a link between antidepressants and MR in MDD.

The group studied MDD patients undergoing treatment with amitriptyline. The sample were randomly split between some being given MR antagonist spironolactone and some a placebo as an adjunct to standard treatment. Hundt et al. found a decrease in the therapeutic effect of amitriptyline in the individuals given spironolactone compared to those given a placebo, who experienced attenuation of clinical symptoms.

Furthermore in 1997 Holsboer emphasised that diffusion of an anti-mineralocorticoid has a negative impact on the efficiency of antidepressants (cited in De Kloet). [5] Otte and colleagues administered MR agonist fludrocortisone in conjunction with antidepressant treatment within a group of 64 patients with MDD; divided into responders to treatment and treatment-resistant. [82] Interestingly they found that responders to treatment experienced a significantly faster response with the add-on of fludrocortisone. They also exhibited lower cortisol levels when compared to the treatment resistant patients following the add-on treatment. The findings fortify conclusions on the importance of MR function in MDD and likewise consolidating the discovery of the association between impaired MR function and treatment resistant depression.

4.2. Up-Regulation of Mineralocorticoid Receptor Expression

Experimental research conducted in the early 1990s initially revealed that one of the first mechanisms of action of antidepressant drugs is the up-regulation of MR in the hippocampus (see Table 3). [117] This increase in MR mRNA was followed by reduced CRH mRNA in the PVN, therefore it was anticipated that an up-regulation of MR expression could produce established homeostasis of the HPA axis and possibly a lot sooner in non-treatment resistant MDD patients. Reul and colleagues supported this view by proposing that MR mRNA up-regulation also increases GR capacity thus assisting the negative feedback mechanism. [48]

Interestingly, Bachmann et al. administered GR antagonists to rats and reported that throughout the entire time course, there was an up-regulation of MR. [118] They postulated that it is possible for the change in MR:GR ratio towards MR predominantly could in fact be responsible for the therapeutic effect of GR antagonists in MDD. Animal studies demonstrate that long-term anti-depressant treatment can up-regulate MR gene expression in the hippocampus, moreover than GR (See Table 3). [117, 119]

Table 3. The Effects of Antidepressants on the GR and the MR Hippocampal mRNA in Rodents

AUTHOR	ANTI-DEPRESSANT/ DAILY DOSE	DURATION OF TREATMENT	HIPPOCAMPAL MR EXPRESSION mRNA	HIPPOCAMPAL GR EXPRESSION mRNA
Brady et al. [117]	IMI 5mg	14 Days	→	→
		8 Weeks	↑	↓
Seckl and Fink [119]	AMI 20mg	2 Days	↑	→
		14 Days	↑	↑
	DMI 10mg	2 Days	→	→
		14 Days	↑	↑
	CIT 20mg	2 Days	→	→
		14 Days	↑	→
Reul et al. [48]	AMI 4.5mg	3-7 Days	↓	↓
		5 Weeks	↑	↑
		7 Weeks	↑	↑
Yau et al. [123]	AMI 10mg	9 Weeks	↑	→
Yau et al. [124]	FLU 10mg	9 Days	↓	↓
	VEN 10mg			
Yau et al. [125]	FLU 10mg	4 Weeks	→	↑
	MOC 5mg		→	→
	DES 10mg		↑	→
	TIA 10mg		→	→
	VEN 10mg		→	→
Heydendael and Jacobson [125]	IMI 20mg	4 Weeks	(CA1 & CA2 of Hippocampus only)↑	(CA1 of Hippocampus only) ↑
		8 Weeks	(CA3 of Hippocampus only) ↓	→
	PHE 25mg	4 Weeks	(CA2, CA3 & Dentate Gyrus of Hippocampus only) ↑	→

IMI = Imipramine, AMI = Amitriptyline, DMI = Desipramine, CIT = Citalopram, FLU = Fluoxetine, MOC = Moclobemide, TIA = Tianeptine, VEN = Venlafaxine, PHE = Phenelzine. Up arrow indicates a significant increase in antidepressant treated rodent cells vs. wild-type; down arrow a significant decrease in antidepressant treated rodent cells vs. wild-type and right arrow indicates no significant difference between antidepressant treated rodent cells compared with wild-type.

However the same effect has not been observed with short-term treatment. In humans anti-depressants also produce differing effects on MR and GR dependent upon the length of administration. Short-term use can down-regulate MR and GR expression, whereas long-term use can up-regulate the receptors. Furthermore, after 6-9 weeks of antidepressant administration MR function can remain increased and GR function returns to normal status. [53] Jahn et el proposed that co-administration of metyrapone, (impedes upon the synthesis of corticosteroids), with antidepressant treatment for MDD can accelerate and provide a better and longer-lasting clinical outcome. [120] It is possible that this is due to metyrapone having an up-regulatory effect on MR expression.

These studies signify the critical position held by MR in the pathophysiology of and recovery from MDD. Moreover findings indicate that enhancement of antidepressant actions benefit from the effective and stimulated performance of MR. In addition the investigations seemingly propose that the initial neuroendocrine reaction to anti-depressants could be the up-regulation of hippocampal MR, which may be necessary to activate a flow of events that lead to clinical effect. Moreover, these findings demonstrate the beneficial outcomes of considering MR function and expression when developing treatment methods for treatment resistant MDD. A further point to make here is that if manipulation of MR accelerates treatment response then this holds vital implications for public health improvements. Explicitly a faster response to treatment may lessen individual distress and the risk of suicide in MDD patients and consequently could lower national costs for mental health care. According to Lai et al. enhanced memory, reduced anxiety and modified stress response to novel situations in mice signifies that MR overexpression could be neuroprotective by possibly augmenting neuronal survival in cognitive and affective disorders. [121] However, further investigation is required before treatments that enhance the function and expression of MR and also those that prevent the synthesis of corticosteroids are safe for clinical application along with routine treatment.

Conclusion

Extensive research has shown that the HPA axis of depressed individuals is hyperactive caused by the impairment of the negative feedback mechanism, which could be attributable to a reduction in MR and GR expression and function. A balanced MR/GR ratio is crucial to an effective HPA axis and any disturbance to this ratio could leave the host vulnerable to MDD. Furthermore, investigations imply that MR function in MDD is overactive, possibly compensating for a disturbed GR function, also that dysfunction of MR is associated with treatment resistant MDD and thus could be an indicator of poor outcome following treatment. MR gene variants could increase vulnerability to MDD therefore further investigation on the role of MR polymorphisms MRI180V and 215G/C is required. The exact mechanisms that underlie the role of MR in MDD are still mysterious; nevertheless MR may perhaps mediate 5HT1A levels, which could be taken advantage of in future treatment strategies. Also, irrespective of the possible aberrant interaction between MR and CRH, the actual communication itself could be exploited for a possible future treatment target. Future therapies for MDD should consider targeting the MR/GR imbalance and instigating increased MR function and/or expression particularly for treatment-resistant MDD and for accelerating treatment response, especially since MR efficiency is essential to the therapeutic effect of antidepressants. *In vitro* study on the effect of antidepressants on MR is required. *In vivo* imaging using positron emission tomography to measure receptor binding would allow the further study of both MR and GR abnormalities in the HPA axis of MDD patients and provide truer enlightenment on their function within this disorder. Efforts are currently being made to produce a suitable radioactive tracer for this research to begin.

Disclosure

The authors have no conflict of interest to disclose.

References

[1] Gillespie CF, Nemeroff CB. Hypercortisolemia and depression. *Psychosomatic Medicine.* 2005;67:S26-S8.

[2] Bao AM, Meynen G, Swaab DF. The stress system in depression and neurodegeneration: Focus on the human hypothalamus. *Brain Research Reviews.* 2008;57(2):531-53.

[3] Pariante CM, Lightman SL. The HPA axis in major depression: classical theories and new developments. *Trends in Neurosciences.* 2008;31(9):464-8.

[4] Cowen PJ. Not fade away: The HPA axis and depression. Psychological Medicine: *A Journal of Research in Psychiatry and the Allied Sciences.* 2010;40(1):1-4.

[5] De Kloet ER, Vreugdenhil E, Oitzl MS, Joels M. Brain corticosteroid receptor balance in health and disease. *Endocrine Reviews.* 1998;19(3):269-301.

[6] Carroll BJ, Cassidy F, Naftolowitz D, et al. Pathophysiology of hypercortisolism in depression. *Acta Psychiatrica Scandinavica.* 2007;115:90-103.

[7] Nemeroff CB, Widerlov E, Bissette G, et al. Elevated Concentrations of CSF Corticotropin-Releasing Factor-Like Immunoreactivity in Depressed-Patients. *Science.* 1984;226(4680):1342-4.

[8] Barden N, Reul J, Holsboer F. Do Antidepressants Stabilize Mood Through Actions on the Hypothalamic-Pituitary-Adrenocortical System. *Trends in Neurosciences.* 1995;18(1):6-11.

[9] Stokes PE. The potential role of excessive cortisol induced by HPA hyperfunction in the pathogenesis of depression. *European Neuropsychopharmacology.* 1995;5:77-82.

[10] Gibbons JL, McHugh PR. Plasma-Cortisol in Depressive-Illness. *Journal of Psychiatric Research.* 1962;1(2):162-71.

[11] Sachar EJ, Hellman L, Fukushim DK, Gallaghe TF. Cortisol Production in Depressive Illness - A Clinical and Biochemical Clarification. *Archives of General Psychiatry.* 1970;23(4):289.

[12] Carpente WT, Bunney WE. Adrenal Cortical Activity in Depressive Illness. *American Journal of Psychiatry.* 1971;128(1):31.

[13] De Kloet ER. Hormones and the stressed brain. Stress: Current *Neuroendocrine and Genetic Approaches.* 2004;1018:1-15.

[14] Seckl JR, Dickson KL, Yates C, Fink G. Distribution of Glucocorticoid and Mineralocorticoid Receptor Messenger-RNA Expression in Human Postmortem Hippocampus. *Brain Research. 1991*;561(2):332-7.
[15] Herman JP, Patel PD, Akil H, Watson SJ. Localization and Regulation of Glucocorticoid and Mineralocorticoid Receptor Messenger-RNAs in the Hippocampal-Formation of the Rat. *Molecular Endocrinology.* 1989;3(11):1886-94.
[16] Bradbury MJ, Akana SF, Dallman MF. Roles of Type-I And Type-II Corticosteroid Receptors in Regulation of Basal Activity in the Hypothalamo-Pituitary-Adrenal Axis During the Diurnal Trough and the Peak - Evidence for a Nonadditive Effect of Combined Receptor Occupation. *Endocrinology.* 1994;134(3):1286-96.
[17] Wellhoener P, Born J, Fehm HL, Dodt C. Elevated resting and exercise-induced cortisol levels after mineralocorticoid receptor blockade with canrenoate in healthy humans. *Journal of Clinical Endocrinology & Metabolism.* 2004;89(10):5048-52.
[18] Reul J, Probst JC, Skutella T, et al. Increased stress-induced adrenocorticotropin response after long-term intracerebroventricular treatment of rats with antisense mineralocorticoid receptor oligodeoxynucleotides. *Neuroendocrinology.* 1997;65(3):189-99.
[19] Dekloet ER. Brain Corticosteroid Receptor Balance and Homeostatic Control. *Frontiers in Neuroendocrinology.* 1991;12(2):95-164.
[20] Holsboer F. The corticosteroid receptor hypothesis of depression. *Neuropsychopharmacology.* 2000;23(5):477-501.
[21] Pariante CM, Miller AH. Glucocorticoid receptors in major depression: Relevance to pathophysiology and treatment. *Biological Psychiatry.* 2001;49(5):391-404.
[22] Nemeroff CB. The corticotropin-releasing factor (CRF) hypothesis depression: New findings and new directions. *Molecular Psychiatry.* 1996;1(4):336-42.
[23] Young EA, Haskett RF, Murphyweinberg V, et al. Loss of Glucocorticoid Fast Feedback in Depression. *Archives of General Psychiatry.* 1991;48(8):693-9.
[24] Reul J, Dekloet ER. 2 Receptor Systems for Corticosterone in Rat-Brain - Microdistribution and Differential Occupation. *Endocrinology.* 1985;117(6):2505-11.
[25] Carroll BJ, Feinberg M, Greden JF, et al. A Specific Laboratory Test for the Diagnosis of Melancholia - Standardization, Validation, and Clinical Utility. *Archives of General Psychiatry.* 1981;38(1):15-22.
[26] Vale W, Spiess J, Rivier C, Rivier J. Characterization of a 41-Residue Ovine Hypothalamic Peptide that Stimulates Secretion of Corticotropin and Beta-Endorphin. *Science.* 1981;213(4514):1394-7.
[27] Ising M, Horstmann S, Kloiber S, et al. Combined Dexamethasone /Corticotropin Releasing hormone test predicts treatment response in major depression--a potential biomarker? *Biological Psychiatry.* 2007;62(1):47-54.
[28] Holsboertrachsler E, Stohler R, Hatzinger M. Repeated Administration of the Combined Dexamethasone-Human Corticotropin Releasing Hormone Stimulation Test During Treatment of Depression. *Psychiatry Research.* 1991;38(2):163-71.
[29] Arana GW, Baldessarini RJ, Ornsteen M. The Dexamethasone Suppression Test for Diagnosis and Prognosis in Psychiatry - Commentary and Review. *Archives of General Psychiatry.* 1985;42(12):1193-204.
[30] Holsboer F, Barden N. Antidepressants and hypothalamic pituitary adrenocortical regulation. *Endocrine Reviews.* 1996;17(2):187-205.

[31] Sawhney G, Goswami U, Chandra R, et al. Selectivity of antidepressant therapy and dexamethasone suppression test in depression: A prospective study. *Indian Journal of Pharmacology*. 2001;33(3):235.

[32] Amsterdam JD, Maislin G, Winokur A, et al. The oCRH stimulation test before and after clinical recovery from depression. *Journal of Affective Disorders*. 14(3):213-22.

[33] Heim C, Mletzko T, Purselle D, et al. The dexamethasone/corticotropin-releasing factor test in men with major depression: Role of childhood trauma. *Biological Psychiatry*. 2008;63(4):398-405.

[34] Kunugi H, Ida I, Owashi T, et al. Assessment of the dexamethasone /CRH test as a state-dependent marker for hypothalamic-pituitary-adrenal (HPA) axis abnormalities in major depressive episode: A multicenter study. *Neuropsychopharmacology*. 2006;31(1):212-20.

[35] Kathol RG, Jaeckle RS, Lopez JF, Meller WH. Patho-Physiology of HPA Axis Abnormalities in Patients with Major Depression - An Update. *American Journal of Psychiatry*. 1989;146(3):311-7.

[36] Molcrani MC, Duval F, Crocq MA, et al. HPA axis dysfunction in depression: Correlation with monoamine system abnormalities. *Psychoneuroendocrinology*. 1997;22(Supplement 1):S63-S8.

[37] Holsboer F, Lauer CJ, Schreiber W, Krieg JC. Altered Hypothalamic-Pituitary-Adrenocortical Regulation in Healthy-Subjects at High Familial Risk for Affective-Disorders. *Neuroendocrinology*. 1995;62 (4):340-7.

[38] Appelhof BC, Huyser J, Verweij M, et al. Glucocorticoids and relapse of major depression (dexamethasone/corticotropin-releasing hormone test in relation to relapse of major depression). *Biological Psychiatry*. 2006;59(8):696-701.

[39] Schule C. Neuroendocrinological mechanisms of actions of antidepressant drugs. *Journal of Neuroendocrinology*. 2007;19(3):213-26.

[40] Greden JF, Gardner R, King D, et al. Dexamethasone Suppression Tests in Anti-Depressant Treatment of Melancholia - The Process of Normalization and Test-Retest Reproducibility. *Archives of General Psychiatry*. 1983;40(5):493-500.

[41] Ribeiro SCM, Tandon R, Grunhaus L, Greden JF. The DST as a Predictor of Outcome in Depression - A Metaanalysis. *American Journal of Psychiatry*. 1993;150(11):1618-29.

[42] Holsboer F, Liebl R, Hofschuster E. Repeated Dexamethasone Suppression Test During Depressive-Illness - Normalization of Test Result Compared with Clinical Improvement. *Journal of Affective Disorders*. 1982;4(2):93-101.

[43] Plotsky PM, Owens MJ, Nemeroff CB. Psychoneuroendocrinology of depression - Hypothalamic-pituitary-adrenal axis. *Psychiatric Clinics of North America*. 1998;21(2):293.

[44] Belmaker RH, Agam G. Mechanisms of disease: Major depressive disorder. New *England Journal of Medicine*. 2008;358(1):55-68.

[45] Linkowski P, Mendlewicz J, Kerkhofs M, et al. 24-Hour Profiles of Adrenocorticotropin, Cortisol, and Growth-Hormone in Major Depressive-Illness - Effect of Antidepressant Treatment. *Journal of Clinical Endocrinology & Metabolism*. 1987;65(1):141-52.

[46] Heuser IJE, Schweiger U, Gotthardt U, et al. Pituitary-adrenal-system regulation and psychopathology during amitriptyline treatment in elderly depressed patients and normal comparison subjects. *American Journal of Psychiatry.* 1996;153(1):93-9.
[47] Reul J, Labeur MS, Grigoriadis DE, et al. Hypothalamic-Pituitary-Adrenocortical Axis Changes in the Rat After Long-Term Treatment with the Reversible Monoamine Oxidase-A Inhibitor Moclobemide. *Neuroendocrinology.* 1994;60(5):509-19.
[48] Reul J, Stec I, Soder M, Holsboer F. Chronic Treatment of Rats with the Antidepressant Amitriptyline Attenuates the Activity of the Hypothalamic-Pituitary-Adrenocortical System. *Endocrinology.* 1993;133(1):312-20.
[49] Nikisch G. Involvement and role of antidepressant drugs of the hypothalamic-pituitary-adrenal axis and glucocorticoid receptor function. *Neuroendocrinology Letters.* 2009;30(1):11-6.
[50] Carvalho LA, Pariante CM. In vitro modulation of the glucocorticoid receptor by antidepressants. Stress-the *International Journal on the Biology of Stress.* 2008;11(6):411-24.
[51] Carvalho LA, Garner BA, Dew T, et al. Antidepressants, but not antipsychotics, modulate GR function in human whole blood: An insight into molecular mechanisms. *European Neuropsychopharmacology.* 2010;20(6):379-87.
[52] Barden N. Regulation of corticosteroid receptor gene expression in depression and antidepressant action. *Journal of Psychiatry & Neuroscience.* 1999;24(1):25-39.
[53] Mason BL, Pariante CM. The effects of antidepressants on the hypothalamic-pituitary-adrenal axis. *Drug News & Perspectives.* 2006;19(10):603-8.
[54] Thakore JH, Barnes C, Joyce J, et al. Effects of antidepressant treatment on corticotropin-induced cortisol responses in patients with melancholic depression. *Psychiatry Research.* 1997;73(1-2):27-32.
[55] Scharnholz B, Weber-Hamann B, Lederbogen F, et al. Antidepressant treatment with mirtazapine, but not venlafaxine, lowers cortisol concentrations in saliva: A randomised open trial. *Psychiatry Research.* 2010;177(1-2):109-13.
[56] Schule C, Baghai T, Laakmann G. Mirtazapine decreases stimulatory effects of reboxetine on cortisol, adrenocorticotropin and prolactin secretion in healthy male subjects. *Neuroendocrinology.* 2004;79(1):54-62.
[57] Heit S, Owens MJ, Plotsky P, Nemeroff CB. Corticotropin-releasing factor, stress, and depression. *Neuroscientist.* 1997;3(3):186-94.
[58] Banki CM, Karmacsi L, Bissette G, Nemeroff CB. CSF corticotropin-releasing hormone and somatostatin in major depression: response to antidepressant treatment and relapse. *Eur Neuropsychopharmacol.* 1992;2(2):107-13.
[59] Pintor L, Torres X, Navarro V, et al. Corticotropin-releasing factor test in melancholic patients in depressed state versus recovery: A comparative study. *Progress in Neuro-Psychopharmacology and Biological Psychiatry.* 2007;31(5):1027-33.
[60] Van Den Eede F, Van den Bossche B, Hulstijn W, et al. Combined dexamethasone/CRF test in remitted outpatients with recurrent major depressive disorder. *Journal of Affective Disorders.* 2006;93(1-3):259-63.
[61] Carvalho LA, Juruena MF, Papadopoulos AS, et al. Clomipramine In Vitro Reduces Glucocorticoid Receptor Function in Healthy Subjects but not in Patients with Major Depression. *Neuropsychopharmacology.* 2008;33(13):3182-9.

[62] Dinan TG, Lavelle E, Cooney J, et al. Dexamethasone augmentation in treatment-resistant depression. *Acta Psychiatrica Scandinavica.* 1997;95(1):58-61.

[63] Bouwer C, Claassen J, Dinan TG, Nemeroff CB. Prednisone augmentation in treatment-resistant depression with fatigue and hypocortisolaemia: A case series. *Depression & Anxiety* (1091-4269). 2000;12(1):44-50.

[64] Belanoff JK, Flores BH, Kalezhan M, et al. Rapid reversal of psychotic depression using mifepristone. *Journal of Clinical Psychopharmacology.* 2001;21(5):516-21.

[65] Barden N. Modulation of glucocorticoid receptor gene expression by antidepressant drugs. *Pharmacopsychiatry.* 1996;29(1):12-22.

[66] Augustyn M, Otczyk M, Budziszewska B, et al. Effects of some new antidepressant drugs on the glucocorticoid receptor-mediated gene transcription in fibroblast cells. *Pharmacological Reports.* 2005;57(6):766-73.

[67] Budziszewska B, Basta-Kaim A, Kubera M, et al. Effect of lipopolysaccharide and antidepressant drugs on glucocorticoid receptor-mediated gene transcription. *Pharmacological Reports.* 2005;57(4):540-4.

[68] Budziszewska B, Jaworska-Feil L, Kajta M, Lason W. Antidepressant drugs inhibit glucocorticoid receptor-mediated gene transcription - a possible mechanism. *British Journal of Pharmacology.* 2000; 130(6):1385-93.

[69] Pariante CM, Pearce BD, Pisell TL, et al. P.1.155 Steroid-independent translocation of the glucocorticoid receptor by the antidepressant, desipramine. *European Neuropsychopharmacology.* 1997;7(Supplement 2):S184-S.

[70] Miller AH, Vogt GJ, Pearce BD. The phosphodiesterase type 4 inhibitor, rolipram, enhances glucocorticoid receptor function. *Neuropsychopharmacology.* 2002;27(6):939-48.

[71] Pariante CM, Hye A, Williamson R, et al. The antidepressant clomipramine regulates cortisol intracellular concentrations and glucocorticoid receptor expression in fibroblasts and rat primary neurones. *Neuropsychopharmacology.* 2003;28(9):1553-61.

[72] Pariante CM, Kim RB, Makoff A, Kerwin RW. Antidepressant fluoxetine enhances glucocorticoid receptor function in vitro by modulating membrane steroid transporters. *British Journal of Pharmacology.* 2003;139(6):1111-8.

[73] Lai M, McCormick JA, Chapman KE, et al. Differential regulation of corticosteroid receptors by monoamine neurotransmitters and antidepressant drugs in primary hippocampal culture. *Neuroscience.* 2003;118(4):975-84.

[74] Semont A, Fache MP, Ouafik L, et al. Effect of serotonin inhibition on glucocorticoid and mineralocorticoid expression in various brain structures. *Neuroendocrinology.* 1999;69(2):121-8.

[75] Joëls M, Karst H, DeRijk R, de Kloet ER. The coming out of the brain mineralocorticoid receptor. *Trends in Neurosciences.* 2008;31(1):1-7.

[76] de Kloet R, DeRijk RH, Meijer OC. Therapy Insight: is there an imbalanced response of mineralocorticoid and glucocorticoid receptors in depression? *Nature Clinical Practice Endocrinology & Metabolism.* 2007;3(2):168-79.

[77] Young EA, Lopez JF, Murphy-Weinberg V, et al. Mineralocorticoid receptor function in major depression. *Archives of General Psychiatry.* 2003;60(1):24-8.

[78] Juruena MF, Pariante CM, Papadopoulos AS, et al. Prednisolone suppression test, but not dexamethasone predicts outcome in treatment resistant depression. *European Neuropsychopharmacology.* 2010;20: S351-S2.

[79] Juruena MF, Cleare AJ, Papadopoulos AS, et al. Different responses to dexamethasone and prednisolone in the same depressed patients. *Psychopharmacology.* 2006;189(2):225-35.

[80] Juruena MF, Cleare AJ, Papadopoulos AS, et al. The prednisolone suppression test in depression: Dose-response and changes with antidepressant treatment. *Psychoneuroendocrinology.* 2010;35(10): 1486-91.

[81] Juruena MF, Pariante CM, Papadopoulos AS, et al. Prednisolone suppression test in depression: prospective study of the role of HPA axis dysfunction in treatment resistance. *British Journal of Psychiatry.* 2009;194(4):342-9.

[82] Otte C, Hinkelmann K, Moritz S, et al. Modulation of the mineralocorticoid receptor as add-on treatment in depression: A randomized, double-blind, placebo-controlled proof-of-concept study. *Journal of Psychiatric Research.* 2010;44(6):339-46.

[83] Heuser I, Deuschle M, Weber B, et al. Increased activity of the hypothalamus-pituitary-adrenal system after treatment with the mineralocorticoid receptor antagonist spironolactone. *Psychoneuroendocrinology.* 2000;25(5):513-8.

[84] DeRijk RH, Wust S, Meijer OC, et al. A common polymorphism in the mineralocorticoid receptor modulates stress responsiveness. Journal of Clinical Endocrinology & Metabolism. 2006;91(12):5083-9.

[85] Muhtz C, Zyriax B-C, Bondy B, et al. Association of a common mineralocorticoid receptor gene polymorphism with salivary cortisol in healthy adults. Psychoneuroendocrinology. 2011;36(2):298-301.

[86] Kuningas M, de Rijk RH, Westendorp RGJ, et al. Mental performance in old age dependent on cortisol and genetic variance in the mineralocorticoid and glucocorticoid receptors. *Neuropsychopharmacology.* 2007;32(6):1295-301.

[87] van Leeuwen N, Kumsta R, Entringer S, et al. Functional mineralocorticoid receptor (MR) gene variation influences the cortisol awakening response after dexamethasone. *Psychoneuroendocrinology.* 2010;35(3):339-49.

[88] Vreeburg SA, Hoogendijk WJG, van Pelt J, et al. Major Depressive Disorder and Hypothalamic-Pituitary-Adrenal Axis Activity Results From a Large Cohort Study. *Archives of General Psychiatry.* 2009;66(6):617-26.

[89] Klok MD, Vreeburg SA, Penninx BWJH, et al. Common functional mineralocorticoid receptor polymorphisms modulate the cortisol awakening response: Interaction with SSRIs. *Psychoneuroendocrinology.* 2011;36(4):484-94.

[90] Berger S, Wolfer DP, Selbach O, et al. Loss of the limbic mineralocorticoid receptor impairs behavioral plasticity. *Proceedings of the National Academy of Sciences of the United States of America.* 2006;103(1):195-200.

[91] Sanchez MM, Young LJ, Plotsky PM, Insel TR. Distribution of corticosteroid receptors in the rhesus brain: Relative absence of glucocorticoid receptors in the hippocampal formation. *Journal of Neuroscience.* 2000;20(12):4657-68.

[92] Rozeboom AM, Akil H, Seasholtz AF. Mineralocorticoid receptor overexpression in forebrain decreases anxiety-like behavior and alters the stress response in mice. *Proceedings of the National Academy of Sciences of the United States of America.* 2007;104(11):4688-93.

[93] Savitz J, Lucki I, Drevets WC. 5-HT1A receptor function in major depressive disorder. *Progress in Neurobiology.* 2009;88(1):17-31.

[94] Dekloet ER, Sybesma H, Reul H. Selective Control by Corticosterone of Serotonin-1 Receptor Capacity in Raphe-Hippocampal System. *Neuroendocrinology.* 1986;42(6):513-21.

[95] Chalmers DT, Kwak SP, Mansour A, et al. Corticosteroids Regulate Brain Hippocampal 5-HT1A Receptor Messenger-RNA Expression. *Journal of Neuroscience. 1993*;13(3):914-23.

[96] Meijer OC, Dekloet ER. Corticosterone Suppresses the Expression of 5-HT1A Receptor Messenger-RNA in Rat Dentate Gyrus. *European Journal of Pharmacology-Molecular Pharmacology Section.* 1994;266(3):255-61.

[97] Meijer OC, Dekloet ER. A Role for the Mineralocorticoid Receptor in a Rapid and Transient Suppression of Hippocampal 5-HT1A Receptor Messenger-RNA by Corticosterone. *Journal of Neuroendocrinology.* 1995;7(8):653-7.

[98] Kuroda Y, Watanabe Y, Albeck DS, et al. Effects of Adrenalectomy and Type-I Or Type-II Glucocorticoid Receptor Activation on 5-HT1A and 5-HT2 Receptor-Binding and 5-HT Transporter Messenger-RNA Expression in Rat-Brain. *Brain Research.* 1994;648(1):157-61.

[99] Welner SA, Demontigny C, Desroches J, et al. Autoradiographic Quantification of Serotonin1A Receptors In Rat-Brain Following Antidepressant Drug-Treatment. *Synapse.* 1989;4(4):347-52.

[100] Drevets WC, Thase ME, Moses-Kolko EL, et al. Serotonin-1A receptor imaging in recurrent depression: replication and literature review. *Nuclear Medicine and Biology.* 2007;34(7):865-77.

[101] Dekloet ER, Rots NY, Vandenberg D, Oitzl MS. Brain Mineralocorticoid Receptor Function. In: deKloet ER, Azmitia EC, Landfield PW, editors. Brain Corticosteroid Receptors - Studies on the Mechanism, Function, and Neurotoxicity of Corticosteroid Action. New York: New York Acad Sciences; 1994. p. 8-21.

[102] Joels M, Dekloet ER. Mineralocorticoid and Glucocorticoid Receptors in the Brain - Implications for Ion Permeability and Transmitter Systems. *Progress in Neurobiology.* 1994;43(1):1-36.

[103] van Eekelen JAM, Rots NY, Sutanto W, et al. Brain corticosteroid receptor gene expression and neuroendocrine dynamics during aging. *The Journal of Steroid Biochemistry and Molecular Biology.* 1991;40(4-6):679-83.

[104] Gesing A, Bilang-Bleuel A, Droste SK, et al. Psychological stress increases hippocampal mineralocorticoid receptor levels: Involvement of corticotropin-releasing hormone. *Journal of Neuroscience.* 2001;21(13):4822-9.

[105] Lai M, Bae SE, Bell JE, et al. Mineralocorticoid receptor mRNA expression is increased in human hippocampus following brief cerebral ischaemia. *Neuropathology and Applied Neurobiology.* 2009;35(2):156-64.

[106] Lopez JF, Chalmers DT, Little KY, Watson SJ. Regulation of serotonin(1A), glucocorticoid, and mineralocorticoid receptor in rat and human hippocampus: Implications for the neurobiology of depression. *Biological Psychiatry.* 1998;43(8):547-73.

[107] Wang SS, Kamphuis W, Huitinga I, et al. Gene expression analysis in the human hypothalamus in depression by laser microdissection and real-time PCR: the presence of multiple receptor imbalances. *Molecular Psychiatry.* 2008;13(8):786-99.

[108] Xing GQ, Russell S, Webster MJ, Post RM. Decreased expression of mineralocorticoid receptor mRNA in the prefrontal cortex in schizophrenia and bipolar disorder. *International Journal of Neuropsychopharmacology.* 2004;7(2):143-53.

[109] Webster MJ, Knable MB, O'Grady J, et al. Regional specificity of brain gluclocorticoid receptor mRNA alterations in subjects with schizophrenia and mood disorders. *Molecular Psychiatry.* 2002;7(9):985-94.

[110] Klok MD, Alt SR, Irurzun Lafitte AJM, et al. Decreased expression of mineralocorticoid receptor mRNA and its splice variants in postmortem brain regions of patients with major depressive disorder. *Journal of Psychiatric Research.* In Press, Corrected Proof.

[111] Herman JP, Schafer MKH, Young EA, et al. Evidence for Hippocampal Regulation of Neuro-Endocrine Neurons of the Hypothalamo Pituitary Adrenocortical Axis. *Journal of Neuroscience.* 1989;9(9):3072-82.

[112] Jacobson L, Sapolsky R. The Role of the Hippocampus in Feedback-Regulation of the Hypothalamic-Pituitary-Adrenocortical Axis. *Endocrine Reviews.* 1991;12(2):118-34.

[113] Gesing A, Bilang-Bleuel A, Linthorst ACE, Reul JMHM. Corticotropin-releasing hormone-mediated rise in rat brain mineralocorticoid receptor density following acute stress: Functional significance. *Society for Neuroscience Abstracts.* 1999;25(1-2):707.

[114] Sutanto W, Rosenfeld P, de Kloet ER, Levine S. Long-term effects of neonatal maternal deprivation and ACTH on hippocampal mineralocorticoid and glucocorticoid receptors. *Brain Res Dev Brain Res.* 1996;92(2):156-63.

[115] Vallee M, Mayo W, Dellu F, et al. Prenatal stress induces high anxiety and postnatal handling induces low anxiety in adult offspring: Correlation with stress-induced corticosterone secretion. *Journal of Neuroscience.* 1997;17(7):2626-36.

[116] Holsboer F. The rationale for corticotropin-releasing hormone receptor (CRH-R) antagonists to treat depression and anxiety. *Journal of Psychiatric Research.* 1999;33(3):181-214.

[117] Brady LS, Whitfield HJ, Fox RJ, et al. Long-Term Antidepressant Administration Alters Corticotropin-Releasing Hormone, Tyrosine-Hydroxylase, and Mineralocorticoid Receptor Gene-Expression in Rat-Brain - Therapeutic Implications. *Journal of Clinical Investigation.* 1991;87(3):831-7.

[118] Bachmann CG, Linthorst ACE, Holsboer F, Reul J. Effect of chronic administration of selective glucocorticoid receptor antagonists on the rat hypothalamic-pituitary-adrenocortical axis. *Neuropsychopharmacology.* 2003;28(6):1056-67.

[119] Seckl JR, Fink G. Antidepressants Increase Glucocorticoid and Mineralocorticoid Receptor Messenger-RNA Expression In Rat Hippocampus Invivo. *Neuroendocrinology.* 1992;55(6):621-6.

[120] Jahn H, Schick M, Kiefer F, et al. Metyrapone as additive treatment in major depression - A double-blind and placebo-controlled trial. *Archives of General Psychiatry.* 2004;61(12):1235-44.

[121] Lai M, Horsburgh K, Bae SE, et al. Forebrain mineralocorticoid receptor overexpression enhances memory, reduces anxiety and attenuates neuronal loss in cerebral ischaemia. *European Journal of Neuroscience.* 2007;25(6):1832-42.

[122] Katz ER, Stowe ZN, Newport DJ, Kelley ME, Pace TW, Cubells JF, et al (2012). Regulation of mRNA expression encoding chaperone and co-chaperone proteins of the

glucocorticoid receptor in peripheral blood: association with depressive symptoms during pregnancy. *Psychol Med* 42(5): 943-956.

[123] Menke A, Arloth J, Putz B, Weber P, Klengel T, Mehta D, et al (2012). Dexamethasone Stimulated Gene Expression in Peripheral Blood is a Sensitive Marker for Glucocorticoid Receptor Resistance in Depressed Patients. *Neuropsychopharmacology* 37(6): 1455-1464.

[124] Menke A, Klengel T, Rubel J, Bruckl T, Pfister H, Lucae S, et al (2013). Genetic variation in FKBP5 associated with the extent of stress hormone dysregulation in major depression. *Genes Brain Behav.* 12(3):289-96.

[125] Heydendael W, Jacobson L (2008) Differential effects of imipramine and phenelzine on corticosteroid receptor gene expression in mouse brain:potential relevance to antidepressant response. *Brain Res.* 1238:93-107.

In: Horizons in Neuroscience Research. Volume 15
Editors: Andres Costa and Eugenio Villalba
ISBN: 978-1-63321-523-8

Chapter 8

Betaine in the Brain

Nafisa M. Jadavji**, PhD
Department of Experimental Neurology; Center for Stroke Research Berlin, Charité Medical University, Berlin; German Center for Neurodegenerative Diseases, Partner Site Berlin, Germany

Abstract

Elevated levels of plasma homocysteine have been negatively linked to the development of cardiovascular disease, impaired neurological function and most recently to neurodegenerative diseases. Homocystinuria, an in born error of metabolism results in significantly high levels of plasma homocysteine. Choline, along with its' metabolite betaine can lower levels of plasma homocysteine by remethylating it back to methionine with the aid of the enzyme betaine homocysteine *S*-methyltransferase (BHMT). Folates can also methylate homocysteine to methionine. Mice with genetic deficiencies in methylenetratrahydrofolate reductase (MTHFR) and methionine synthase reductase (MTRR) enzymes involved in folate and homocysteine metabolism have reduced levels of betaine in brain tissue. Interestingly, when patients suffering from homocystinuria are treated with betaine there is a reversal of most negative symptoms including reducing plasma homocysteine levels, however neurological symptoms still remain. This chapter will discuss interactions of betaine with folate metabolism, implications of reduced betaine levels in brain function and importance of this nutrient in normal and disease brain function.

Introduction

Homocysteine is a non-protein amino acid and is generated during the metabolism of methionine. Elevated levels of plasma homocysteine (>15 μM) have been linked to increased risk of developing cardiovascular disease [1] and have been associated with impaired neurological function and most recently neurodegenerative diseases [2], specifically mild

* E-mail: nafisa.jadavji@charite.de.

cognitive impairment [3], dementia and Alzheimer's disease [4], however, this remains link unclear [5,6]. Plasma homocysteine levels can be influenced by age, gender, menopausal status and physiological determinants [7]. It is interesting to note that homocysteine levels rise as we age [7]. These data suggest that choline, betaine, folates and homocysteine play an important role in normal brain function.

Homocysteine Metabolism

Homocysteine can be metabolized by three pathways highlighted in Figure 1. Choline along with its metabolite betaine along with enzymes choline dehydrogenase (CHDH) and betaine homocysteine *S*-methyltransferase (BHMT) can remethylate homocysteine back to methionine [8]. Folic acid can be reduced by methylenetetrahydrofolate reductase (MTHFR) to generate 5-methyltetrahydrofolate (5-mTHF) and this can also remethylate homocysteine back to methionine, Methionine synthase reductase (MTRR) is an enzyme involved in folate metabolism, along with vitamin B12 it activates methionine synthase (MTR) which provides catalyzes the methylation reaction of homocysteine to methionine. Both choline and folate pathways are linked to methylation through homocysteine metabolism. Specifically, when methioinine is activated it is converted into *S*-adenosylmethioine (SAM), which is a global methyl donor [9]. The third pathway is transulphuration, which is irreversible and uses the enzyme cystathionine β-synthase (CBS) and generating cysteine [10].

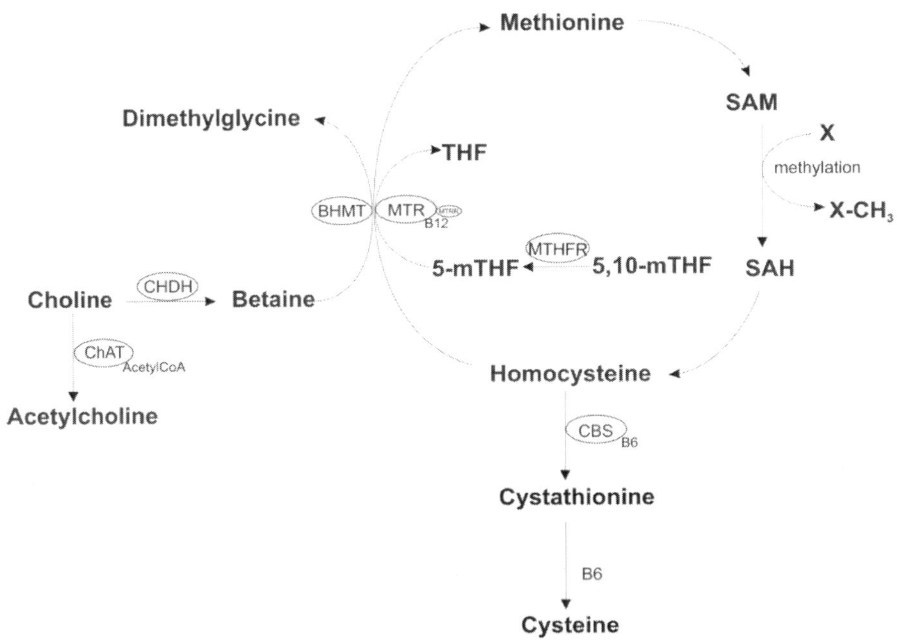

Figure 1. Homoysteine Metabolic Pathway. Abbreviations: Acetyl-CoA – Acetyl coenzyme A; BHMT – betaine homocysteine methyltransferase; B12 – vitamin B12; B6 – vitamin B6; CBS – ChAT – choline acetyltransferase; CHDH – choline dehydrogenase; MTHFR – methylenetetrahydrofolate reductase; MTHR – methyltetrahydrofolate; mTHF– methyltetrahydrofolate; MTR – methionine synthase; MTRR – methionine synthase reductase; SAM – S-adenosylmethionine; SAH S-adenosylhomocysteine.

Homocystinuria

Homocystinuria is an born error of metabolism which results in significantly elevated levels of plasma homocysteine >50μM [11–13]. Genetic deficiencies in enzymes involved in folate metabolism, such as MTHFR and MTRR can lead to homocystinuria [14]. Patients present with developmental delays, vascular disease, gait abnormalities, and seizures [15]. When these patients are treated with betaine, as well as other B-vitamins (e.g. folates) they show some improvements of symptoms [12,13]. The treatment of homocystinuria with betaine will be discussed later in the chapter.

Choline and Betaine

Choline and betaine can be obtained from our diet. Eggs, beef, liver, soya bean are good sources of choline whereas wheat, spinach and shellfish are good sources of betaine [16]. Choline can also be synthesized *de novo* from tissue, specifically from phosphatidlycholine which is a phospholipid. Betaine is oxidized by choline dehydrogenase (CHDH) from choline (Figure 1) this occurs primarily in the liver and kidney [16]. Betaine is a methyl donor for the homocysteine to methinione reaction and is therefore linked to folate metabolism [17]. Interestingly, during times of choline deficiency there is a decrease in betaine levels and more 5-mTHF is used for homocysteine remethylation [16].

Choline and Betaine in the Brain

Choline is transported through the blood brain barrier by a specific carrier. Transport of choline into the brain is highest during development [16]. Choline plays a significant role in brain development and normal brain function during adulthood [18]. Supplementation of choline during pregnancy enhances hippocampal neurogenesis, as well as the size of cholinergic neurons and Ach storage [16]. Choline is precursor of acetylcholine (Ach), a neurotransmitter in the central and peripheral nervous system (Figure 1) and plays a role in learning and memory [19]. In addition, choline is involved in lipid metabolism in the brain [16]. Interestingly, there are very low levels of BHMT in the brain [20], so betaine cannot metabolize homocysteine in the brain [8]. Furthermore, betaine levels in the brain under normal conditions are low (~0.1U/mg) in procine samples [20], ~1 mmol/kg H_20 in rats [21] and 15nmol/g in mice [22]. Interestingly, the elderly and patients with Alzheimer's disease have reduced levels free choline in the brain [16].

Reduced levels of Ach have been observed in mice with genetic deficiencies in folate metabolism [22,23], specifically deficiencies in MTHFR and MTRR. These mice also have impairments in memory function [22, 23]. Severe MTHFR deficient mice, an animal model for homocystinuira, have significantly reduced levels of betaine along with elevated levels of glycerophosphocholine (GPC) [22]. Elevated levels of glycerophosphocholine (GPC) may indicate a breakdown of phospholipids in order to increase levels of choline for acetylcholine synthesis [22]. MTRR knockout mice are embryonic lethal, so a gene trap mouse model was created [24]. *Mtrr*$^{gt/gt}$ mice have significantly elevated levels of betaine in plasma and reduced levels of betaine in the brain, specifically in the cerebellum and hippocampus [25]. The decrease in betaine levels in both the MTHFR and MTRR deficient mice maybe a result of betaine being converted back to choline or a preference in choline reuiqredfor Ach synthesis, this can be supported by the increase in GPC [22].

Treatment with Betaine

Animal studies show that maternal betaine supplementation prior to pregnancy until weaning showed a reduction in homocysteine concentration of *Mthfr*$^{-/-}$ offspring from 2.66nmol/mg to 1.99nmol/mg in brain tissue and 1.907 to 1.255nmol/mg in liver tissue. Betaine supplemented *Mthfr*$^{-/-}$ mice also had increased body and brain weight in comparison non-supplemented mice [26]. While there was some relief of the negative symptoms involved in a severe MTHFR deficiency, there is not a total amelioration. Morphological changes within the cerebellum and hippocampus were not completely restored in betaine supplemented *Mthfr*$^{-/-}$ mice.

In a study investigating the role of one-carbon metabolism associated with cognitive impairment, betaine was associated with better performance of construction, sensiomotor speed and executive function in the Dutch elderly individuals. Furthermore patients that were treated with a combination of vitamin B12 and folic acid showed the greatest improvement in memory function [27]. Betaine has been used to treat patients with homocysintuira, this is usually beneficial, however, varies from patient to patient. Overall, betaine therapy does not normalize plasma homocysteine [28,29], although some studies have shown a reduction in plasma homocysteine levels after treatment with betaine [30,31]. Treatment with betaine in homocystinuric patients has shown to result in biochemical improvements [28,31], normal brain structure and myelination in MRI scans [30,31] and developmental progress as well as decrease in mortality rate [32], however, neurological symptoms do not always disappear [33].

Betaine, as treatment for homocystinuria is limited in its effects. A combination of choline, betaine, B-vitamins, as well as SAM may be more benefical for effective treatment of homocystinuira as well as for individuals with elevated levels of homocysteine. Furthermore, increasing methionine uptake into the brain may help alleviate the negative symptoms associated with elevated levels of homocysteine.

References

[1] Blom, H. J. and Smulders, Y. (2011) Overview of homocysteine and folate metabolism. With special references to cardiovascular disease and neural tube defects. *J. Inherit. Metab. Dis*. 34, 75–81.

[2] Herrmann, W. and Obeid, R. (2011) Homocysteine: a biomarker in neurodegenerative diseases. *Clin. Chem. Lab. Med.* 49, 435–41.

[3] De Jager, C. a, Oulhaj, A., Jacoby, R., Refsum, H. and Smith, a D. (2012) Cognitive and clinical outcomes of homocysteine-lowering B-vitamin treatment in mild cognitive impairment: a randomized controlled trial. *Int. J. Geriatr. Psychiatry* 27, 592–600.

[4] Ravaglia, G., Forti, P., Maioli, F., Martelli, M., Servadei, L., Brunetti, N., Porcellini, E. and Licastro, F. (2005) Homocysteine and folate as risk factors for dementia and Alzheimer disease. *Am. J. Clin. Nutr.* 82, 636–43.

[5] West, R. K., Beeri, M. S., Schmeidler, J., Mitchell, D. B., Carlisle, K. R., Angelo, G., Mavris, R., Langhoff, E., Rosendorff, C. and Silverman, J. M. (2011) Homocysteine and cognitive function in very elderly nondemented subjects. *Am. J. Geriatr. Psychiatry* 19, 673–7.

[6] Ho, R. C. M., Cheung, M. W. L., Fu, E., Win, H. H., Zaw, M. H., Ng, A. and Mak, A. (2011) Is high homocysteine level a risk factor for cognitive decline in elderly? A systematic review, meta-analysis, and meta-regression. *Am. J. Geriatr. Psychiatry* 19, 607–17.

[7] Bolander-Gouaille, C. and Bottiglieri, T. (2007) *Related Vitamins and Neuropsychiatric Disorders Vitamins, Related Disorders, Neuropsychiatric Second.,* Springer-Verlag France, Berlin.

[8] Caudill, M. A. (2009) Folate and choline interrelationships: metabolic and potential health implications. In *Folate in Health and Disease* (Bailey, L. B., ed.) 2nd Editio., pp 449–465, CRC Press, Boca Raton.

[9] Chiang, P., Gordon, R., Tal, J., Zeng, G., Doctor, B., Pardhasaradhi, K. and McCann, P. (1996) *S-Adenosylmethionine and methylation.* FASEB J. 10, 471–480.

[10] Orendác, M., Zeman, J., Stabler, S. P., Allen, R. H., Kraus, J. P., Bodamer, O., Stöckler-Ipsiroglu, S., Kvasnicka, J. and Kozich, V. (2003) Homocystinuria due to cystathionine beta-synthase deficiency: novel biochemical findings and treatment efficacy. *J. Inherit. Metab.* Dis. 26, 761–73.

[11] Vilaseca, M. a, Vilarinho, L., Zavadakova, P., Vela, E., Cleto, E., Pineda, M., Coimbra, E., Suormala, T., Fowler, B. and Kozich, V. (2003) CblE type of homocystinuria: mild clinical phenotype in two patients homozygous for a novel mutation in the MTRR gene. *J. Inherit. Metab. Dis.* 26, 361–9.

[12] Zavadakova, P., Fowler, B., Zeman, J., Suormala, T., Pristoupilová, K., Kozich, V. and Zavad'áková, P. (2002) CblE type of homocystinuria due to methionine synthase reductase deficiency: clinical and molecular studies and prenatal diagnosis in two families. *J. Inherit. Metab. Dis.* 25, 461–76.

[13] Zavadáková, P., Fowler, B., Suormala, T., Novotna, Z., Mueller, P., Hennermann, J. B., Zeman, J., Vilaseca, M. A., Vilarinho, L., Gutsche, S., et al. (2005) cblE type of homocystinuria due to methionine synthase reductase deficiency: functional correction by minigene expression. *Hum. Mutat.* 25, 239–47.

[14] Watkins, D. and Rosenblatt, D. S. (2011) PART 17 : VITAMINS Inherited Disorders of Folate and Cobalamin Transport and Metabolism. In *Metabolic and Molecular Basis of Inherited Disease* (Valle, D., Beaudet, A. L., Vogelstein, D., Kinzler, K. W., Antonarakis, S. E., and Ballabio, A., eds.), McGraw-Hill, New York.

[15] Schoonderwaldt, H. C., Boers, G. H., Cruysberg, J. R., Schulte, B. P., Slooff, J. L. and Thijssen, H. O. (1981) Neurologic manifestations of homocystinuria. *Clin. Neurol. Neurosurg.* 83, 153–62.

[16] Ueland, P. M. (2011) Choline and betaine in health and disease. *J. Inherit. Metab. Dis.* 34, 3–15.

[17] Ueland, P. M., Holm, P. I. and Hustad, S. (2005) Betaine: a key modulator of one-carbon metabolism and homocysteine status. *Clin. Chem. Lab. Med.* 43, 1069–75.

[18] Zeisel, S. (2004) Nutritional importance of choline for brain development. *J. Am. Coll. Nutr.* 23, 621–626.

[19] Gandhi, C. C., Kelly1, R. M., Wiley, R. G. and Walsh, T. J. (2000) Impaired acquisition of a Morris water maze task following selective destruction of cerebellar purkinje cells with OX7-saporin. Behav. *Brain Res.* 109, 37–47.

[20] Sunden, S. L. F., Renduchintala, M. S., Park, E. I., Miklasz, S. D. and Garrow, T. A. (1997) *Betaine – Homocysteine Methyltransferase Expression in Porcine and Human Tissues and Chromosomal Localization of the Human Gene* 1 345, 171–174.

[21] Lien, Y. H. (1995) Role of organic osmolytes in myelinolysis. A topographic study in rats after rapid correction of hyponatremia. *J. Clin. Invest.* 95, 1579–86.

[22] Jadavji, N. M., Deng, L., Leclerc, D., Malysheva, O., Bedell, B. J., Caudill, M. a and Rozen, R. (2012) Severe methylenetetrahydrofolate reductase deficiency in mice results in behavioral anomalies with morphological and biochemical changes in hippocampus. *Mol. Genet. Metab.* 106, 149–59.

[23] Chan, a, Tchantchou, F., Graves, V., Rozen, R. and Shea, T. B. (2008) Dietary and genetic compromise in folate availability reduces acetylcholine, cognitive performance and increases aggression: critical role of S-adenosyl methionine. *J. Nutr. Health Aging* 12, 252–61.

[24] Elmore, C. L., Wu, X., Leclerc, D., Watson, E. D., Bottiglieri, T., Krupenko, N. I., Krupenko, S. A., Cross, J. C., Rozen, R., Gravel, R. A., et al. (2007) Metabolic derangement of methionine and folate metabolism in mice deficient in methionine synthase reductase. *Mol. Genet. Metab.* 91, 85–97.

[25] Jadavji, N. M., Bahous, R. H., Deng, L., Malysheva, O. V, Grand'Maison, M., Bedell, B. J., Caudill, M. A. and Rozen, R. (2014) Mouse model for deficiency of methionine synthase reductase exhibits short-term memory impairment and disturbances in brain choline metabolism. *Biochem. J.*

[26] Schwahn, B. C., Laryea, M. D., Chen, Z., Melnyk, S., Pogribny, I., Garrow, T., James, S. J. and Rozen, R. (2004) Betaine rescue of an animal model with methylenetetrahydrofolate reductase deficiency. *Biochem. J.* 382, 831–40.

[27] Eussen, S. J. P. M., Ueland, P. M., Clarke, R., Blom, H. J., Hoefnagels, W. H. L., van Staveren, W. a and de Groot, L. C. P. G. M. (2007) The association of betaine, homocysteine and related metabolites with cognitive function in Dutch elderly people. *Br. J. Nutr.* 98, 960–8.

[28] Strauss, K. a, Morton, D. H., Puffenberger, E. G., Hendrickson, C., Robinson, D. L., Wagner, C., Stabler, S. P., Allen, R. H., Chwatko, G., Jakubowski, H., et al. (2007) Prevention of brain disease from severe 5,10-methylenetetrahydrofolate reductase deficiency. *Mol. Genet. Metab.* 91, 165–75.

[29] Holme, E., Kjellman, B. and Ronge, E. (1989) Betaine for treatment of homocystinuria caused by methylenetetrahydrofolate reductase deficiency. *Arch. Dis. Child.* 64, 1061–4.

[30] Engelbrecht, V., Rassek, M., Huismann, J. and Wendel, U. (1997) MR and proton MR spectroscopy of the brain in hyperhomocysteinemia caused by methylenetetrahydrofolate reductase deficiency. *AJNR. Am. J. Neuroradiol.* 18, 536–9.

[31] Al Tawari, A. a, Ramadan, D. G., Neubauer, D., Heberle, L. C. and Al Awadi, F. (2002) An early onset form of methylenetetrahydrofolate reductase deficiency: a report of a family from Kuwait. *Brain Dev.* 24, 304–9.

[32] Fattal-Valevski, a., Bassan, H., Korman, S. H., Lerman-Sagie, T., Gutman, a. and Harel, S. (2000) Methylenetetrahydrofolate Reductase Deficiency: Importance of Early Diagnosis. *J. Child Neurol.* 15, 539–543.
[33] Wendel, U. and Bremer, H. J. (1984) Betaine in the treatment of homocysfinuria reductase deficiency *. *Eur J Pediatr* 142, 147–150.

In: Horizons in Neuroscience Research. Volume 15
Editors: Andres Costa and Eugenio Villalba
ISBN: 978-1-63321-523-8

Chapter 9

Clinical Outcome Prediction Using Bayesian Regression and Artificial Neural Networks

***Benjamin W. Y. Lo*[1*], *Yusuke Nishimura*[2] *and Hitoshi Fukuda*[3]**
[1]Divisions of Neurosurgery & Critical Care Medicine,
Department of Clinical Epidemiology and Biostatistics,
University of Toronto & McMaster University, Ontario, Canada
[2]Department of Neurosurgery, Nagoya University, Nagoya, Japan
[3]Department of Neurosurgery, Kurashiki Central Hospital, Kurashiki, Okayama, Japan

Abstract

The goal of this chapter is to use a combination of traditional and Bayesian logistic regression, as well as artificial neural networks for clinical outcome prediction. Using the example of ruptured brain aneurysms, this chapter provides novel insights on specific brain-body interactions that are essential in influencing outcome in patients with ruptured brain aneurysms. Thus far, there is scarce existing literature on such important relationships. The multicenter Tirilazad database (3550 patients) was used to create this clinical outcome prediction model in order to elucidate significant brain-body interactions. Traditional logistic regression main effects model included five statistically significant single prognostic variables, namely, neurological grade, age, diastolic blood pressure, time to surgery and stroke. Logistic regression models found the significance of liver disease and hypertension in development of brain swelling, and the negative consequences of seizures in those with history of myocardial infarction and post admission fever worsening neurological outcome. Bayesian logistic regression using informative prior probability likelihoods detected a number of prognostic factors that may play a role in multiplicative interaction effects, as well as possible heterogeneity in patient subgroups. An artificial neural networks model, with similar model discrimination values as traditional logistic regression model, was created to explore the influence of non-linear relationships on the importance of prognostic factors.

[*] Corresponding author: Email: benjamin.lo@alumni.utoronto.ca.

Part One – Introduction

Ruptured Brain Aneurysms and Aneurysmal Subarachnoid Hemorrhage

Patients with ruptured brain aneurysms and associated subarachnoid hemorrhage have a mortality rate of at least 45% in the first month after rupture [1-7]. Neurological damage can be from the primary injury of the aneurysm rupture itself and a number of secondary injurious processes that can further worsen the affected individual's neurological state. These secondary processes can be related to the nervous system, such as rebleeding from the ruptured aneurysm, brain swelling, occurrence of a delayed second stroke, brain blood vessel vasospasm leading to strokes, seizures and overaccumulation of brain-spinal fluid causing hydrocephalus [1-7]. In addition, other body organ systems can be affected, such as myocardial infarction, fever and overaccumulation of fluid in the lungs causing pulmonary edema. Together, these processes can lead to long term disability. Types of disability include physical, neurocognitive and psychological impairment.

The Tirilazad Database

Tirilazad was a 21-aminosteroid compound produced by Pharmacia and Upjohn, Kalamazoo, Michigan. It was originally investigated as a free radical scavenger for the potential treatment of cerebral vasospasm [8-12]. Between 1991 and 1997, 3550 patients from 162 centers in 21 countries in Europe, Australia, North America and Africa participated in five randomized, double-blind, placebo-controlled trials of Tirilazad [8-12].

Included patients were adults, with evidence of aneurysmal subarachnoid hemorrhage. The patients received either placebo or active agent (in intravenous doses of 0.6, 2, 6 or 15 mg/kg/day depending on the trial) from the third to tenth day after subarachnoid hemorrhage onset [8-12]. Excluded patients were: (1) those with traumatic subarachnoid hemorrhage, (2) infectious mycotic aneurysm, (3) severe underlying medical illness including serious cardiovascular disease, such as myocardial infarction within 6 months, uncontrolled hypertension, serious cardiac arrhythmia or congestive heart failure, (4) pregnant or lactating patients, (5) patients taking corticosteroids or calcium channel blockers other than nimodipine, (6) known intolerance to calcium channel blockers, and (7) patients whose aneurysms were treated with Gulielmi or other detachable coils [8-12].

Patients in the placebo and treatment arms were treated similarly. Over 85% of patients underwent surgical clipping (rather than coiling), and half of them were operated within the first 48 hours. Baseline demographics in both arms were balanced in terms of sex, age, number of pre-existing medical conditions, mean time to treatment, mean admission systolic blood pressure, admission neurological grade, ruptured aneurysm location, admission amount of blood. Potential confounders were accounted for in statistical analysis. Percentages of patients experiencing neurological and systemic disabilities were similar in different treatment groups [8-12].

Only one percent of patients was lost to follow up. The primary outcome measures were the patient's Glasgow Outcome Score at 3 months after aneurysm rupture. Glasgow Outcome Score is a 5-point scale, defined as: 5 – good recovery with normal life activities despite minor deficits, 4 – moderate disability being disabled but independent, 3 – severe disability being conscious but dependent, 2 – persistent vegetative state, and 1 – death [13].

Two systematic reviews [14,15] on the five trials of Tirilazad found no substantial heterogeneity among these trials and no significant differences in the: (1) number of patients who died between treatment and placebo arms, (2) number of adverse events between the two arms, and (3) outcomes among patients who were administered different dosages of Tirilazad.

Overview of Chapter

Part Two demonstrates creation of a clinical outcome prediction model using traditional logistic regression. In our clinical example of aneurysmal subarachnoid hemorrhage, it provides novel insight on clinical prediction factors, as well as significant brain-body interactions. Multivariable binary logistic regression model will be used. Novel significant brain-body axes will be noted. They play key roles in the clinical deterioration of patients with ruptured brain aneurysms. Based on the authors' clinical expertise in neurosurgery and critical care medicine, these epidemiologic findings will be correlated to underlying pathophysiologic mechanisms.

Part Three demonstrates how Bayesian analysis and artificial neural networks can be used in a complementary way to explore findings from traditional logistic regression. Bayesian regression makes use of informative prior probability likelihoods based on the authors' shared clinical experience. The entire patient population as well as subgroups will be investigated to note any differences in odds ratios of individual prognostic variables among these groups. Furthermore, a novel artificial neural networks model will be designed to examine whether its results will complement or differ from those generated by traditional logistic regression, with consideration of possible non-linear interactions.

Part Four summarizes chapter findings and discusses future directions.

Part Two – Clinical Outcome Prediction Using Traditional Logistic Regression

Objectives

In this part, we aim to create a clinical outcome prediction model using traditional binary logistic regression. In our clinical example of aneurysmal subarachnoid hemorrhage, we aim to investigate the potential factors for outcome prediction in patients with ruptured brain aneurysms.

Methods

Using the Tirilazad database, the identified independent variables included:

1. patient's neurological grade at hospital admission,

2. age (in years),
3. sex,
4. patient's weight (kg),
5. aneurysm size (mm),
6. presence of vasospasm at the time of hospital admission as noted on the admission angiogram (admission vasospasm),
7. presence of hemorrhage within the brain's ventricles (intraventricular hemorrhage),
8. presence of blood clot within the brain at the time of hospital admission (intracerebral hematoma),
9. location of aneurysm (anterior or posterior circulation in the brain),
10. thickness of subarachnoid hemorrhage (thin as 1 mm or less of blood clot, thick as more than 1 mm of blood clot),
11. systolic blood pressure on admission (mmHg),
12. diastolic blood pressure on admission (mmHg),
13. temperature on admission (degree Celsius),
14. occurrence of fever one week after hospital admission,
15. presence of prior episode of subarachnoid hemorrhage,
16. history of migraines,
17. history of hypertension,
18. history of angina,
19. history of myocardial infarction,
20. history of diabetes mellitus,
21. history of hepatic disease,
22. history of thyroid disease,
23. time to surgical treatment (hours),
24. treatment arm (investigational agent Tirilazad versus placebo),
25. development of hydrocephalus,
26. development of brain swelling (cerebral edema),
27. occurrence of stroke post admission,
28. development of vasospasm during course of treatment,
29. development of severe vasospasm requiring balloon angioplasty to dilate constricted vessels,
30. development of seizures requiring use of anti-seizure medication, and
31. development of fluid accumulation in the lungs (pulmonary edema) after admission.

The dependent variable used for analysis is the dichotomized Glasgow Outcome Score at three months. Good outcome represents functional independence (GOS 5 and GOS 4). Poor outcome represents functional dependence (GOS 3), persistent vegetative state (GOS 2) or death (GOS 1).

Using IBM SPSS Version 19.0™, independent variables were entered into univariate models in which chi square analyses were used to investigate for associations with poor neurologic outcome. Two-way interaction terms that included brain and body (systemic) prognostic factors were examined. Variables that reached a probability of 0.10 were then entered into a multivariable binary logistic regression model. Significance of this model was set at a probability of 0.05. In both our univariate and multivariable binary logistic models, there were at least 10 to 15 subjects per independent variable. In the multivariable logistic regression model, both independent risk factors as well as two-way interaction terms were investigated for their cumulative predictor effects for poor neurologic outcome. In addition, the following diagnostics were performed:

1. correlation matrix to rule out multicollinearity among predictor variables,
2. Hosmer and Lemeshow test to ensure good data fit to model and model calibration,
3. c statistic and classification to examine model discrimination, and
4. split sample analysis (70% training and 30% testing) to examine model generalizability [16].

Results

Correlation matrix of the included 31 independent variables was examined to ensure that there were no correlated variables. Single prognostic variables found to be statistically significant in univariate analyses included the following: (1) neurological grade, (2) age, (3) time to surgical treatment, (4) presence of fever one week post admission, (5) aneurysm size, (6) angiographic vasospasm on admission, (7) presence of intraventricular hemorrhage, (8) development of cerebral edema, (9) presence of intracerebral hematoma, (10) history of hypertension, (11) history of myocardial infarction, (12) development of stroke post admission, (13) history of previous episode of subarachnoid hemorrhage, (14) location of aneurysm, (15) development of vasospasm, and (16) development of seizures requiring use of anti-epileptic medications.

Interaction terms found to be statistically significant in univariate analyses included the following: (1) fever by aneurysm size, (2) fever by intraventricular hemorrhage, (3) fever by seizures, (4) intraventricular hemorrhage by aneurysm size, (5) hepatic disease by brain edema, (6) hypertension by brain edema, (7) hepatic disease by hydrocephalus, (8) hydrocephalus by aneurysm size, (9) myocardial infarction by blood clot thickness, (10) seizures by myocardial infarction, and (11) seizures by aneurysm size.

The above significant single prognostic variables and interaction terms were entered into a multivariable binary logistic regression model. The final model comprised 5 significant single prognostic variable terms and 4 significant interaction terms.

Statistically significant terms in the final model are shown below.

The Omnibus Tests of Model Coefficients indicated that this model was statistically significant (chi square = 887.9, $p < 0.001$). This model showed a large significant reduction in -2 log likelihood (-2LL from 2240 to 1699, $p < 0.001$). Nagelkerke R Square statistic indicated that approximately 47% of the variance in poor neurologic outcome could be predicted from the combination of the significant terms. In terms of model discrimination (Figure 1), the c statistic was 0.87 (95% confidence interval of 0.86-0.89).

Classification Table indicated that 84% of subject outcomes were predicted correctly. In terms of model calibration, the Hosmer and Lemeshow test showed a chi square of 11.38, $p = 0.181$, suggesting that the model appeared to fit the data reasonably well. Generalizability of model findings was supported by split sample validation (70% training, 30% testing), whereby the following criteria were met: (1) no multicollinearity (standard errors less than 2 for significant terms), (2) classification accuracy (84.5%) was greater than 1.25 times chance accuracy (82.5%), (3) individual relationship Wald statistic significance less than alpha level of 0.05, and (4) classification accuracy rate for training sample was 84.5%, whereas classification accuracy rate for testing sample was 82.5%, both satisfying minimum requirement for holdout sample (0.9*84.5%=76%).

Variable	β coefficient	Standard Error	Odds Ratio ($e^{β}$)	95% Confidence Interval	Significance (p value)
Neurological Grade	0.72	0.06	2.06	1.83-2.32	<0.001
Age (per year) [For every Δ5 years]	0.05 [0.25]	0.005	1.06 [1.28]	1.05-1.07 [1.22-1.42]	<0.001
Diastolic Blood Pressure (mmHg) [For Δ5 mmHg]	0.01 [0.05]	0.004	1.01 1.05	1.00-1.02 [1.05-1.10]	0.02
Time to Surgery (hour)	0.01	0.004	1.01	1.00-1.02	0.02
Stroke	1.39	0.33	4.03	2.11-7.69	<0.001
Seizures by Fever on Day 8	0.87	0.13	2.39	1.86-3.06	<0.001
Brain Edema by Hepatic Disease	1.70	0.80	5.47	1.13-26.46	0.03
Brain Edema by Hypertension	0.98	0.26	2.66	1.59-4.45	<0.001
Seizures by Myocardial Infarction	1.12	0.41	3.05	1.35-6.87	0.007

Interpretation of Results

There has been scarce epidemiologic literature to attempt to characterize complex brain-body interactions in those with ruptured brain aneurysms. Our analyses included both non-treatment and treatment related prognostic factors, as well as significant two-way interactions that captured clinically relevant brain-body associations in aneurysmal subarachnoid hemorrhage. Traditional significant prognostic variables including stroke, neurological grade and age were confirmed through the main effects logistic regression model.

In patients with ruptured brain aneurysms, excessive activation of the sympathetic nervous system as well as systemic inflammatory response can lead to deleterious effects on the internal regulatory mechanisms of multiple organ systems [17-22]. These body systems include the cardiovascular, pulmonary, temperature regulation, immune and neuro-gut axes. Overstimulation of the insular cortex and amygdala of the dominant brain can trigger both seizures and cardiac arrhythmias. This epidemiologic study confirms the negative effects of the combination of seizures and history of cardiovascular disease (in the form myocardial injury) on neurologic outcome (OR 3.05, 95% CI 1.35-6.87, p=0.007).

Excessive activation of the sympathetic nervous system also leads to a surge of catecholamine release. This catecholamine surge acts on the postganglionic fibers to the heart and blood vessels. In turn, autonomic control over blood pressure regulation and vascular resistance is lost [1-7,17-22]. Disrupted blood brain barrier is the end result as fluid and proteins leak into worsening areas of brain edema [1-7,17-22]. Epidemiologically, our database captured the interdependent brain-body association of the development of brain edema in those with hypertension. Specifically, in those with poorly controlled blood pressure who develop brain swelling as a result of ruptured brain aneurysms, the odds ratio of poor neurologic outcome is 2.66 (95% CI 1.59-4.45, p<0.001).

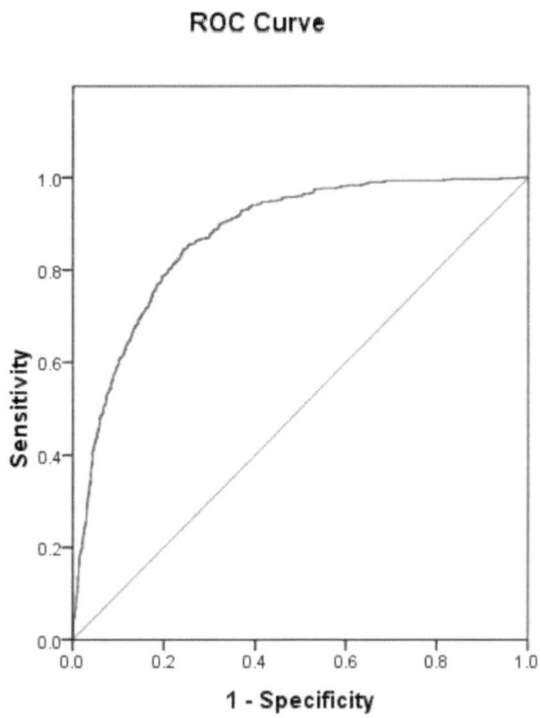

Figure 1. ROC curve to analyze binary logistic regression model discrimination. Output figure generated by IBM SPSS Version 19.0™ (Armonk, NY).

Fevers markedly increase cerebral metabolic rate and can worsen secondary injurious processes in the brain, such as brain swelling [1-7,17-22]. They must be recognized in a timely fashion and treated aggressively. Early onset fevers can be secondary to dysfunction of the temperature regulation centers in the hypothalamus [21]. Late onset fevers are more likely to be infectious, but can include fevers secondary to drugs and pulmonary embolism [17-22]. In those with fever at one week after aneurysm rupture and seizures, the odds ratio for poor neurologic outcome can be as high as 2.39 fold (95% CI 1.86-3.06, p<0.001).

Original insights are given in this chapter to attempt to delineate the significance of hepatic component of the neuro-gut axis in influencing neurologic outcome in aneurysmal subarachnoid hemorrhage. Through our epidemiologic data, we propose that liver disease can lead to significantly worse neurologic outcome via the development of brain swelling. Scarce published literature on this area captures the incidence of hepatic failure at 1.9% after aneurysmal subarachnoid hemorrhage [1,17]. Yet, the incidence of hepatic enzyme elevation is common in patients with ruptured brain aneurysms, and can be as high as 24% [1,17]. Possible liver dysfunction related mechanisms on neurologic outcome include: (1) altered blood flow to the brain, (2) release of chemicals which are toxic to the nervous system, (3) altered blood clotting cascade, and (4) overall systemic inflammatory states [1, 17].

In systemic inflammatory states after aneurysm rupture, inflammatory markers are released (such as cytokines, leukotrienes, prostaglandins and endothelins) [1-7, 17-22]. These chemicals interfere with the liver's ability to remove neurotoxic substances from blood [1-7, 17-22]. In turn, these neurotoxic substances (such as glutamate, manganese, mercaptans and ammonia) can leak across the disrupted blood brain barrier [1-7, 17-22]. This is especially common in those with alcoholic liver disease [1-7, 17-22]. The end result is a viscous cycle of neuron swelling and death, marked increase in brain blood flow (cerebral hyperemia) and brain swelling (cerebral edema) [1-7, 17-22]. In those with chronic liver disease, there are usually states of decreased cerebral blood flow because of blood shunting into the liver's portal circulation [1-7, 17-22]. However, rupture of brain aneurysms can predispose these patients to decompensated states with acute marked increases in brain blood flow (luxuriant perfusion) and decreased differences in the oxygen content between the arterial blood vessels and jugular veins [1-7, 17-22]. As shown in this epidemiologic study, development of brain swelling in the face of liver disease may predispose the aneurysmal subarachnoid hemorrhage patient to poor neurologic outcome by 5.5 fold, and can be as high as 26 fold (OR 5.47, 95% CI 1.13-26.46, p=0.03).

Limitations and Future Directions

Further delineation can be done in future studies to investigate the impact of other body systems, such as the kidney and immune systems, on neurologic outcome. Furthermore, the Tirilazad database did not capture some important variables, such as smoking and alcohol consumption.

In the next section of this chapter, informative prior probability likelihoods will be incorporated in Bayesian logistic regression analysis to investigate their effects on the odds ratios of prognostic variables on the entire group and subgroups of Tirilazad trial patients. In addition, an artificial neural networks model will be designed to examine whether its results will complement or differ from those generated by traditional logistic regression model, with consideration of possible non-linear interactions.

Part Three – Comparative Analyses Using Bayesian Regression and Artificial Neural Networks

Objectives

In this part, we aim to perform a comparative analysis of prognostic factors in aneurysmal subarachnoid hemorrhage using Bayesian logistic regression and artificial neural networks, with results generated from frequentist logistic regression. Informative prior probability likelihoods will be incorporated in Bayesian logistic regression analysis to investigate their effects on the odds ratios of prognostic variables on the entire group and subgroups of Tirilazad trial patients. In addition, an artificial networks model will be designed to examine whether its results will complement or differ from those generated by frequentist logistic regression model, with consideration of possible non-linear interactions.

Methods

Our analyses made use of the Tirilazad database, as well as the same independent and dependent variables as those from the previous section. To explore clinical prognostic factors for aneurysmal subarachnoid hemorrhage, the following statistical analyses were performed:

1. Bayesian logistic regression with WinBUGS 1.4.3, on both the entire patient population and 5 subgroups of 710 patients. Informative priors, odds ratios for each independent variable were estimated to lie between 0.01 and 100, with mean of 1 and precision of 0.189. Two chains of initial values were used, with 1000 iterations as burn-in and 100,000 total iterations. Convergence diagnostics were examined [23-28].
2. Artificial neural networks with IBM SPSS Version 19.0™, using 3 partitions for split sample validation technique (60% training, 20% testing and 20% holdout). Model discrimination was examined with the c statistic. The most robust clinical prediction model was presented with estimates of the relative importance of each prognostic variable [29-32].

Results

Bayesian Logistic Regression

Bayesian logistic regression made use of informative priors, with odds ratios for each independent variable estimated to lie between 0.01 and 100, with mean of 1 and precision of 0.189. The final Bayesian logistic regression model took into account Markov Chain Monte Carlo (MCMC) sampling. That is, Markov chains were used to generate Monte Carlo samples from probability distributions. In order to ensure that the resultant chain was a representative sample from the distributions, two chains of initial values were used with 1000 iterations used as burn in, with a total of 100 000 iterations. Kernel density probability distributions were smooth in contour. They represented posterior probability distributions for beta coefficients of clinical prognostic variables. Brooks-Gelman-Rubin (BGR) diagnostics showed that the two chains converged, thus, ensuring that sampling was from the same distribution as convergence occurred. Auto correlation plots showed that beta coefficients for age, diastolic blood pressure, systolic blood pressure, temperature and weight did not demonstrate rapid drop off, demonstrating correlation of simulation values to previous values for beta coefficients of these variables. Comparison of Bayesian and frequentist logistic regression results is demonstrated below.

Comparison of Bayesian and Frequentist Logistic Regression Results

Neurological Grade

Bayesian Logistic Regression:	
ALL (1-3552):	Exp(B)1.84, 1.72-1.98
GROUP 1 (1-710):	1.83, 1.54-2.18
GROUP 2 (711-1420):	1.77, 1.53-2.07
GROUP 3 (1421-2130):	2.03, 1.72-2.41
GROUP 4 (2131-2840):	2.08, 1.74-2.51
GROUP 5 (2841-3550):	1.91, 1.63-2.25
Frequentist Logistic Regression:	2.06, 1.83-2.32, **$p < 0.001$**

Age (years)

Bayesian Logistic Regression:	
ALL (1-3552):	Exp(B)1.04, 1.03-1.05
GROUP 1 (1-710):	1.04, 1.02-1.06
GROUP 2 (711-1420):	1.03, 1.02-1.05
GROUP 3 (1421-2130):	1.04, 1.03-1.07
GROUP 4 (2131-2840):	1.05, 1.03-1.07
GROUP 5 (2841-3550):	1.05, 1.03-1.07
Frequentist Logistic Regression:	1.06, 1.05-1.07, **$p < 0.001$**

Diastolic blood pressure on admission (mmHg)

Bayesian Logistic Regression:	
ALL (1-3552):	Exp(B)1.00, 1.00-1.01
GROUP 1 (1-710):	1.01, 1.00-1.03
GROUP 2 (711-1420):	1.00, 0.98-1.02
GROUP 3 (1421-2130):	1.01, 1.00-1.03
GROUP 4 (2131-2840):	1.00, 0.98-1.01
GROUP 5 (2841-3550):	0.99, 0.97-1.01
Frequentist Logistic Regression:	1.00, 0.99-1.01, $p=0.42$

Systolic blood pressure on admission (mmHg)

Bayesian Logistic Regression:	
ALL (1-3552):	Exp(B)1.00, 1.00-1.01
GROUP 1 (1-710):	0.99, 0.98-1.00
GROUP 2 (711-1420):	1.01, 1.00-1.02
GROUP 3 (1421-2130):	1.01, 1.00-1.02
GROUP 4 (2131-2840):	1.01, 1.00-1.02
GROUP 5 (2841-3550):	1.02, 1.00-1.03
Frequentist Logistic Regression:	1.00, 0.99-1.01, $p=0.27$

Temperature on admission (degrees Celsius)

Bayesian Logistic Regression:	
ALL (1-3552):	Exp(B)1.11, 0.96-1.26

GROUP 1 (1-710):	1.03, 0.83-1.25
GROUP 2 (711-1420):	1.08, 0.82-1.38
GROUP 3 (1421-2130):	1.14, 0.90-1.44
GROUP 4 (2131-2840):	1.24, 0.99-1.69
GROUP 5 (2841-3550):	1.25, 0.96-1.68
Frequentist Logistic Regression:	1.08, 0.93-1.24, p=0.32

Time to surgical treatment (hours)

Bayesian Logistic Regression:	
ALL (1-3552):	Exp(B)1.00, 1.00-1.00
GROUP 1 (1-710):	1.00, 1.00-1.01
GROUP 2 (711-1420):	1.00, 1.00-1.00
GROUP 3 (1421-2130):	1.00, 1.00-1.00
GROUP 4 (2131-2840):	1.00, 0.99-1.00
GROUP 5 (2841-3550):	1.00, 1.00-1.00
Frequentist Logistic Regression:	1.01, 1.00-1.01, **p=0.03**

Weight (kg)

Bayesian Logistic Regression:	
ALL (1-3552):	Exp(B)1.00, 1.00-1.00
GROUP 1 (1-710):	1.00, 0.99-1.02
GROUP 2 (711-1420):	0.97, 0.95-0.99
GROUP 3 (1421-2130):	1.00, 1.00-1.02
GROUP 4 (2131-2840):	0.99, 0.98-1.01
GROUP 5 (2841-3550):	0.98, 0.97-1.00
Frequentist Logistic Regression:	0.99, 0.98-1.00, p=0.07

Presence of fever one week post admission

Bayesian Logistic Regression:	
ALL (1-3552):	Exp(B)1.10, 0.92-1.33
GROUP 1 (1-710):	1.20, 0.76-1.89
GROUP 2 (711-1420):	1.00, 0.67-1.50
GROUP 3 (1421-2130):	0.99, 0.65-1.53
GROUP 4 (2131-2840):	1.64, 1.07-2.53
GROUP 5 (2841-3550):	0.82, 0.54-1.26
Frequentist Logistic Regression:	1.94, 1.51-2.49, **p<0.001**

Aneurysm size (mm)

Bayesian Logistic Regression:	
ALL (1-3552):	Exp(B)1.34, 1.10-1.64
GROUP 1 (1-710):	1.31, 0.78-2.19
GROUP 2 (711-1420):	1.02, 0.67-1.55
GROUP 3 (1421-2130):	1.34, 0.83-2.16
GROUP 4 (2131-2840):	1.18, 0.73-1.90
GROUP 5 (2841-3550):	**2.20, 1.41-3.41**

Frequentist Logistic Regression: 1.37, 1.04-1.80, **p=0.02**

History of angina

Bayesian Logistic Regression:

ALL (1-3552):	Exp(B)1.13, 0.69-1.82
GROUP 1 (1-710):	1.14, 0.38-3.31
GROUP 2 (711-1420):	1.53, 0.40-5.69
GROUP 3 (1421-2130):	0.85, 0.29-2.28
GROUP 4 (2131-2840):	1.10, 0.43-2.79
GROUP 5 (2841-3550):	3.14, 0.72-14.06
Frequentist Logistic Regression:	1.13, 0.58-2.21, p=0.72

Presence of angiographic vasospasm on admission

Bayesian Logistic Regression:

ALL (1-3552):	Exp(B)1.54, 1.16-2.05
GROUP 1 (1-710):	0.68, 0.31-1.41
GROUP 2 (711-1420):	1.50, 0.81-2.74
GROUP 3 (1421-2130):	**2.52, 1.34-4.78**
GROUP 4 (2131-2840):	**2.58, 1.31-5.10**
GROUP 5 (2841-3550):	1.22, 0.62-2.39
Frequentist Logistic Regression:	1.54, 1.04-2.30, **p=0.03**

Presence of intraventricular hemorrhage on admission

Bayesian Logistic Regression:

ALL (1-3552):	Exp(B)1.50, 1.25-1.81
GROUP 1 (1-710):	1.24, 0.81-1.91
GROUP 2 (711-1420):	**2.27, 1.50-3.44**
GROUP 3 (1421-2130):	**1.88, 1.22-2.92**
GROUP 4 (2131-2840):	1.16, 0.74-1.81
GROUP 5 (2841-3550):	1.26, 0.83-1.92
Frequentist Logistic Regression:	1.33, 1.02-1.73, **p=0.04**

Development of cerebral edema

Bayesian Logistic Regression:

ALL (1-3552):	Exp(B)1.86, 1.42-2.42
GROUP 1 (1-710):	1.96, 1.16-3.32
GROUP 2 (711-1420):	1.29, 0.71-2.34
GROUP 3 (1421-2130):	**2.78, 1.44-5.42**
GROUP 4 (2131-2840):	1.98, 0.97-4.03
GROUP 5 (2841-3550):	**2.71, 1.40-5.28**
Frequentist Logistic Regression:	1.84, 1.29-2.62, **p<0.001**

History of diabetes mellitus

Bayesian Logistic Regression:

ALL (1-3552):	Exp(B)1.45, 0.94-2.24

GROUP 1 (1-710):	0.81, 0.30-2.08
GROUP 2 (711-1420):	2.28, 0.84-6.27
GROUP 3 (1421-2130):	1.36, 0.47-3.78
GROUP 4 (2131-2840):	2.32, 0.81-6.71
GROUP 5 (2841-3550):	1.71, 0.62-4.82
Frequentist Logistic Regression:	1.00, 0.99-1.01, p=0.42

Presence of intracerebral hematoma on admission

Bayesian Logistic Regression:	
ALL (1-3552):	Exp(B)1.35, 1.09-1.65
GROUP 1 (1-710):	0.92, 0.53-1.56
GROUP 2 (711-1420):	1.23 0.79-1.91
GROUP 3 (1421-2130):	1.78, 1.10-2.88
GROUP 4 (2131-2840):	1.03, 0.61-1.73
GROUP 5 (2841-3550):	1.97, 1.25-3.12
Frequentist Logistic Regression:	1.39, 1.04-1.87, **p=0.03**

History of hepatic disease

Bayesian Logistic Regression:	
ALL (1-3552):	Exp(B)1.17, 0.75-1.79
GROUP 1 (1-710):	1.39, 0.46-3.78
GROUP 2 (711-1420):	0.84, 0.26-2.39
GROUP 3 (1421-2130):	1.18, 0.43-2.95
GROUP 4 (2131-2840):	1.29, 0.49-3.30
GROUP 5 (2841-3550):	1.08, 0.40-2.74
Frequentist Logistic Regression:	1.76, 0.90-3.45, p=0.10

Development of hydrocephalus

Bayesian Logistic Regression:	
ALL (1-3552):	Exp(B)1.68, 1.45-1.95
GROUP 1 (1-710):	1.40, 0.97-1.99
GROUP 2 (711-1420):	1.90, 1.36-2.66
GROUP 3 (1421-2130):	1.49, 1.04-2.13
GROUP 4 (2131-2840):	**2.09, 1.48-2.96**
GROUP 5 (2841-3550):	1.56, 1.09-2.21
Frequentist Logistic Regression:	0.95, 0.73-1.23, p=0.68

History of hypertension

Bayesian Logistic Regression:	
ALL (1-3552):	Exp(B)1.62, 1.38-1.89
GROUP 1 (1-710):	1.74, 1.22-2.48
GROUP 2 (711-1420):	1.80, 1.26-2.57
GROUP 3 (1421-2130):	1.43, 0.98-2.09
GROUP 4 (2131-2840):	1.52, 1.06-2.18
GROUP 5 (2841-3550):	**2.00, 1.36-2.94**

Frequentist Logistic Regression:	1.37, 1.05-1.79, **p=0.02**

History of myocardial infarction

Bayesian Logistic Regression:	
ALL (1-3552):	Exp(B)1.76, 1.06-2.89
GROUP 1 (1-710):	2.10, 0.76-5.75
GROUP 2 (711-1420):	0.92, 0.15-4.40
GROUP 3 (1421-2130):	**2.52, 1.02-6.39**
GROUP 4 (2131-2840):	2.32, 0.81-6.90
GROUP 5 (2841-3550):	1.00, 0.15-6.07
Frequentist Logistic Regression:	2.07, 1.01-4.27, **p=0.05**

Occurrence of stroke post admission

Bayesian Logistic Regression:	
ALL (1-3552):	Exp(B) 6.14, 5.08-7.44
GROUP 1 (1-710):	**10.74, 6.86-17.13**
GROUP 2 (711-1420):	**7.20, 4.63-11.4**
GROUP 3 (1421-2130):	5.38, 3.48-8.46
GROUP 4 (2131-2840):	**7.70, 4.92-12.30**
GROUP 5 (2841-3550):	5.40, 3.50-8.48
Frequentist Logistic Regression:	4.03, 2.11-7.69, **p<0.001**

Development of pulmonary edema post admission

Bayesian Logistic Regression:	
ALL (1-3552):	Exp(B)1.00, 0.77-1.29
GROUP 1 (1-710):	1.26, 0.81-1.94
GROUP 2 (711-1420):	1.12, 0.51-2.35
GROUP 3 (1421-2130):	0.77, 0.39-1.45
GROUP 4 (2131-2840):	1.13, 0.62-2.03
GROUP 5 (2841-3550):	**2.74, 1.07-7.11**
Frequentist Logistic Regression:	0.83, 0.56-1.23, p=0.37

History of migraines

Bayesian Logistic Regression:	
ALL (1-3552):	Exp(B)0.74, 0.61-0.89
GROUP 1 (1-710):	0.72, 0.49-1.05
GROUP 2 (711-1420):	0.71, 0.47-1.07
GROUP 3 (1421-2130):	0.86, 0.55-1.30
GROUP 4 (2131-2840):	0.81, 0.53-1.22
GROUP 5 (2841-3550):	0.85, 0.52-1.38
Frequentist Logistic Regression:	0.93, 0.70-1.25, p=0.65

Presence of previous episode of subarachnoid hemorrhage

Bayesian Logistic Regression:	
ALL (1-3552):	Exp(B)1.66, 1.30-2.10

GROUP 1 (1-710): 1.61, 0.85-2.94
GROUP 2 (711-1420): **2.01, 1.17-3.42**
GROUP 3 (1421-2130): **2.23, 1.26-3.90**
GROUP 4 (2131-2840): 1.32, 0.73-2.35
GROUP 5 (2841-3550): 1.91, 1.12-3.23
Frequentist Logistic Regression: 1.60, 1.07-2.39, **p=0.02**

Location of aneurysm

Bayesian Logistic Regression:
ALL (1-3552): Exp(B)1.00, 0.96-1.01
GROUP 1 (1-710): 1.00, 0.97-1.03
GROUP 2 (711-1420): 1.25, 0.77-1.99
GROUP 3 (1421-2130): 1.04, 0.62-1.71
GROUP 4 (2131-2840): 1.12, 0.68-1.84
GROUP 5 (2841-3550): 1.48, 0.78-2.78
Frequentist Logistic Regression: 1.72, 1.18-2.51, **p=0.005**

Thickness of subarachnoid hemorrhage

Bayesian Logistic Regression:
ALL (1-3552): Exp(B)1.00, 0.99-1.02
GROUP 1 (1-710): 1.00, 0.98-1.02
GROUP 2 (711-1420): **1.64, 1.13-2.40**
GROUP 3 (1421-2130): **2.60, 1.75-3.94**
GROUP 4 (2131-2840): **2.82, 1.91-4.24**
GROUP 5 (2841-3550): **2.64, 1.78-3.98**
Frequentist Logistic Regression: 1.02, 0.99-1.05, p=0.13

Sex

Bayesian Logistic Regression:
ALL (1-3552): Exp(B)1.01, 0.83-1.24
GROUP 1 (1-710): 3.25, 0.04-234.39
GROUP 2 (711-1420): 2.35, 0.03-165.2
GROUP 3 (1421-2130): 1.27, 0.81-2.02
GROUP 4 (2131-2840): 0.92, 0.64-1.35
GROUP 5 (2841-3550): 1.22, 0.84-1.78
Frequentist Logistic Regression: 0.73, 0.51-1.04, p=0.09

History of thyroid disease

Bayesian Logistic Regression:
ALL (1-3552): Exp(B)1.10, 0.82-1.46
GROUP 1 (1-710): 1.09, 0.62-1.88
GROUP 2 (711-1420): 1.21, 0.69-2.11
GROUP 3 (1421-2130): 1.13, 0.53-2.30
GROUP 4 (2131-2840): 0.96, 0.41-2.11
GROUP 5 (2841-3550): 1.21, 0.54-2.66

Frequentist Logistic Regression: 0.98,0.62-1.54, p=0.92

Development of vasospasm during course of treatment

Bayesian Logistic Regression:

ALL (1-3552):	Exp(B)1.73, 1.47-2.04
GROUP 1 (1-710):	1.70, 1.16-2.50
GROUP 2 (711-1420):	1.80, 1.25-2.60
GROUP 3 (1421-2130):	1.82, 1.24-2.64
GROUP 4 (2131-2840):	1.17, 0.79-1.73
GROUP 5 (2841-3550):	**2.44, 1.65-3.61**
Frequentist Logistic Regression:	1.58, 1.21-2.06, **p<0.001**

Treatment Group

Bayesian Logistic Regression:

ALL (1-3552):	Exp(B)1.00, 0.96-1.04
GROUP 1 (1-710):	1.06, 0.97-1.16
GROUP 2 (711-1420):	1.05, 0.96-1.15
GROUP 3 (1421-2130):	0.88, 0.72-1.07
GROUP 4 (2131-2840):	1.18, 1.00-1.38
GROUP 5 (2841-3550):	0.95, 0.83-1.08
Frequentist Logistic Regression:	0.74, 0.48-1.13, p=0.17

Development of severe vasospasm requiring balloon angioplasty

Bayesian Logistic Regression:

ALL (1-3552):	Exp(B)1.34, 0.95-1.89
GROUP 1 (1-710):	1.71, 0.88-3.29
GROUP 2 (711-1420):	0.42, 0.10-1.50
GROUP 3 (1421-2130):	1.44, 0.68-3.03
GROUP 4 (2131-2840):	**2.27, 1.21-4.29**
GROUP 5 (2841-3550):	1.20, 0.24-5.56
Frequentist Logistic Regression:	1.31, 0.77-2.20, p=0.32

Development of seizures requiring use of anti-epileptic medications

Bayesian Logistic Regression:

ALL (1-3552):	Exp(B)1.25, 1.06-1.47
GROUP 1 (1-710):	**2.11, 1.27-3.64**
GROUP 2 (711-1420):	1.47, 1.02-2.11
GROUP 3 (1421-2130):	1.61, 1.04-2.65
GROUP 4 (2131-2840):	1.19, 0.81-1.77
GROUP 5 (2841-3550):	1.42, 1.00-2.01
Frequentist Logistic Regression:	1.68, 1.28-2.21, **p<0.001**

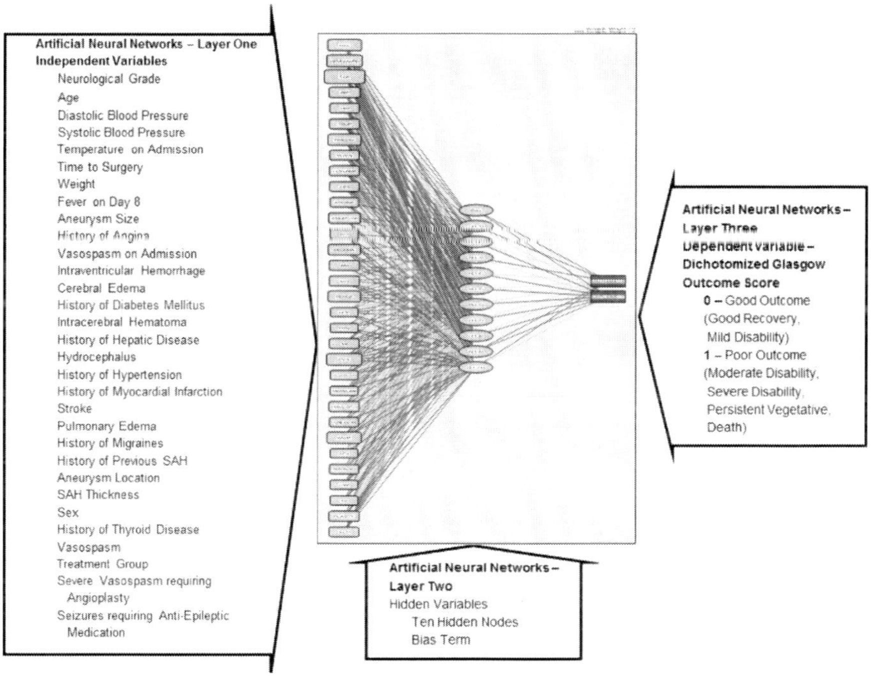

Figure 2. Artificial neural networks output using the dichotomized Glasgow Outcome Score with insets for each layer. Output figure generated by IBM SPSS Version 19.0™ (Armonk, NY).

Artificial Neural Networks

The most robust artificial neural networks for clinical prediction in aneurysmal subarachnoid hemorrhage was found using the multilayer perceptron model with 3 partitions (training sample 60%, testing sample 20%, holdout sample 20%), as shown in Figure 2.

The network architecture had 1 hidden layer and 10 latent variables. The sigmoid function was specified as the activation function from input to hidden layers, and from hidden to output layers. Training criteria made use of periodic synaptic weight optimization, with the error optimization function using gradient descent.

Training time required 53 seconds. Model classification showed 84% correct predictions in the training sample, 81% correct predictions in the testing sample and 85% correct predictions in the holdout sample. C statistic was used to examine discrimination, with area under the ROC curve being 0.870, for both favourable and unfavourable outcomes (Figure 3). The predicted-by-observed chart showed ability of the model to predict both favourable and unfavourable outcomes (Figure 4).

Variables in order of magnitude of relative importance are shown as follows:

1. age - 0.132,
2. neurological grade - 0.082,
3. occurrence of stroke post admission - 0.079,
4. thickness of subarachnoid hemorrhage - 0.069,
5. time to surgical treatment - 0.060,

6. history of hepatic disease - 0.054,
7. development of cerebral edema - 0.049,
8. presence of angiographic vasospasm on admission - 0.044,
9. diastolic blood pressure - 0.038,
10. development of severe vasospasm requiring angioplasty - 0.034,

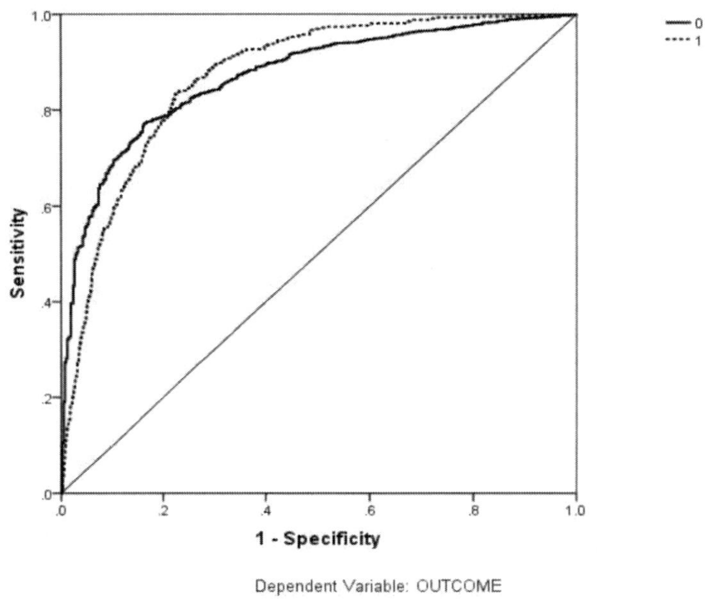

Figure 3. ROC curve to analyze model performance. Output figure generated by IBM SPSS Version 19.0™ (Armonk, NY).

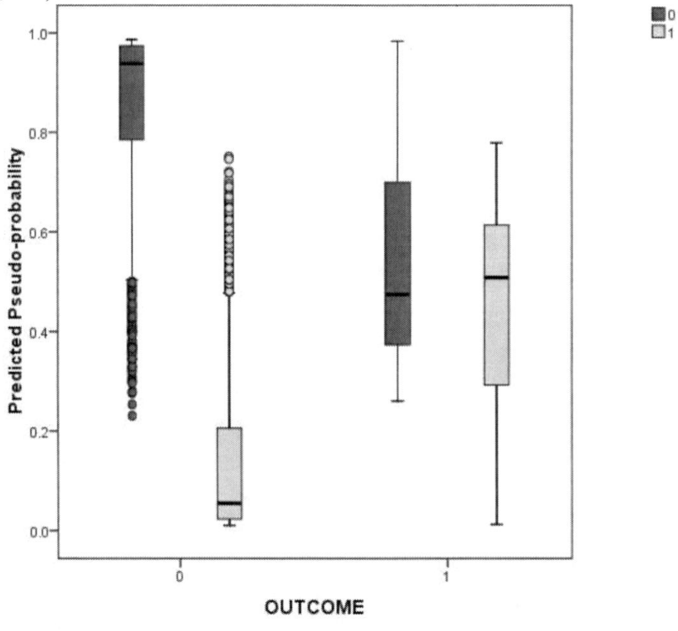

Figure 4. Predicted-by-Observed Chart showing ability to predict both favourable and unfavourable outcomes. Output figure generated by IBM SPSS Version 19.0™ (Armonk, NY).

11. presence of previous episode of subarachnoid hemorrhage - 0.032,
12. history of diabetes mellitus - 0.028,
13. temperature on admission - 0.027,
14. presence of intracerebral hematoma on admission - 0.027,
15. presence of fever one week post admission - 0.026,
16. development of seizures requiring use of anti-epileptic medications - 0.025,
17. history of myocardial infarction - 0.024,
18. aneurysm size - 0.023,
19. patient's weight - 0.021,
20. systolic blood pressure on admission - 0.021,
21. history of thyroid disease - 0.019,
22. location of aneurysm - 0.018,
23. history of angina - 0.014,
24. presence of intraventricular hemorrhage on admission - 0.011,
25. history of migraines - 0.009,
26. treatment group - 0.008,
27. development of vasospasm during course of treatment - 0.007,
28. development of pulmonary edema post admission - 0.006,
29. history of hypertension - 0.006,
30. sex - 0.006, and
31. development of hydrocephalus - 0.004.

Interpretation of Results

Comparative analyses of clinical prognostic factors for aneurysmal subarachnoid hemorrhage using Bayesian regression and artificial neural networks demonstrated that these techniques complemented results generated by frequentist logistic regression.

Bayesian Logistic Regression Analysis

Beta coefficients for prognostic variables were similar in magnitude using frequentist multivariable logistic regression and Bayesian logistic regression with informative prior likelihoods, using the entire patient population group. However, dissimilar odds ratios for patient subgroups were noted, with increased magnitudes of odds ratios for the following independent variables: (1) aneurysm size, (2) presence of angiographic vasospasm on admission, (3) presence of intraventricular hemorrhage on admission, (4) development of cerebral edema, (5) development of hydrocephalus, (6) history of hypertension, (7) history of myocardial infarction, (8) occurrence of stroke post admission, (9) development of pulmonary edema post admission, (10) previous episode of subarachnoid hemorrhage, (11) thickness of subarachnoid hemorrhage, (12) sex, (13) development of vasospasm during course of treatment, (14) development of severe vasospasm requiring balloon angioplasty, and (15) development of seizures requiring use of anti-epileptic medications.

The aforementioned observations point to different effect sizes across patient subgroups, suggesting possibility of heterogeneity in patient characteristics, multiplicative interactions between prognostic variables, as well as non-linear relationships.

In this section, Bayesian analyses carry advantages of enabling the investigator to: (1) incorporate informative prior likelihoods based on known knowledge, (2) examine effect sizes across patient subgroups, (3) hypothesize on possibility of divergent subgroups, (4) examine variability in treatment responses, and (5) examine multiplicative interaction effects between prognostic variables.

Artificial Neural Networks

The most robust artificial neural networks for clinical prediction in aneurysmal subarachnoid hemorrhage was found using the multilayer perceptron model with sigmoid activation functions, mean square error and gradient descent, with 10 latent variables in one layer. Model classification showed large percentage (84%) of correct predictions in the training sample, which is retained across the testing and holdout samples. Model discrimination value was demonstrated with area under the ROC curve being 0.870, for both favourable and unfavourable outcomes.

The ten predictors with the highest normalized importance values include age, neurological grade, occurrence of stroke post admission, thickness of subarachnoid hemorrhage, time to surgical treatment, history of hepatic disease, development of cerebral edema, presence of angiographic vasospasm on admission, diastolic blood pressure and development of severe vasospasm requiring angioplasty.

In comparison with frequentist logistic regression presented in the previous section, both models had similar model discrimination values. In general, the significant variables generated with logistic regression were similar to those from artificial neural networks. The effects of non-linear relationships were noted in the clinically robust artificial neural networks model. This resulted in a different ordering of the variables' relative importance estimates when compared with their beta coefficients from traditional logistic regression. For example, age exhibited the highest relative importance in the artificial neural networks model as the likelihood of unfavourable outcome markedly increased with age in a non-linear fashion.

In this section, artificial neural networks were shown to have the advantages of being able to take into account non-linear relationships, as well as model discrimination values that were comparable to those generated with frequentist logistic regression. However, they were shown to have the disadvantages of not being able to provide details on the various higher order interactions and specific non-linear relationships for variables. In addition, it is difficult to clinically interpret the variables' relative importance estimates.

Limitations and Future Directions

Insight gained from Bayesian analyses and artificial neural networks models may be used by the investigator in order to improve the design of the traditional logistic regression model. These techniques, therefore, play a complementary role to frequentist methods. More sophisticated models can be explored with development of more advanced statistical softwares that can both specify and process high order interactions and non-linear relationships in Bayesian analyses and artificial neural networks. In addition, Bayesian principles can be integrated into artificial neural networks which allow investigators to incorporate prior likelihood probabilities while designing elements of a particular artificial neural network.

Part Four - Conclusion

This chapter represents the first paper to analyze both non-treatment and treatment related prognostic factors, as well as two way brain-body interactions. It provided a number of important novel findings, which are key in elucidating the mechanisms of neurological deterioration in aneurysmal subarachnoid hemorrhage patients. This is also the first paper to elucidate epidemiologic links of brain interactions across multiple organ systems, namely: (1) the neuro-gut and cardiovascular axes in pathogenesis of brain edema, which can be deleterious in those with ruptured brain aneurysms, as well as (2) the cardiovascular and temperature regulation axes in combination with seizures, which can also be deleterious in those with ruptured brain aneurysms.

This chapter also used both Bayesian regression and artificial neural networks in an innovative way to complement results from frequentist logistic regression models. Using informative prior likelihoods, both entire group and subgroup analyses were performed. It is hypothesized that subgroup heterogeneity, multiplicative interactions and non-linear relationships can affect the magnitudes of odds ratios across patient subgroups. The effects of such non-linear relationships were noted in the clinically robust artificial neural networks model.

Future Directions

The techniques of frequentist logistic regression, Bayesian regression and artificial neural networks were applied to the Tirilazad database. Since the conduct of the Tirilazad trials, there have been advancements in both medical and surgical treatment of cerebral aneurysms. These include aneurysm coiling and neurocritical care of aneurysmal subarachnoid hemorrhage patients. In addition, the Tirilazad database did not capture variables such as smoking, alcohol consumption, and other systematic complications including renal and hematologic complications.

The prognostication system described in this chapter made use of a combination of techniques that are generalizable and applicable to development of clinical prediction rules in other areas of medicine. Most importantly, this chapter has detailed novel brain-body associations which significantly influence clinical outcome in aneurysmal subarachnoid hemorrhage. Recognition and treatment of these factors are essential in optimizing outcome and preventing deterioration in those with ruptured brain aneurysms.

Acknowledgments

Dr. Wan Yee Yin reviewed this manuscript. Dr. R. Loch Macdonald provided us with the Tirilazad database. The authors of this chapter do not have any financial relationship with the pharmaceutical company, Pharmacia & Upjohn, USA, manufacturer of the trial agent, Tirilazad.

References

[1] Bederson, JB. AANS Publications Committee. Subarachnoid hemorrhage: pathophysiology and management. Park Ridge, Ill.: American Association of Neurological Surgeons, 1997.

[2] Layon, AJ; Gabrielli, A; Friedman, W. *Textbook of neurointensive care*. Philadelphia: Saunders, 2004.

[3] Lee, K. The neuroICU book. New York: McGraw-Hill Medical, 2012.

[4] Mashour, G; Farag, E. *Case studies in neuroanesthesia and neurocritical care*. Cambridge; New York: Cambridge University Press, 2011.

[5] Matta, B; Menon, D; Smith, M. *Core topics in neuroanaesthesia and neurointensive care*. Cambridge: Cambridge University Press, 2011.

[6] Torbey, M. *Neurocritical care*. Cambridge; New York: Cambridge University Press, 2010.

[7] Wijdicks, E. *The practice of emergency and critical care neurology*. New York: Oxford University Press, 2010.

[8] Haley, EC, Jr; Kassell, NF; Alves, WM; Weir, BK; Hansen, CA. Phase II trial of tirilazad in aneurysmal subarachnoid hemorrhage. A report of the Cooperative Aneurysm Study. *J Neurosurg*. 1995, 82, 786–90.

[9] Kassell, NF; Haley, ECJ; Apperson-Hansen, C; Alves, WM. Randomized, double-blind, vehicle-controlled trial of tirilazad mesylate in patients with aneurysmal subarachnoid hemorrhage: a cooperative study in Europe, Australia, and New Zealand. *J Neurosurg*. 1996, 84, 221–8.

[10] Haley, EC, Jr; Kassell, NF; Apperson-Hansen, C; Maile, MH; Alves, WM. A randomized, double-blind, vehicle-controlled trial of tirilazad mesylate in patients with aneurysmal subarachnoid hemorrhage: a cooperative study in North America. *J Neurosurg*. 1997, 86, 467–74.

[11] Lanzino, G; Kassell, NF. Double-blind, randomized, vehicle-controlled study of high-dose tirilazad mesylate in women with aneurysmal subarachnoid hemorrhage. Part II. A cooperative study in North America. *J Neurosurg*. 1999, 90, 1018–24.

[12] Lanzino, G; Kassell, NF; Dorsch, NW; et al. Double-blind, randomized, vehicle-controlled study of high-dose tirilazad mesylate in women with aneurysmal subarachnoid hemorrhage. Part I. A cooperative study in Europe, Australia, New Zealand, and South Africa. *J Neurosurg*. 1999, 90, 1011–7.

[13] Jennett, B; Bond, M. Assessment of outcome after severe brain damage: a practical scale. *Lancet.* 1975, 480-4.

[14] Jang, Y; Ilodigwe, D; Macdonald, R. Metaanalysis of tirilazad mesylate in patients with aneurysmal subarachnoid hemorrhage. *Neurocrit Care*. 2009, 10(1), 141-7.

[15] Zhang, S; Wang, L; Liu, M; Wu, B. Tirilazad for aneurysmal subarachnoid haemorrhage (Review). *Cochrane Database Syst Rev*. 2010 Feb 17.

[16] Hosmer, D; Lemeshow, S; Sturdivant, R. Applied logistic regression. Hoboken, NJ: Wiley, 2013.

[17] Butterworth, RF. The liver-brain axis in liver failure: neuroinflammation and encephalopathy. *Nat Rev Gastroenterol Hepatol*. 2013 Jul 2.

[18] Lannes, M; Teitelbaum, J; del, Pilar, Cortes, M; et al. Milrinone and homeostasis to treat cerebral vasospasm associated with subarachnoid hemorrhage: the Montreal Neurological Hospital protocol. *Neurocrit Care*. 2012 Jun, 16(3), 354-62.

[19] Lee, VH; Oh, JK; Mulvagh, SL; Wijdicks, EF. Mechanisms in neurogenic stress cardiomyopathy after aneurysmal subarachnoid hemorrhage. *Neurocrit Care*. 2006, 5(3), 243-9.

[20] Mrozek, S; Vardon, F; Geeraerts, T. Brain temperature: physiology and pathophysiology after brain injury. *Anesthesiol Res Pract*. Epub 2012 Dec 26.

[21] Schneider, HJ; Kreitschmann-Andermahr, I; Ghigo, E; et al. Hypothalamopituitary dysfunction following traumatic brain injury and aneurysmal subarachnoid hemorrhage: a systematic review. *JAMA*. 2007 Sep 26, 298(12), 1429-38.

[22] Lo, BW; Macdonald, RL; Baker, A; Levine, MA. Clinical outcome prediction in aneurysmal subarachnoid hemorrhage using Bayesian neural networks with fuzzy logic inferences. Comput Math Methods Med. 2013, 2013:904860. Epub 2013 Apr 10.

[23] Berry, D. Statistics: a Bayesian perspective. Belmont, CA: Duxbury Press, 1996.

[24] Iversen, G. Bayesian statistical inference. Beverly Hills: Sage Publications, 1984.

[25] Kery, M; Schaub, M. Bayesian population analysis using WinBUGS: a hierarchical perspective. Waltham, MA: Academic Press, 2012.

[26] Kruschke, J. Doing Bayesian data analysis: a tutorial with R and BUGS. Burlington, MA: Academic Press, 2011.

[27] Ntzoufras, I. Bayesian modeling using WinBUGS. Hoboken, NJ: Wiley, 2009. Spiegelhalter D, Abrams K, Myles J. Bayesian approaches to clinical trials and health-care evaluation. Chichester: Wiley, 2002.

[28] Woodworth, G. Biostatistics: a Bayesian introduction. Hoboken, New Jersey: Wiley-Interscience, 2004.

[29] Garson, GD. Neural networks: an introductory guide for social scientists. London; Thousand Oaks, CA: Sage, 1998.

[30] Haykin, S. Neural networks: A comprehensive foundation. New York: Macmillan, 1994.

[31] IBM SPSS Neural Networks 20. Upper Saddle River, NJ: Prentice Hall, 2011.

[32] Samarasinghe, S. Neural networks for applied sciences and engineering: from fundamentals to complex pattern recognition. Boca Raton, FL: Auerback, 2007.

In: Horizons in Neuroscience Research. Volume 15
Editors: Andres Costa and Eugenio Villalba

ISBN: 978-1-63321-523-8

Chapter 10

Pathological Changes of Event-Related Brain Oscillations in Parkinson's Disease

Juliana Dushanova
Institute of Neurobiology, Bulgarian Academy of Sciences, Bulgaria

Abstract

Brain event-related oscillations provide temporal and spatial codes for information processing and give a useful tool for detecting subtle abnormalities in cognitive processes. They are one of the promising candidates explaining the neural mechanisms at central nervous system in Parkinson's patients (PP) and healthy subjects (HS). This study reports auditorily elicited brain oscillatory responses in association with sensorimotor and cognitive processes. Brain oscillations at specific frequencies and times provide a window into the organization and transfer of information underlying such processes in the human brain. Brain event-related oscillatory responses evoked from given auditory stimuli (low-800, high tone type-1000 Hz) were quantified by means of event-related desynchronization / synchronization (ERD/ERS) technique for PP and HS. The differences between the ERD/ERS responses at different frequency bands (delta 1.5–4, theta 4–7, alpha 7–14, beta1 13-20, beta2 20-32, gamma 32-50 Hz) for PP and HS during auditory discrimination tasks are of relevance to the understanding of these cortical functions. Widespread delta- and theta- ERS, and alpha-ERD were significantly pronounced in HS in response to either tone during sensory information processing (0-250 ms after the stimulus onset). More prominent fronto-parietal delta-ERS was observed in response to a low than to a high tone, with predominantly theta- and alpha-ERD in HS following either tone during the cognitive processing (250-600 ms after the stimulus onset). Beta1-, beta2- and gamma-ERD were revealed everywhere in HS after either tone within both intervals except for frontal beta1- and gamma-ERS after a high tone during the cognitive processing. When comparing the PP and HS during the sensory processing, PP delta- and theta- ERS were less pronounced after either tone with exception of frontal higher delta-ERS and theta-ERD at the left motor area in response to a high tone. Widespread alpha-ERS was found only in PP during the sensory processing in response to either tone. PP delta-ERD after a low tone and delta-ERS after a high tone, theta- , alpha-ERD and parietal theta-ERS following either tone were revealed during cognitive processing. Only frontal theta- and alpha-ERD were more prominent for PP than for HS. Beta1-ERD was less pronounced in PP within both intervals after either tone. Beta2-ERS was observed only in PP, but the most prominent beta2-ERS followed the high tone during cognitive processing. Gamma-ERD was observed in PP during sensory processing except for fronto-parietal ERS after a low tone. During cognitive processing, either tone, prominent gamma-ERS was revealed everywhere for PP. Deviations of the ERD/ERS at specific frequencies for the PP compared with HS during sensorimotor and cognitive processing are clear evidence for disturbances in the temporal and regional integration of these frequency components and the relationships between cortical and basal ganglia circuits in parkinsonism.

1. Introduction

Parkinson's disease (PD) is a common and disabling disorder of movement characterized by poverty, slowness and impaired scaling of voluntary movements (akinesia and bradykinesia), muscle rigidity, and tremor of the limbs at rest. PD is caused by a disruption of dopaminergic neurotransmission in the basal ganglia, which serve as an integrative centre for the sensory and cognitive processing of information and as a mutual link between this processing and disturbed motor performance. The basal ganglia and the cerebellum transmit information via the thalamus to the cerebral cortex in order to regulate movement. The neurotransmitter changes affect the output of the striatum into the globus pallidus as well as into the thalamus and cerebral cortex beyond. Alterations of the basal ganglia with proven neuronal degenerative disorders of dopaminergic neurons and a reduction in activity in frontostriatal neural circuitry have been suggested to play a role in the executive dysfunction of PD (Taylor et al., 1990; Innis et al., 1993; Lewis et al., 2003; Owen, 2004; Leblois et al., 2006; Anik et al., 2007). The slowed information processing, insufficient encoding strategies and planning, and attentional set-shifting are related to memory deficits and cognitive impairment in PD (Daum et al., 1995; Pillon et al., 1997; Knoke et al., 1998; Robertson & Empson, 1999; Sawamoto et al., 2002; Cools, 2006). Neuropsychological studies of PD patients report cognitive deficits even during the early stages of the disease (van Spaendonck et al., 2006). The primary working memory deficit in PD is associated with impaired free recall performances (Higginson et al., 2003).

Many electroencephalographic (EEG) studies on PD have used the event-related potential (ERP) method, where the early modal dependent and obligatory N1 and P2 components permit analysis of sensory events while the later N2 and P3 potentials reflect the cognitive processes involving the assessment of stimuli, decision making, strategy selection and recognition memory. ERP investigations have shown P3 predominantly with prolonged latencies and/or diminished amplitudes for Parkinsonian patients (PP) when compared to healthy subjects (HS) (Evarts et al., 1981; Tachibana et al., 1992; Philipova et al., 1997; Wascher et al., 1997; Minamoto et al., 2001; Antal et al., 2002; Wang et al., 2002). Such results have been interpreted as electrophysiological signs of cognitive slowing with respect to stimulus classification and attentional processing (Robertson & Empson, 1999).

One valuable means of assessing deviations from the normal state in PD is to study oscillatory brain processes. In the ERP method, however, the functional significance of the responses in different frequency bands is lost. More clarification could be expected when attentional processes in PD during a representation of discrimination tasks (Vieregge et al., 1994) are examined using event-related desynchronization/synchronization (ERD/ERS) method. In the early stages, PD also affects cognitive functions (Cools, 2006). Cognitive processes require transient integration between different brain areas. Hence, dynamic links are formed, mediated by the ERS or ERD of neuronal assemblies. ERD is defined as a relative decrease in the power of a certain frequency band during stimulus processing, while ERS is a relative increase in the power of the same frequency (Pfurtscheller & Klimesch, 1991). The ERD/ERS method has been used to study auditory and visual working memory encoding and categorization processes in PP; studies indicate less theta-ERS and upper alpha-ERD reflected disturbance of both the basal ganglia activity as well as activity related to their thalamo-cortical neuronal nets at frontal electrode locations (Schmiedt et al., 2005; Ellfolk et al., 2006). The encoding of auditory stimuli elicits alpha- and theta-ERS, while memory retrieval during the presentation of a target stimulus elicits theta-ERS and alpha-ERD (Karrasch et al., 1998, 2004; Krause et al., 1996, 1999). Oscillations in the beta frequency band are associated with cognitive control of behaviour or "executive functions" (Pfurtscheller & Lopes da Silva, 1999; Engel et al., 2001). By means of an auditory stop-signal task, the differential participation of beta subbands in voluntary motor control can be revealed: ERD in the 20–30 Hz band is related to initiation of movement, while ERS in a low frequency beta band (12–16 Hz) is exclusively linked to the stopping of planned action (Pfurtscheller & Lopes da Silva, 1999; Engel et al., 2001). One proposed hypothesis for these observations is that lower-frequency beta subbands represent inhibitory components of cognitive control and are more generalized, while higher frequency beta subbands take part in response choice and activation and are more specialized in terms of both function and cortical distribution.

Some recent investigations (Basar, 2001; Ozgoren et al., 2005; Sutoh et al., 2000; Gurtubay et al., 2001) propose that beta and gamma cortical rhythms may serve cognitive processes such as linking perception to action or movement planning (Donoghue et al., 1998). Research both on animals and humans has suggested that gamma-frequency activity also plays an important role in attention as well as working and long-term memory (Herrmann et al., 2004). Current investigations using intracranial and high-density electro- and magnetoencephalographic recordings explore the involvement of gamma-band synchronization in various cognitive paradigms in humans (Engel et al., 2001; Basar et al., 2001; Herrmann et al., 2004; Pantev, 1995; Tallon-Baudry et al., 1996; Farmer, 1998; Fries et al., 2001). Other works associate the changes in EEG spectral power in the gamma frequency band to interactions between the cortex and basal ganglia (Gatev & Wichmann, 2008). Additionally, akinesia in PP has been related in some studies to abnormally increased beta (15–30 Hz) and decreased gamma (35–80 Hz) synchronous oscillatory activity in the basal ganglia (Weinberger et al., 2006). Other results suggest that resting tremor in PD is associated with an altered balance between beta and gamma oscillations in the motor circuits of the subthalamic nucleus (STN) and is exhibited as increased oscillatory activity in the low gamma frequency range (35–55 Hz) during periods with stronger tremor (Weinberger et al., 2008). Therapeutic doses of dopaminergic medication in PP attenuate the beta band power in the STN, giving rise to the hypothesis that the beta prominence is pathological in PD (Cassidy et al., 2002; Kühn et al., 2006; Levy et al., 2001; 2002; Priori et al., 2002). Treatment of PP with dopaminergic therapy leads to increased gamma band activity in the basal ganglia and thus to improvement in motor performance (Brown et al., 2001), hence the suggestion that synchronization of the activity of populations of basal ganglia neurons in the gamma band may facilitate motor processing (Brown, 2003). Investigations based on local field potentials recorded from the STN in PP show increased power in the beta range (13–35 Hz) while the patient is at rest (Cassidy et al., 2002; Levy et al., 2002; Brown et al., 2001; Priori et al., 2004). This suggests that there is excessive synchrony in the basal ganglia networks in PD and some of the clinical signs of the disease, it is proposed, stem from this abnormal synchrony between basal ganglia and cortical circuits (Brown, 2003; Marsden et al., 2001).

The aim of the present study was to investigate the functional relationships between oscillatory EEG-dominant components with ERD/ERS method for PP and HS during auditory discrimination tasks within two poststimulus intervals of 0–250 and 250–600 ms. We first focused on time-frequency analysis of delta, theta and alpha rhythms, the appearance of which is well established in PD and is thought to reflect the degree of cortical activation during the information processing. We later shift our attention to the beta and gamma bands, where our aim is to assess the differences between PP and HS in these frequency bands and check an assumption that some PP clinical symptoms stem from excessive synchrony between the basal ganglia and cortical circuits. This investigation of the oscillatory processes and ERD/ERS in HS and PP could contribute to clarification of the disturbances of the neurophysiological mechanisms of this disease.

2. Methods

2.1. Experimental Procedure

We investigated eleven voluntary patients with a mean age of 61 ± 12.2 years (±s.d.; 7 males, 4 females) with a diagnosis of idiopathic Parkinson's disease for no longer than 2.8 years, assessed by a neurologist at the University Neurological Hospital, with score of I on the Hoehn–Yahr scale of motor function (Hoehn & Yahr, 1967). Patients receiving levodopa (L–dopa) drugs (Sinemet) were included in order to reduce the heterogeneity in the medication. During the experimental session, all patients were in off-phase of the medication. None of the patients had dementia, depression, a presence of atherosclerosis, attendant neurological complications or pronounced tremor. The same number of healthy volunteers was included as aged-matched healthy controls with a mean age of 59.5 ± 9.5 years. Screening confirmed that subjects were free of past or current psychiatric and neurological disorders. All participants were right handed and without deficits in hearing. Handedness was assessed by a questionnaire adapted from the Edinburgh Handedness Inventory (Oldfield, 1971). The study was performed with the approval of the local ethics committee. The subjects were introduced to the nature of the investigation and their informed written consent was obtained according to the declaration of Helsinki.

Each subject was comfortably seated in an ergonomically designed chair inside a Faraday cage, monitored by a Canon Video system. The experimental design included a binary sensory-motor reaction task. Each sensory-motor series consisted of 50 computer generated low frequency (LT – 800 Hz) and 50 high frequency (HT – 1000 Hz) acoustic stimuli with an intensity of 60 dB, duration of 50 ms, and an inter-stimulus interval of 2.5–3.5 s presented to the subject in a randomized order. PP and HS were asked to press a key with the index finger of each hand and make rapid and accurate choice responses with the left hand to the high frequency (HT) or with the right hand to the low frequency (LT) tone. The movement performance from the stimulus presentation to the onset of voluntary force production (onset of reaction time) and from the stimulus presentation to the force peak (force peak latency) were measured by a force transducer.

An electroencephalogram (EEG) was recorded from Fz, Cz, Pz, C3' and C4' (10/20, system), using Ag/AgCl Nihon-Kohden electrodes with a reference to both processi mastoidei and a ground electrode, placed on the forehead. An oculogram (EOG) was recorded from m. orbicularis oculi dex. We placed two EOG electrodes next to the eyes to register eye movements. EEG and EOG data were recorded using a Nihon-Kohden EEG-4314F (cut-off frequencies of 0.3–70 Hz) and recorded together with markers of the movement performance as a force profile and a surface electromyographic activity pattern of the first dorsal interosseus muscles (bandpass filtered 0.03–500 Hz). The signals were digitized on-line (10 bit A/D converter, 256 samples/s). The data recordings for the sensorimotor task were synchronized to the marker of the stimulus onset (-0.2 s before and 0.8 s after the stimulus). Only recordings that were artifact-free with respect to event-related potentials were processed. We applied a Chebyshev Type II bandpass filter (1-70 Hz) and second-order notch filter at frequency 50 Hz (AC) component. We defined an independent reference interval in order to quantify the changes in the time-frequency energy density of the signal. We used stimulus-nonrelated subepochs within the resting condition series, distant enough (-1.4 s) from the stimulus onset, not including event-related properties, and exceeding the period of the lowest frequency studied in the signal (1.5 Hz, 0.67 s). We preselected trials by applying a bootstrap estimation within the reference period and a false discovery rate correction for multiple comparisons (0.05) to the available data across the indexes corresponding to time and number of the trial, eliminating the need for a strict assumption of ergodicity (Durka et al., 2004).

2.2. Analysis

The time-frequency analysis (TF) represented the power of a continuous EEG signal as a function of both time and frequency (Matlab®, Mathworks, Inc.). For time amplitude-frequency distributions, the filtered signal was analyzed with a sliding-window fast Fourier transform with length 200 ms and step 10 ms. The amplitude was computed for every time window t and frequency bin f by the real and imaginary Fourier coefficients. The amplitude modulations obtained for each frequency band for each subject in a group were added across trials in order to compare amplitude changes in the post-stimulus intervals with respect to pre-stimulus interval reference amplitudes, i.e., to derive ERD/ERS. This method characterizes the relative amplitude decrement/increment of the given frequency during the post-stimulus period in relation to pre-stimulus amplitude modulation of the same frequency (Pfurtscheller G, Klimesch, 1991). This resulted in ERD/ERS values which could then be presented as percentage changes with respect to the reference interval. Negative values indicate a relative power decrease (ERD), whereas positive values point to a relative power increase (ERS). Relatively, sensory processing takes place during the first post-stimulus interval (T1: 0–250 ms) and cognitive processing during the second post-stimulus interval (T2: 250–600 ms), defined as beginning when a tone ends. The peak amplitude modulations, defined in 10 ms bins, were specified as dominating components. Afterwards, the high amplitudes in each frequency band, respectively, were added over trials and across subjects to compare their amplitude changes in the post-stimulus intervals with those in the reference interval. Thus, we calculated the ERD/ERS of delta ($\delta \sim$ 1.5–4 Hz), theta ($\theta \sim$ 4.1–7 Hz) and alpha ($\alpha \sim$ 7.1–13 Hz) waves as percentage power differences in each frequency band compared to the reference interval for both post-stimulus intervals. We also defined the ERD/ERS of beta ($\beta_1 \sim$ 13.1–20 Hz), ($\beta_2 \sim$ 20.1–32 Hz) and gamma ($\gamma \sim$ 32.1–50 Hz) frequency rhythms during the post-stimulus intervals T1 and T2.

We employed the detection of the temporally dynamic processes similar to the approach applied by Foffani et al. (Foffani et al., 2005) that describes the behaviour of β_1 -, β_2 -, γ - ERS rhythms in zones which vary both in amplitude and frequency. Each zone *i* is characterized by a value *ERSi* (*t*) and a frequency value *Fi* (*t*), both dependent on time, which separately describe event-related synchronization and corresponding frequency modulations for the beta1, beta2 and gamma rhythms in two post-stimulus time windows. The significance of the observed ERS values was tested for each frequency band and EEG channel using a permutation test, including corrections for multiple comparisons between time points for time course analysis (Mason & Newton, 1990). The latencies of ERS and corresponding frequency (relative to stimulus onset) were measured as the last zero-crossing before a significant modulation, after subtracting the baseline mean ERS value. Although the ERS clearly occurred, the relationship between the ERS peaks and the maximum of the average event-related synchronization (AERS) was not always evident. Since the AERSs of different channels were not identical, an exact coincidence between the peaks times was not observed. The probabilities for the amplitudes and latencies were not uniform and the activity distribution was clearly not Gaussian. We estimated the latency shift between the largest peaks for each pair of channels.

We performed statistical analyses of the ERD/ERS for the post-stimulus intervals and assessed the statistical difference between the groups (PP and HS) for each tone type and interval by means of a bootstrap nonparametric procedure (Mason & Newton, 1990). The characteristics were grouped by tone, interval, patients and healthy controls and analyzed by means of a permutation test for multiple comparisons (Mason & Newton, 1990). The computed random distribution for interval was analyzed with a nonparametric test (Kruskal-Wallis [KW] test, $p < 0.05$) for pairs comparison of the scalp leads between patient and control group. This procedure reduces the influence of random variations in experimental conditions between trials. The ERD/ERS analysis served to identify the most robust differences between groups and was generally done for the two time windows. The parameters of the movement performance (onset of reaction time, force peak latency and error of performance) were processed statistically by Mann–Whitney U test.

3. Experimental Results

The grand average ERD/ERS values as a function of time and frequency band at the frontal, central, parietal, left and right motor areas were used for assessment of group means with the standard errors (±SE) for the post-stimulus intervals T1 (0–250 ms) and T2 (250–600 ms) (Figures 1, 2).

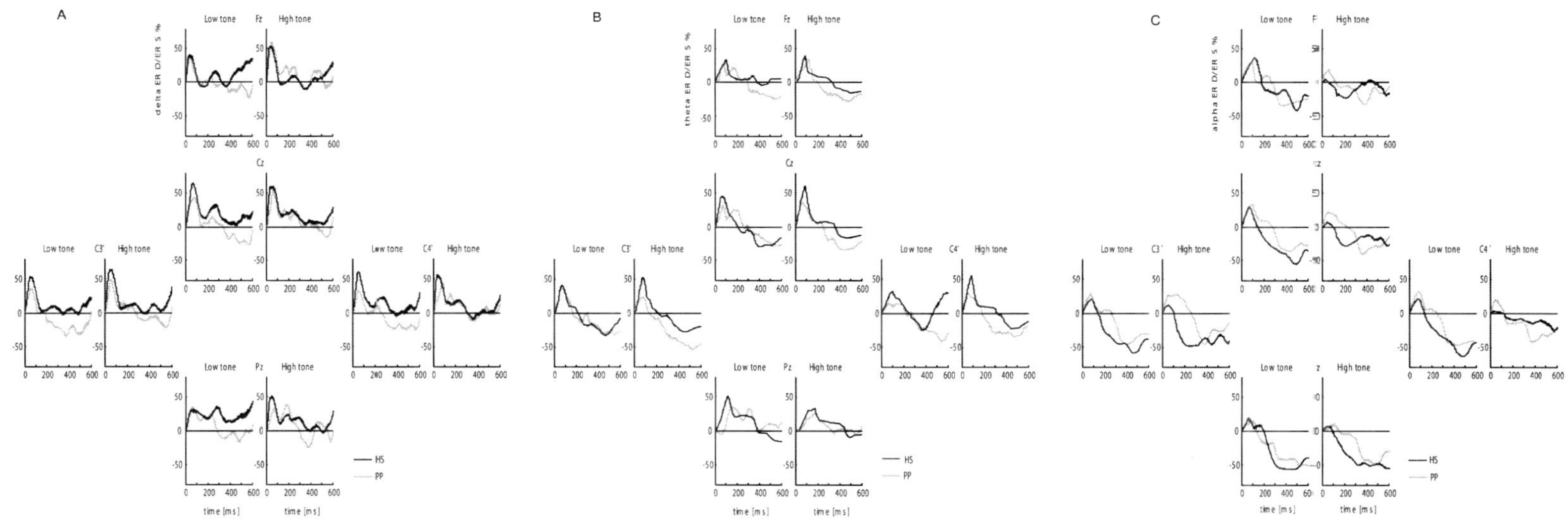

Figure 1. Grand average ERD/ERS over all CS and PP trials in response to the low (left) and high tone (right) were displayed as a function of time (0–600 ms) at channels Fz, Cz, C3’, C4’, Pz for A) δ-, B) θ- , C) α –frequencies.

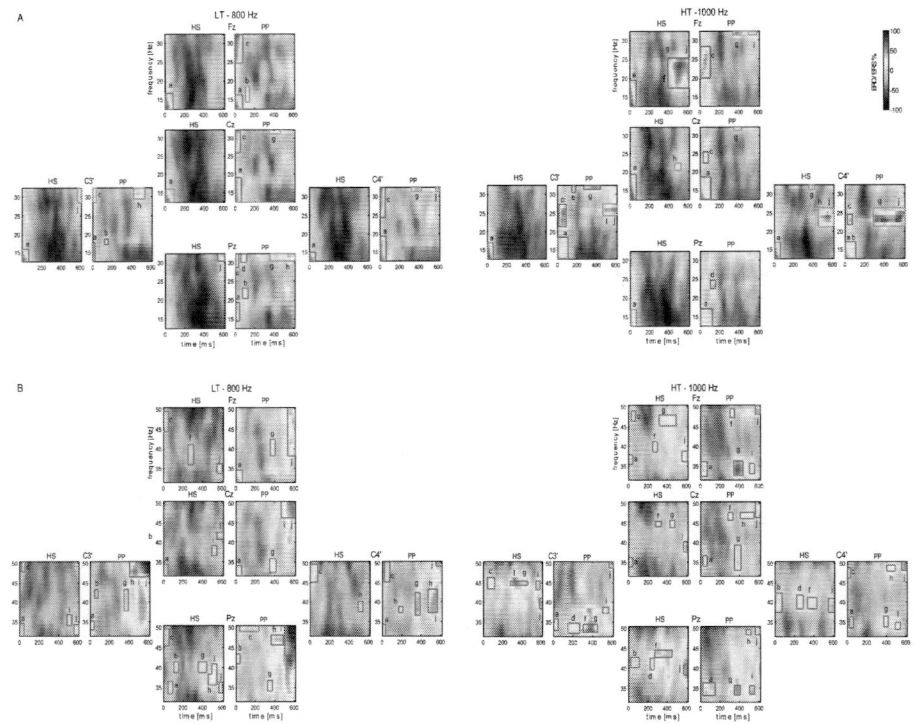

Figure 2. Beta and Gamma band plots of ERD/ERS following stimulus onset. (A) β-ERD/ERS over all HS and PP trials in response to LT (left) and HT (right) displayed as a function of time and frequency for each channel. Important change maxima are relatively indicated by lower case letters, marking consecutive sub-intervals shown in Tables 1, 2: **a**, **b** for β1 and **c**, **d**, **e** for β2 during T1; **f** for β1 and g, h, i, j for β2 during T2 (Table 4). The boxes indicate significant group differences in the same time sub-intervals. (B) γ-ERD/ERS over all HS and PP trials in response to LT (left) and HT (right) displayed as a function of time for all channels. The relevant change maxima are highlighted by lower case letters, marking consecutive sub-intervals a, b, c for T1 (shown in Table 3) and f, g, h, j for T2 (shown in Table 5).

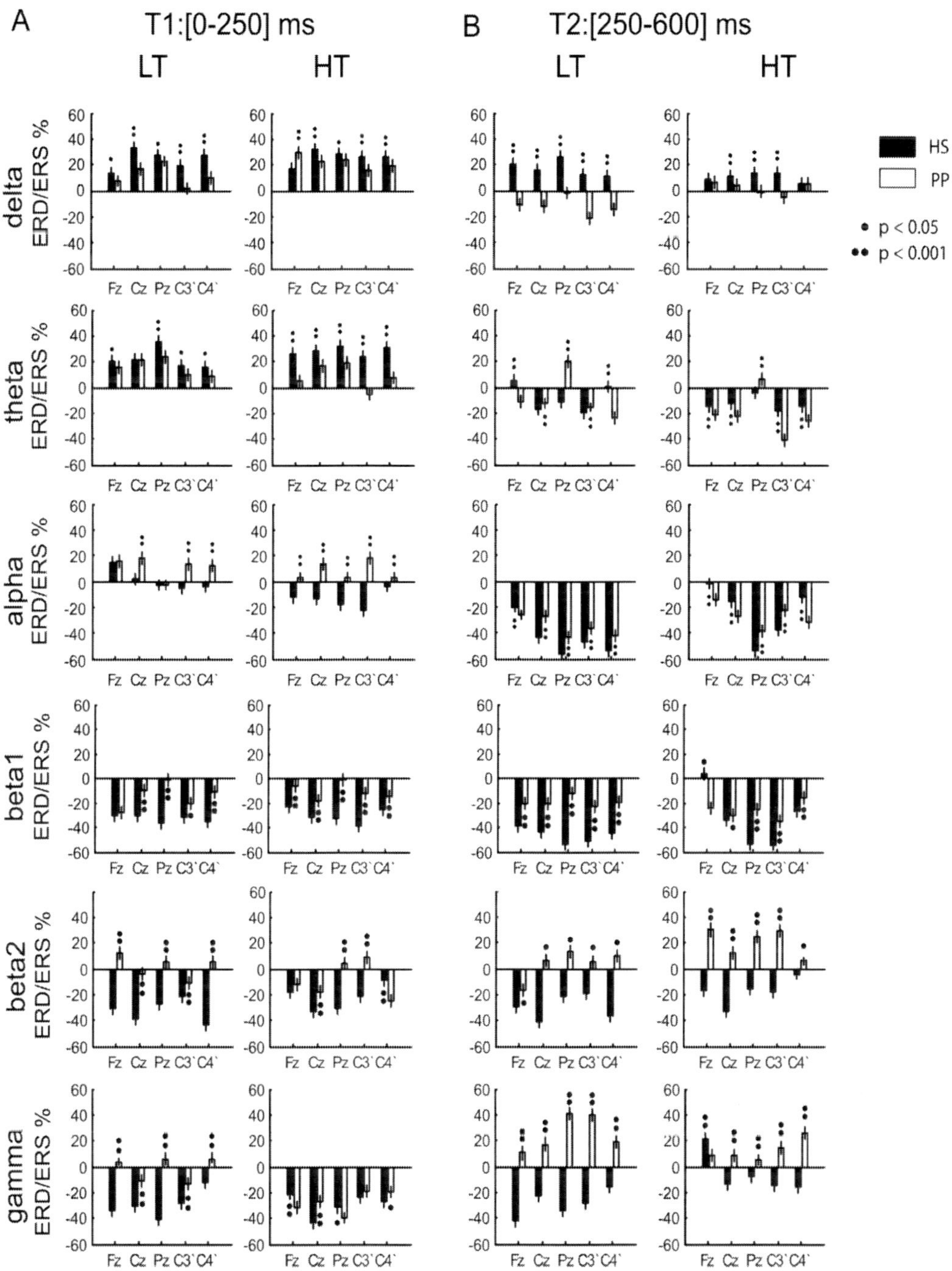

Figure 3. Group means (±SE) and statistical results of δ-, θ-, α-, β1-, β2-, γ-ERD/ERS over all HS and PP trials after LT (left) and HT (right) for the early T1 (A) and late time period T2 (B). The significant difference in the ERD/ERS of HS (black) and PP (white) are presented for each pair channels and marked by * ($p<0.05$) or ** ($p<0.001$, Kruskal-Wallis test).

The statistical group comparison for pair channels are shown graphically to illustrate the delta-, theta-, alpha-, beta-, gamma- ERD/ERS following the low frequency tone type (Figure

3, LT) and high frequency tone type (Figure 3, HT) for the early (0–250 ms) and late (250–600 ms) post-stimulus intervals.

Early (0–250 ms) Post-stimulus Interval

The patterns of δ-ERD/ERS were different between the groups for both intervals in response to both tone types (Dushanova et al., 2009). In the early post-stimulus interval, central δ-ERS amplitude responses were most pronounced in the HS after both tones (Cz, Figures 1A, 3A, 1st row) and in the PP at the frontal side for the high tone (Figure 3A, HT) and parietal area for the low tone (Figure 3A, LT). The least pronounced δ-ERS were those appearing at the frontal side in the HS and at the left motor area in PP following both tones (Figure 3A). The comparison by the bootstrap procedure of δ-ERS between the two groups after the low frequency tone determined that the control δ-ERS was significantly higher than that of the PP for all channels (Fz, Pz, $p < 0.05$; Cz, C3', C4', $p < 0.001$). Both groups displayed δ-ERS following the high frequency tone (Figure 3A, HT). This was significantly higher for the HS than the PP at centro-parietal, left and right motor areas (Cz, C3', C4' $p < 0.001$; Pz, $p < 0.05$). The PD patients' frontal δ-ERS, however, was greater than that of the HS ($p < 0.001$).

In the first post-stimulus interval following both tone types, the θ-ERS responses were most prominent at parietal electrodes for both groups (Figures 1B, 3A, 2nd row). Following the low tone, the θ-ERS elicited was significantly higher for HS than for PP at frontal, left and right motor areas ($p < 0.05$), parietal side ($p < 0.001$), and centrally was non-significantly different (Cz, $p > 0.05$; Figure 3A, LT, 2nd row). Following the high frequency tone, PP produced a significantly lower θ-ERS than the HS at right motor area, fronto-central and parietal sides ($p < 0.001$; Figure 3A, HT). The left motor area showed a pronounced θ-ERS for the HS and a weak θ-ERD in the PP following the high frequency tone.

The PP showed fronto-central, left and right motor α-ERS responses following the LT, while the HS had synchronization only at the fronto-central sides but α-ERD at the parietal side, right and left motor areas (Figures 1C, 3A, LT, 3rd row). The alpha frequency band in the first interval after the LT showed an enhanced central α-ERS for the PP in comparison with the HS ($p < 0.001$; Figure 3A, LT). The right and left motor areas manifested different reversal alteration as α-ERS for the PP and weakly elicited α-ERD for the HS. Following the high tone, we detected significantly different processes, α-ERS in the PP contrasted with α-ERD in the HS at all electrodes, most prominently at the left motor side (Figure 3A, HT).

Despite the early short-term β1 synchrony during the first period, both groups exhibited mean β1 ERD following either tone type (Figure 3A, 4th row), which was significantly greater for HS in comparison with PP in all channels ($p < 0.001$, bootstrap, KW test). The prolonged β2 synchronized bursts for PP during T1 had an effect on the β2 band behavior during the entire early time period (T1, Figure 3A, 5th row). The mean β2 ERS were prominent only in PP at frontal-parietal and right motor areas after LT and at parietal and left motor areas following HT during T1 (Figure 3A). The results for γ closely resembled the β2-frequency band behaviour. The mean γ ERS were more prominent than those for β2 in PP. During the sensory processing (T1), PP showed mean γ ERS responses at fronto-parietal and right motor areas after LT, but not after HT (Figure 3A, 6th row).

Table 1. Mean ERS β1 bursts (%) across the trials with mean frequency peaks F (Hz) for HS and PP after LT and HT presented in short–term zones D (ms) during early T1 period

Channels Subjects Tone(period)	Fz ERS(%)/ F/ t (±SE) D	Cz ERS(%)/ F/ t (±SE) D	Pz ERS(%)/ F/ t (±SE) D	C3' ERS(%)/ F/ t (±SE) D	C4' ERS(%)/F /t (±SE) D
			HS		
LT (T1)	64.1 (±1.2)/ 13 Hz 40 (±7.3) ms [8–72] ms (a)	50.1 (±5.8) /13 Hz 36 (±6.9) ms [8–64] ms (a)	18.8 (±0.5)/13 Hz 20 (±5.2) ms [8–32] ms (a)	41 (±3.3)* /13 Hz 36 (±28) ms [8–80] ms (a)	44.9(±5.2)*/13 Hz 32.0 (±6.5)** ms [8–72] ms (a)
HT (T1)	68.6(±10.2)/13.9(±0.3)Hz [13–15] Hz 36 (±6.9) ms [8–64] ms (a)	62.7(±8.5)*/13.7(±0.3)Hz [13–15] Hz 32 (±6.5)** ms [8–56] ms (a)	30.49(±3.3)*/13Hz 28(±6.1)** ms [8–48] ms (a)	54.7(±8.3)*/13.5(±0.3)Hz [13–15] Hz 32(±6.5) ms [8–56] ms (a)	50.4(±6.6)*/13 Hz 32(±6.5) ms [8–56] ms (a)
Subjects Tone(period)			PP		
LT (T1)	69.4(±0.9)*/ 13 Hz 12 (±4)** ms [8–80] ms (a)	57.2(±3)*/13 Hz 16 (±4.6)** ms [8–64] ms (a)	–	36.8(±2.8)/13 Hz 12 (±4)** ms [8–16] ms (a)	33.9(±5)/13Hz 48 (±6.5) ms [24–72] ms (a)
	15.8(±0.2)/14(±0.6) Hz 124 (±5.2) ms [112–136] ms (b)	–	–	19.3(±1.3)/18 Hz 148 (±5.2) ms [136–160] ms (b)	–
HT (T1)	–	34.4(±3.8)/13.8(±0.3) Hz 53.6 (±11.2) ms [8–104] ms (a)	21.3(±0.8)/ 13 Hz 46 (±8.7) ms [8–88] ms (a)	35(±1.9)/ 13 Hz 20(±5.2)** ms [8–32] ms (a)	40.2(±2.8)/13 Hz 24 (±5.7) ** ms [8–40] ms (a)
	–	–	–	–	25(±1.1)/16 Hz 92 (±5.2) ms [80–112] ms (b)

Note: Bold-marked mean ERS β1 bursts (%) (±SE) are maximum across the channels for each condition and subject separately for LT(T1), HT(T1); **F** (Hz) – mean frequency peaks (±SE) of the maximum ERS β1 bursts across trials; **t** (ms) – mean times (±SE) of maximum ERS β1 bursts across trials during T1 with respect to stimulus onset; **D** (ms) – time duration of these short–term zones with ERS β1 bursts; [1] * =a significant difference in ERS between the groups HS, PP for each channel and tone separately, marked the higher value ($p<0.001$, KW test); [2] ** =a significant difference in **t** between the groups for each channel and tone separately, marked the shorter value ($p<0.001$, KW test). Lower case letters in Table 1 marked consecutive sub-intervals with ERS β1 bursts: **a**, **b** during T1.

Table 2. Mean ERS β2 bursts (%) across the trials with mean frequency peaks F (Hz) for HS and PP after LT and HT presented in short–term zones D (ms) during early T1 period (same format as Table 1)

Channels Subjects tone(period)	Fz ERS(%)/ F/ t (±SE) D	Cz ERS(%)/F/ t (±SE) D	Pz ERS(%)/ F/ t (±SE) D HS	C3’ ERS(%)/ F/ t (±SE) D	C4’ ERS(%)/F /t (±SE) D
LT (T1)	–	–	–	–	–
HT (T1)	–	–	–	–	–
Subjects tone(period)			PP		
LT (T1)	61.1 (±1.1)/ 32 Hz 40 (±5.7) ms [24–56] ms (c)	30.6(±4.7)/32Hz 48 (±5.7) ms [32–64] ms (c)	23.2(±1.7)/25.2(±1.6)Hz 64 (±9.2) ms [8–120] ms (c, d)	22.4(±3.7)/29.7(±0.3)Hz 32 (±4.6) ms [24–40] ms (c)	60.3(±1.1)/32Hz 12 (±4) ms [8–16] ms (c)
HT (T1)	41.2(±2)/24.3(±0.5)Hz 60 (±8.8) ms [32–88] ms (c)	29.8(±1.1)/23Hz 64 (±4.6) ms [56–72] ms (c)	28(±2.7)/24.2(±1.1)Hz 121.6 (±9.1) ms [64–160] ms (d)	46(±2.8)/23.9(±0.4) Hz 64 (±6.5) ms [40–88] ms (c)	38.1(±1.2)/22Hz 56 (±4.6) ms [48–72] ms (c)
	–	–	–	19.3(±0.8)/32 Hz 168 (±4.6) ms [160–176] ms (e)	–

Note: Lower case letters in Table 2 marked consecutive sub-intervals with ERS β2 bursts: **c**, **d**, **e** during T1.

Table 3. Mean ERS γ bursts (%) across the trials with mean spectral peaks F (Hz) for HS and PP after LT and HT presented in short–term zones D (ms) during early T1period (same format as Table 1)

Channels Subjects tone(period)	Fz ERS(%)/ F/ t (±SE) D	Cz ERS(%)/ F/ t (±SE) D	Pz ERS(%)/ F/ t (±SE) D HS	C3' ERS(%)/ F/ t (±SE) D	C4' ERS(%)/F /t (±SE) D
LT (T1)	22.7(±2)/ 47 Hz (c) 20 (±5.2)** ms [8–32] ms	19.8(±2.8)/ 33 Hz (a) 12 (±4)** ms [8–16] ms	20.4(±1.1)/39.4(±1.3)Hz [35–48] Hz (a, b, c) 86.4 (±9.9) ms [24–144] ms	18.9(±1.4)/41(±5.2) Hz [32,50] Hz (a, c) 28 (±9.5)** ms [8–48] ms	30.7(±1.9)/47.6(±0.4)Hz [46–49] Hz (c) 40 (±7.3)** ms [8–72] ms
HT (T1)	18.4(±1)/42.2(±2.5)Hz [36,37,47,48] Hz (a, b) 28 (±6.1)** ms [8–48] ms	19.4(±1.4)/36 Hz (a) 20 (±5.2)** ms [8–32] ms	19.1(±0.5)/40.8(±0.2)Hz 72 (±8.6)** ms (b) [24–120] ms	33.8(±2)/43.8(±0.4) Hz 120 (±12.6) ms (c) [8–250] ms	30.1(±2.3)*/39.9(±0.5)Hz 44 (±7.7) ms (b) [8–80] ms
	–	–	32.2(±4)/ 42.6(±0.4) Hz 208 (±8.6) ms [160–250] ms (d)	–	32.8(±3.3)/41.5(±0.2) Hz 216(±8) ms [176–250] ms (d)
Subjects tone(period)			PP		

Table 3. Continued

LT (T1)	49(±5.5)*/ 32 Hz (a) 44 (±7.7) ms [8–80] ms	36.4(±4)*/32.3(±0.7)Hz(a) 36 (±6.9) ms [8–64] ms	32.5(±2.1)*/45.5(±1.1)Hz [42–50] Hz (c) 56 (±8.6)** ms [8–104] ms	28.2(±3.4)*/37.5(±1.6)Hz [34,35,43] Hz (a) 36 (±6.9) ms [8–64] ms	32.4(±4.3)/43.2(±2.2) Hz [32,46,48,49,50]Hz (a,d) 56 (±8.6) ms [8–104] ms
	–	–	21.9(±0.8)/47.8(±1.4)Hz [33,49,50] Hz (c) 196 (±9.5) ms [136–250] ms	–	30.1(±2.7)/37.9(±0.1) Hz 192 (±8) ms (b) [152–232] ms
HT (T1)	49.7(±3.5)*/34 Hz (a) 40 (±7.3) ms [8–72] ms	27.6(±2.4)*/34.9(±0.1)Hz 44 (±7) ms (a) [16–72] ms	34.3(±3)*/ 34.1 (±0.1) Hz 112 (±11.8) ms (a) [16–208] ms	29.5(±2.6)/34.4(±0.2) Hz 36 (±7)** ms (a) [8–64] ms	21.1(±2)/41(±3.5)Hz [34,36,47] Hz (a) 20 (±5.2)** ms [8–32] ms
	–	–	–	29.9(±1.5)/33.4(±0.1) Hz 188 (±10.1) ms (d) [120–250] ms	–

Note: Lower case letters in Table 3 marked consecutive sub-intervals with ERS γ bursts: **a**, **b**, **c** during T1.

Table 4. Mean ERS β2 bursts (%) across the trials with mean frequency peaks F (Hz) for HS and PP after LT and HT presented in short–term zones D (ms) during late T2 period (same format as Table 1)

Channels Subjects tone(period)	Fz ERS(%)/ F/ t (±SE) D	Cz ERS(%)/F/ t (±SE) D	Pz ERS(%)/ F/ t (±SE) D HS	C3' ERS(%)/ F/ t (±SE) D	C4' ERS(%)/F /t (±SE) D
LT (T2)	–	–	25.8(±1.6)/32Hz	19.6(±1.3)/30.5(±0.2)Hz	–

			584 (±6.5) ms [560–600] ms (j)	588 (±6.1) ms [576–600] ms (j)	
HT (T2)	43.9(±3.1)/23.1(±0.1)Hz 516 (±11.5) ms [424–600] ms (g–j)	18.1(±0.6)/21Hz 488(±5.7) ms [472–504] ms (h)	–	–	28.2(±2.3)/ 32 Hz 384 (±6.5)** ms [360–408] ms (g)
	-	–	–	–	40.6(±3.1)/23.4(±0.4)Hz 532 (±10.6) ms [456–600] ms (h–j)
Subjects tone(period)			PP		
LT (T2)					
	–	15.5(±0.1)/32Hz 392 (±4.6) ms [384–400] ms (g)	17.9(±1)/ 32 Hz 376 (±4.6) ms [368–384] ms (g)	–	30.7(±2.1)/32 Hz 368 (±8.6) ms [320–416] ms (g)
HT (T2)	–	–	24.7(±1.7)/30.9(±0.1)Hz 528 (±10.8) ms [448–600] ms (h)	26(±1.5)/29.3(±0.5)Hz 492 (±9.5) ms [432–552] ms (h)	–
	36.7(±2.9)/ 32Hz 388(±8.9)** ms [336–440] ms (g)	23.8(±1.3)*/32Hz 376 (±7.3)** ms [344–408] ms (g)	25.3(±0.7)/30(±0.03)Hz 452 (±14.8) ms [296–600] ms (g-j)	34.8(±2)/32Hz 352 (±11.8) ms [256–448] ms (g)	60.6(±2)*/22.2(±0.1)Hz 444 (±15.1) ms [280–600] ms (g–j)
	26.1(±1.0)/ 32Hz 548 (±9.5) ms [488–600] ms (i)	–	–	44.8(±2.5)/25Hz 532 (±10.6) ms [456–600] ms (i, j)	–

Note: Lower case letters in Table 4 marked consecutive sub-intervals with ERS β2 bursts: **g**, **h**, **i**, **j** during T2.

Table 5. Mean ERS γ bursts (%) across the trials with mean spectral peaks F (Hz) for HS and PP after LT and HT presented in short–term zones D (ms) during late T2 period (same format as Table 1)

Channels Subjects tone (period)	Fz ERS(%)/ F/ t (±SE) D	Cz ERS(%)/ F/ t (±SE) D	Pz ERS(%)/ F/ t (±SE) D HS	C3' ERS(%)/ F/ t (±SE) D	C4' ERS(%)/F /t (±SE) D
LT (T2)	16.9(±0.5)/ 38 Hz 260 (±4)** ms [250–264] ms (f)	21.8(±1.2)/40 (±0.5) Hz [38, 42] Hz 548 (±9.5) ms [488–600] ms (i-j)	36.8(±1.6)/38.8(±0.4)Hz [35–41] Hz 464 (±14.2) ms [320–600] ms (g-j)	24.3(±1.6)/34.1(±0.3)Hz [33–35] Hz 552.6 (±9.9) ms [496–600] ms (i-j)	16.6(±1.1)/39Hz 524(±5.2) ms [512-536] ms (h)
	24.6(±2)/ 35 Hz 568 (±6.5) ms [544–592] ms (j)	–	–	–	–
HT (T2)	36.3(±1.4)/43(±0.6)Hz [37,38,40,46,47] Hz 432 (±15.7) ms [250–600] ms (f, g, j)	25.6(±2.2)/45.9(±0.6) Hz [45, 49] Hz 312 (±7.3)** ms [280–344] ms (f)	53.7(±2)/ 43.2(±0.2) Hz 432(±15.7) ms [250–600] ms (f, j)	55(±3)/ 44.1(±0.3) Hz [39–45] Hz 432 (±15.7) ms [250–600] ms (f, g, i, j)	32.4(±1.9)/40.7(±0.4) Hz [39–44] Hz 376 (±13.1)** ms [250–496] ms (f)
	–	39(±3)/ 45 Hz 444 (±7.7) ms [408–488] ms (g)	–	–	46(±5.7)/39.3(±0.6) Hz 568 (±8) ms [528–600] ms (j)
	–	48(±7)/ 38.6 (±0.2) Hz 572 (±7.7) ms [536–600] ms (j)	–	–	–

Subjects tone(period)			PP		
LT (T2)	39(±2.6)*/42.7(±0.9)Hz [34–42] Hz 468 (±14.0) ms [328–600] ms (g-j)	48.8(±4.2)*/44.3(±1.2)Hz [35–50] Hz 468 (±14) ms [328–600] ms (g-j)	68.9(±5.4) */ 45(±0.7) Hz [35–50] Hz 432 (±15.7) ms [250–600] ms (g-j)	82.1(±5.2) */47(±0.7) Hz [36–50] Hz 468 (±14.0) ms [328–600] ms (g-j)	54.3(±2.9)*/41.6(±0.7)Hz [38–50] Hz 452 (±14.8) ms [296–600] ms (g-j)
HT (T2)	58(±3.5)*/40.3(±1.1)Hz [34,48,49] Hz 444 (±15.1)ms [280–600] ms (f, g, j)	43(±1.5)/44.9(±0.7)Hz, [47,50, 35,38,39] Hz 432 (±15.7) ms [250–600] ms (f, g, i, j)	52(±2.1)/ 37.8(±0.9) Hz [34,35, 41,45,49,50] Hz 436 (±15.5) ms [264–600] ms (g, i, j)	55.5(±3.2)/36.5(±0.9) Hz [33, 50] Hz 432 (±15.7) ms [250–600] ms (f, g, i, j)	41.6(±2.3)/42.4(±1.2)Hz [34–36, 49,50] Hz 448 (±15) ms [288–600] ms (g, h, i, j)

Note: Lower case letters in Table 5 marked consecutive sub-intervals with ERS γ bursts: **f**, **g**, **h**, **i**, **j** during T2.

During the early post-stimulus interval T1, ERD/ERS β1 patterns appeared in the lower frequency portions of this band for each electrode and group (Dushanova et al., 2010). The maximum synchronized β1 bursts across the channels were localized over fronto-central sides for both groups after LT, but had significantly higher amplitude and shorter durations in PP than HS ($p < 0.001$, bootstrap, KW test; Figure 2A, Table 1, a). Later β1 ERS bursts were found only in PP over the frontal side and left motor area (Figure 2A, Table 1, LT (T1), b). Synchronized β1 bursts were centered on right motor side for PP and fronto-central sides for HS, following HT (Figure 2A, Table 1, HT (T1), a, b). In PP, the peaks were of a significantly lower amplitude and peaked later at centro-parietal sides than in HS (Table 1, HT (T1), a).

During T1, ERS β2 bursts were present only in PP after either tone (Figure 2A, Table 2, LT(T1), HT(T1)). Their maximum amplitude across the scalp was localized over frontal side for either tone, right and left motor areas respectively for LT and HT (Table 2, **c**). They peaked earlier at the right than at the left motor area, fronto-central leads (Table 2, **c**), and parietal side (Table 2, **d**) for either tone, but in higher β2 frequency range after LT than after HT.

The scalp γ burst topography was localized on frontal area for PP following either tone (Figure 2B, Table 3, LT(T1), HT(T1), a) and for HS – on right motor area after LT and left motor area after HT during T1 (Table 3, **c**). The ERS γ bursts in PP peaked later than in HS at the fronto-central for either tone, and at right and left motor areas after LT (Table 3, a, b, c). They were of significantly greater amplitude and more prolonged duration in PP than in HS at the fronto-central and parietal sides after either tone ($p < 0.001$, Table 3). Later γ burst ERS was also found during T1 over right motor and parietal areas in PP after LT (Table 3, b, c), and over left motor area in PP, but parietal and right motor sides in HS, after HT (Table 3, d).

Late (250–600 ms) Post-stimulus Interval

The HS maintained δ-ERS at all electrodes during the late post-stimulus interval T2 following the low frequency tone, while in the PP the early post-stimulus δ-ERS was reversed to become δ-ERD in the late post-stimulus interval (Figure 1A, 3B, 1st row, LT). The highest δ-ERS was located at parietal side for the HS, whereas the PP had a less enhanced parietal δ-ERD. The PP showed the most enhanced δ-ERD at the left motor area (Figure 3B, LT). Following the high frequency tone, δ-ERS was elicited at all electrodes in the HS (Figure 3B, HT). The PP group, in comparison with the HS, showed a less pronounced central δ-ERS (Cz, $p < 0.001$) and specific δ-ERD at parietal and left motor areas ($p < 0.001$).

The comparison of the groups during the late period following the low frequency tone showed θ-ERD in both groups at most electrodes with the following exceptions. PP recorded a large parietal θ-ERS while a less prominent θ-ERS appeared in HS at frontal and right motor areas (Figure 1B, 3B, 2nd row, LT). The most pronounced θ-ERDs for HS were at central and left motor areas ($p < 0.001$). The signal at the parietal area was characterized by a very prominent θ-ERS in the patients but by θ-ERD in the control group. In response to the high frequency tone, we found a significant θ-ERD for the PP as compared with the HS at fronto-central, left and right motor areas ($p < 0.001$; Figure 3B, HT). The signal at the parietal area was characterized in a similar manner to that elicited by a low tone, but with a less pronounced θ-ERS in the PP and smaller θ-ERD in the HS.

Alpha-ERD differences between the groups were observed after low as well as high frequency tones during the late period (Figures 1C, 3B, 3rd row). After the low tone, the HS α-ERD means were significantly more pronounced than those of the PP at centro-parietal sides, left and right motor areas ($p < 0.001$; Figure 3B, 3rd row, LT). The PP displayed a higher frontal α-ERD than that in the HS ($p < 0.001$). Alpha-ERD differences were observed between the groups following the high tone (Figure 3B, HT). The α-ERD signals of HS at parietal and left motor areas were more pronounced than in the PP ($p < 0.001$). The PP fronto-central and right motor α-ERD signals had greater respective magnitudes than those in HS ($p < 0.001$).

During the second period, both groups exhibited mean β1 ERD following either tone type and interval (Figure 3B, 4th row), which was significantly greater for HS in comparison with PP in all channels, except frontal mean β1 ERS the control group following HT during T2 (Figure 3B, 4th row; Pz, C3', C4', $p < 0.001$; Cz, $p < 0.05$). The comparison of the groups also showed mean β2 ERD in HS and β2 ERS in PP during T2 following either tone in all channels except for the frontal area, which showed mean β2 ERD in both groups after LT, significantly more prominently in HS (LT, $p < 0.05$; Figure 3B, 5th row). The γ ERD in HS and γ ERS in PP were observed in all channels after either tone during the cognitive processing (T2), except frontal γ ERS after HT, which was more pronounced in HS than in PP ($p < 0.001$, Figure 3B, 6th row).

During the late post-stimulus interval T2, frontal synchronized β1 peaks at 20 Hz were extracted only in HS after HT and had a prolonged latency of 176 ± 11.5 ms (Figure 2A, HT (T2), f: 516 (±11.5) ms).

During T2, the maximum value of β2 bursts across all recorded areas was centered on right motor area in PP (Figure 2A, Table 4, g) but on parietal side in HS following LT (Table 4, LT(T2), **j**). More widely distributed, prolonged β2 ERS bursts were pronounced over all locations for PP following HT, but limited to frontal-central and right motor areas for HS (Table 4, HT(T2)). The maximum scalp β2 burst was localized at the right motor area in PP and frontal area in HS after HT (Table 4, g - j). At the right motor area, spectral peaks of low β2 exhibited significantly more exaggerated and prolonged bursts in PP than in HS ($p < 0.001$, Table 4, g - j). The frontal synchronized high frequency β2 bursts appeared in PP during two subintervals, the first one with a significantly shorter latency than the low frequency β2 bursts in HS ($p < 0.001$, Table 4, g). In PP, synchronized bursts of high β2 were generated on the left motor area (Table 2, g) earlier than larger synchronized bursts at low β2 (Table 4, i, j).

During T2, synchronized frontal γ bursts were extracted from both groups after either tone (Figure 2B, Table 5, f) and peaked later in PP. The scalp γ bursts were with significantly higher amplitudes in PP than the equivalent responses from HS ($p < 0.001$, Table 5).

Parkinson's patients also showed a longer reaction time onset, but the difference between the two groups was not significant: 440.5 ± 135.8 ms for HS and 508.4 ± 148.5 ms for PP in response to a low tone (mean ± s.d., $p > 0.05$), 455.7 ± 134.6 ms for HS and 500.4 ± 146.5 ms for PP in response to a high tone ($p > 0.05$). Parkinson's patients had significantly longer force peak latency (FPL) in response to either tone type. The FPLs were 672.8 ± 154.7 ms for HS and 919.96 ± 163.7 ms for PP in response to the low tone ($p < 0.02$). The FPLs were 690.7 ± 148.7 ms for HS and 934.9 ± 160.2 ms for PP in response to the high tone ($p < 0.05$). The mean errors (false and missing responses) were 4.5 and 5.1 for PP, respectively, in response to the low and high tone. The mean errors were 3.5 and 4.4 for HS, respectively, in response to the low and high tone. The differences of errors between the two groups were not significant ($p > 0.05$).

4. Discussion

The data obtained confirmed that event-related oscillatory responses in different frequency bands vary with sensory and cognitive processes. We found functional differences between event-related oscillatory activity for cognitive and sensory-motor information processing, and a clear distinction between PP and HS in both the stimuli encoding (0–250 ms) and cognitive processing (250–600 ms) intervals. Attended stimuli produced theta response synchronizations in both groups, more markedly in HS, in the first period up to 250 ms after stimulation. Enhanced theta waves in the early period (up to 250 ms) of visual and auditory stimuli have also been described by Basar (1980), Schurmann and Basar-Eroglu (1994). Theta frequency rhythms are dominant oscillations within the hippocampal formation, which is of crucial importance for the encoding of new information (Klimesch, 1997; Klimesch et al., 2005). In the late post-stimulus period functionally related to cognitive processing, θ-ERD response was predominant, excluding the parietal θ-ERS in PP in response to both frequency tones and the frontal θ-ERS in HS in response to the low frequency tone.

Prominent differences in the α-ERD/ERS responses between the groups were observed during the first time period of 0–250 ms. A widely distributed fronto-central α-ERS manifested in PP in response to both tone types during the first 250 ms after stimulus was absent in HS. In the second period after stimulus absence, α-ERD was found in both groups in response to both tone types. This was generally more prominent in HS, with the exception of frontal side in response to either a high or low frequency tone. Alpha-ERD was more prominent at central and right motor areas in response to the high frequency tone in PP.

Theta and alpha frequency ERD/ERS were significantly different between subject groups. It is known that the oscillatory alterations to θ-ERS are related to memory encoding (Klimesch et al., 2001; Jensen & Tesche, 2002).

Alpha-ERS most probably demonstrates active working memory or attentional processes (Klimesch, 1997; Jensen et al., 2002), whereas α-ERD is functionally related to mental activity (Basar, 1980) and reflects memory search processes (Klimesch, 1997; Klimesch et al., 2005; Pesonen et al., 2006). The recognition of auditory stimuli elicits widespread α-ERD responses (Krause et al., 1994). It is accepted that alpha oscillations are mainly generated by cortico-cortical and thalamo-cortical neuronal networks (Lopes da Silva et al., 1980; Schmiedt et al., 2005; Ellfolk et al., 2006). This fact, together with the changes in the metabolic patterns of thalamic, premotor and prefrontal cortex, parieto-occipital regions, etc., that occur in PP (Fukuda et al., 2001) could explain the abnormality of early time period α-ERS in the PP compared to the HS. Observed slight activity of the basal ganglia–thalamic and cerebellar–thalamic pathways might be implicated in the development of parkinsonian symptoms (Rolland et al., 2007).

Schmiedt et al. (2005) also found differences between PP and HS in the θ- and α-frequency ERD/ERS responses during working memory encoding but in a visual working memory paradigm. We cannot draw direct parallels between their results and ours in the present study because of the different stimulus modality. The early and late δ post-stimulus activities were enhanced in HS. The late period, related to cognitive information processing, exhibited δ-ERS in HS and δ-ERD in PP at most electrodes in response to a low frequency tone, and at parietal and left motor areas in response to the high frequency tone. Many authors agree that the main power of P300 is in the delta range (Demiralp et al., 1999; Karakas et al., 2000; Klimesch et al., 2000; Klimesch et al., 2006). The lower δ-ERS in PP in the late post-stimulus period, which becomes δ-ERD in some recordings, could explain the lower P3 amplitude observed in PP (Philipova et al., 1997). Our patients were medicated by L-dopa drugs and this medication may have had some effect on the present findings. One of the models (Leblois et al., 2006) supposed that high dopamine depletion could modify the network dynamic state from an imbalance between the feedbacks and lead to synchronous oscillations driven by a hyperdirect loop appearing in basal ganglia after inactivation of the striatum.

A reduction in this α-ERD/ERS abnormality and a consequent improvement in PP performance during working memory tasks have been found as the result of L-dopa (Lewis et al., 2003; Marini et al., 2003; Shohamy et al., 2005; Devos et al., 2004). Nevertheless, we found some differences between the two groups. The memory related and stimulus categorized ERD/ERS responses at all these frequencies reflected different underlying neuropathological and cognitive changes in this neurodegenerative disease. Theta activity is suggested to be mostly engaged in memory operations (Klimesch et al., 1996; Karakas et al., 2000; Jensen et al., 2002b) and this θ pathological synchronized enhancement in PD could explain the cognitive dysfunction commonly occurring even in the early stages of Parkinson's disease (Lewis et al., 2003). We found specific significant differences at left motor area as θ-ERS in the HS and θ-ERD in the PP during sensory-motor processing (early period) following the high frequency tone. We also detected different processes of θ-ERS for the PP and ERD for the HS in the parietal lead during the cognitive information processing (late period) following both tones, which reflects different task-related activation of the associative posterior cortex.

These findings are probably due to the auditory cortex being located in the dorsal and lateral part of the superior temporal gyrus as well as in the inferior parietal lobule (Konig et al., 2005). The absence of α-ERD at the frontal electrode locations in the patients with PD indicated that the PP, compared with HS, used different cognitive strategies for stimulus response processing which are normally implemented by fronto-striatal circuits (Krause, 2006). The late higher fronto-central α-ERD in PP accompanied by a lower P3 component amplitude, especially in the fronto-central sides, reflects a disturbance in the frontal regulation of attentional processes as well a disturbance of the basal ganglia activity and their related thalamo-cortical neuronal nets (Stam et al., 1993; Piccirilli et al., 1989; Schmiedt et al., 2005).

In PP, we found hemispheric lateralization for sensory and cognitive processing concerning θ-ERD/ERS at left and right motor areas as well as a significantly higher α-ERS at left compared to right motor area. This finding corresponds with the results of Magnani et al., 1998, Defebvre et al., 1996. These authors suggested that other cortical areas may be activated both to compensate for a dysfunction of motor preparation and to increase the level of cortical activity necessary for the realization of the movement. Another possible explanation is that this hemispheric lateralization is connected with auditory attention and hemispheric differences in the processing of high and low frequencies (Ivry & Robertson, 1998).

Post-stimulus β1 ERD was elicited from both groups during sensory (T1) and cognitive information (T2) processing, though this was significantly more pronounced in HS in response to both tone types at all electrodes. The greater β1 ERD in HS can be explained by the increased excitability level of the neurons (Pfurtscheller & Lopes da Silva, 1999; Brown & Marsden, 1998). Late post-stimulus frontal β1 ERS (T2) was evident only in HS following HT. This HS ERS, comprising components in the band between 13 and 20 Hz, may represent an inhibited frontal cortical network, at least under certain circumstances (Pfurtscheller & Lopes da Silva, 199; Engel et al., 2001).

A frontal β2 ERD was maintained in both groups during the cognitive information processing (T2) following LT, though this was weaker in PP. β2 ERS was only observed in PP. These were weakly elicited during the sensory stimuli processing (T1) and appeared at fronto-parietal and left motor areas (LT: Fz, C3', Pz; HT: C3', Pz). β2 ERS in PP was more prominent during cognitive processing (T2) after either tone type, but particularly so following HT. The β2 change reversals compared to β1 which we observed for the PD patients support the hypothesis of Marceglia et al. (2009), that two distinct information channels in the cortico-basal ganglia–thalamo-cortical loop, involved in motor and non-motor information processing, are formed in the parkinsonian brain. The frontal β synchronization at 20–30 Hz arises both from communication with, and also from within, the STN (Williams et al., 2003). The β synchrony has been ascribed predominantly to a lack of dopaminergic activity in the striatum which, together with the STN, is the recipient of cortical input to the basal ganglia (Fogelson et al., 2006; Williams et al., 2002). Studies with unmedicated PD patients have revealed prominent oscillations in 'basal ganglia β frequency band' (Weinberger et al., 2006; Kühn et al., 2006; Priori et al., 2004; Fogelson et al., 2006). The engagement of the basal ganglia in β band synchronization is found when there is acute or chronic dopaminergic hypoactivity, and while primarily associated with bradykinesia and rigidity, it has also been associated with impairments to complex movements and motor related cognitive behaviour because of the widespread basal ganglia connectivity with the cerebral cortex (Terman et al., 2002). Further, the pathological β synchrony in the cerebellum might lead to a purer breakdown of simple motor tasks because of more focal cerebellothalamic projections into the cerebral cortex that are concentrated on the primary motor cortex (Leblois et al., 2007). A relative functional division between activities in the β band might be supported by the evidence for different patterns of pharmacological sensitivity (Priori et al., 2004) and cortico-subthalamic coupling (Fogelson et al., 2006). The dopaminergic drug treatment suppressed mainly β1 synchrony, graded by the amount of drug-induced suppression in the STN (Kühn et al., 2006; Wang et al., 2005) and cerebral cortex, correlating with the level of improvement in bradykinesia and rigidity but not in parkinsonian rest tremor (Weinberger et al., 2006; Silberstein et al., 2005), the latter of which probably has an independent pathophysiological substrate (Rivlin-Etzion et al., 2006).

Our group of patients showed a significantly reduced γ–ERD compared with HS over central and left motor areas, and only PP showed γ–ERS over fronto-parietal and right motor areas following LT during the sensory stimuli processing (T1). A widespread γ–ERS appeared during later cognitive processing (T2), and then only in PP, following either tone type, with the exception of a more prominent frontal γ–ERS in HS following HT. In our study, we observed switches between cortical activity in the β2 and γ band oscillations. Hence we concluded that a reduction in β2-band synchronized activity allows higher frequency oscillatory activity in the γ range leading to its synchronization. The observed energy changes in the β2 and γ bands indicate that an increase in one is accompanied by a decrease in the other. These T2 changes in PP were more pronounced in the motor cortex than in the parietal and even frontal cortex data. In the parkinsonian state, there was a tendency towards increased synchronized higher frequency fluctuations, specifically in the motor cortex, where instances of peaks were found after both tone types. Except for the β2 band series of data during cognitive processing (T2) after HT, the difference between magnitude of the peaks in the frontal, parietal and contralateral motor areas did not reach significance. Recent data demonstrate that the disruptions of the beta and gamma range cortical rhythms are based on the disturbed temporal relationship between cortical oscillatory activity and basal ganglia activity in Parkinsonism (Gatev & Wichmann, 2008). This finding is also in agreement with studies of PP following dopaminergic medication, which promoted synchronized oscillatory activity at higher frequencies (γ) predominantly at the level of the frontal cortex and striatum (Levy et al., 2001; 2002; Brown et al., 2001; Williams et al., 2002; Leblois et al., 2007).

In recent MEG investigations of various cognitive and sensory tasks (Kaiser et al., 2003; Lutzenberger et al., 2002) the reported γ band activity over the higher sensory areas has not shown a sustained activation, but rather, a peaking activity. In our sensori-motor study, these transient responses were functionally dissociable between the two groups. We observed stimulus-specific γ band activity components over the fronto-parietal cortex, but this was differently manifest in each group and varied over the time course. The topography was compatible with the notion of an auditory dorsal space processing stream involving the posterior temporal, parietal and superior frontal cortex (Rauschecker, 1998; Arnott et al., 2004). If the cognitive processing (T2) γ band activity components represent similar anticipatory activations both for LT and HT, one might assume that the same cortical networks should underlie the same stimulus representations. However, while all components were mainly localized over fronto-central areas, there was some variation between the conditions, showing significant effects on the parietal γ components for LT, but not for parietal γ activity for HT. This suggests that networks encoding the stimulus features are not fixed, and may vary with task demands.

The assessing EEG stimulus-specific oscillatory activity yielded insights into the temporal dynamics of sound processing in short-term memory. Contrasting oscillatory γ activity between the two stimuli, such as between LT ERS and HT ERD during the sensory processing (T1) in PP, as well as between LT ERD and HT ERS during the cognitive processing (T2) in HS, revealed stimulus-specific γ activity behavior in the 30–50 Hz range over the HS's frontal and PP's fronto-parietal cortex. This suggests that γ band activity reflects the general involvement of cortical networks in particular tasks but may index the specific content of short-term memory in each group.

The pronounced and well-synchronized γ burst in HS was present in the very short-term phases around 25–60 ms after the stimulus onset, with spectral peaks ranging from 30 to 45 Hz (Gurtubay et al., 2001). Despite this short-term high synchrony in HS, the common γ behavior during sensory stimuli processing (T1) was desynchronization. However, the PP processes were with higher short-term energy, which is a prerequisite for all maintenance-related processes, and thus defined a persistent synchronization during sensory stimuli processing, mainly at fronto-parietal and right motor areas following LT. It is clear that there is a difference between groups in the early and well-synchronized response that is basically a sensory phenomenon important to preparing the brain for the subsequent processing. This evidence suggests that γ oscillations may be modulated by attentional processes. Several cognitive paradigms for the auditory system have shown early spectral peak responses in the γ band between 30 and 40 Hz at around 25 ms after stimulus onset that last for about 100 ms (Pantev, 1995; Arnott et al., 2004).

Later peaking activity has been recorded in the 200–400 ms interval following an experimental task. The latency and scalp topography vary according to the type of stimulus, indicating task-dependent local network activation. The significantly varying magnitudes of differentiation demonstrated that the topography of stimulus-specific γ-band activity is also task-dependent within the groups (Tiitinen et al., 1993). We found restricted energy changes over all recorded areas in HS, but not in the frontal area after HT. The significant differences between groups in T2 recorded after both stimuli could be due to memory retrieval processes that are activated during the performance of the paradigms. The lower energy in HS during cognitive processing (T2) could be related to fewer attentional processes required to eventually perform a task. The relative strength of differentiation in the γ-band may suggest that performance depends on the different group's ability to retain a representation in memory of the relevant stimulus feature and thus to be able to neglect the irrelevant stimuli. The acquisition and retention of sound frequency information was accompanied by frontal gamma band activity components (Karakaş & Başar, 1998). The high-frequency stimuli were accompanied by more exaggerated, well–synchronized frontal γ band components in performing the tasks. Memory for low versus high frequency tones selectively enhanced oscillatory activity for the posterior versus the frontal components, thus directly demonstrating differentiation of the group's modulation of cortical activity by task demands.

A mechanism that underlies many of the immediately aforementioned cognitive functions is the match of sensory information with memory contents (Kaiser et al., 2009a; 2009b; Visscher et al., 2007). The 'early' γ-band activity occurring 150 ms after stimulus presentation reflects such a match with memory. The 'late' γ activity, which typically emerges with a latency of more than 150 ms, is a temporal signature of utilization processes such as response selection or context updating. We also found a later (250–400 ms) ERS γ response, following only HT in HS, over the frontal location, where this activity peaked in the 33–45 Hz range. In PP, the specific β2 and γ bursts (30–38 Hz) exhibited maximal scalp projection covering areas to the left and right of the motor areas, and with frontal, central or parietal participation that depended on the stimuli. This oscillatory burst reflects a later stimulus context process, although it has also been associated with the motor responses later in the task (Brown, 2003; Kaiser et al., 2009b). The β2/γ oscillation in the groups points to a direct relation to aspects of post-discrimination processes related to the P300 wave (Haig et al., 2000). This oscillatory burst (letters **g-j** for intervals from 320 to 550 ms) also showed a variable relationship to attention, as it was significantly different during the HT and LT task. The results also showed that EEG activity in the frontal, parietal and motor cortex is significantly different between groups, not only in temporal variations (always with a delay in PP) but also in frequency shifts (β2/γ ERD in HS compared to the ERS in PP). Although these shifts do not follow a simple pattern, they are significantly different from HS, raising the possibility that the interactions between basal ganglia activity and cortical rhythms are functionally relevant. Therefore, the normal higher frequency relationships between cortical and basal ganglia activity are strongly altered in the parkinsonian state (Gatev & Wichmann,

2008). The shifts of β/γ patterns occurring in the groups are probably associated with specific types of basal ganglia events related to transitions between cortical idling and more active states (Williams et al., 2003).

Conclusion

Our investigation further demonstrates the close relationship between physiological abnormalities in PD and disturbances in the EEG frequency characteristics. The results of this investigation in PD patients of both sensory and cognitive processing of auditory stimuli suggests that PD should be characterized by multiple impairments in oscillatory networks, which in turn indicates the presence of task-specific disturbances in the temporal and regional integration of all frequency components.

References

Anik, Y., Iseri, P., Demirci, A., Komsuoglu, S., Inan, N. (2007) Magnetization transfer ratio in early period of Parkinson disease. *Acad Radiol* 14(2), (2007 Feb), pp. 189–192, ISSN 1076-6332.

Antal, A., Kéri, S., Dibó, G., Benedek, G., Janka, Z., Vécsei, L., Bodis-Wollner, I. (2002) Electrophysiological correlates of visual categorization, pp. evidence for cognitive dysfunctions in early Parkinson's disease. *Brain Res Cogn Brain Res* 13(2), (2002 Apr), pp. 153–158, ISSN 0926-6410.

Arnott, S. R., Binns, M. A., Grady, C. L., Alain, C. (2004) Assessing the auditory dualpathway model in humans. *NeuroImage* 22(1), (2004 May), pp. 401-408, ISSN 1053-8119.

Basar, E. (1980) EEG brain dynamics. Relations between EEG and brain evoked potentials. Elsevier, Amsterdam .

Başar, E., Basar-Eroglu, C., Karakas, S., Schurmann, M. (2001) Gamma, alpha, delta, and theta oscillations govern cognitive processes. *Int J Psychophysiol* 39(2-3), (2001 Jan) pp. 241-248, ISSN 0167-8760.

Brown, P. & Marsden, C. D. (1998) What do the basal ganglia do? Lancet 351(9118), (1998 Jun 13), pp. 1801-1804, ISSN 0140-6736.

Brown, P., Oliviero, A., Mazzone, P., Insola, A., Tonali, P. & Di Lazzaro, V. (2001) Dopamine dependency of oscillations between subthalamic nucleus and pallidum in Parkinson's disease. *J Neurosci* 21 (3), (2001 Feb 1) pp. 1033-1038, ISSN 0270-6474.

Brown, P. (2003) Oscillatory nature of human basal ganglia activity: relationship to the pathophysiology of Parkinson's disease. *Mov Disord* 18(4), (2003 Apr), pp. 357-363, ISSN 0885-3185.

Cassidy, M., Mazzone, P., Oliviero, A., Insola, A., Tonali, P., Di Lazzaro, V., Brown, P. (2002) Movement-related changes in synchronization in the human basal ganglia. Brain 125(Pt 6), (2002 Jun), pp. 1235-1246, ISSN 0006-8950.

Cools R (2006) Dopaminergic modulation of cognitive functionimplications for L-DOPA treatment in Parkinson's disease. *Neurosci Biobehav Rev* 30(1), (2005 Jun 1), pp. 1–23, ISSN 0149-7634.

Gurtubay, I. G., Alegre, M., Labarga, A., Malanda, A., Iriarte, J., Artieda, J. (2001) Gamma band activity in an auditory oddball paradigm studied with the wavelet *Clin Neurophysiol* 112(7), (2001 Jul), pp. 1219-1228, ISSN 1388-2457.

Daum, I., Schugens, M. M., Spieker, S., Poser, U., Schönle, P. W., Birbaumer, N. (1995) Memory and skill acquisition in Parkinson's disease and frontal lobe dysfunction. *Cortex* 31(3), (1995 Sep), pp. 413–432, ISSN 0010-9452.

Defebvre, L, Bouniez, JL, Destke, A, Guieu, JD (1996) Movement related ERD pattern preceding voluntary movement in untreated Parkinson's disease. *J Neurol Neurosurg Psychiatry* 60(3), (1996 Mar), pp. 307–312, ISSN 0022-3050.

Demiralp, T., Ademoglu, A., Schürmann, M., Bas¸ar-Eroglu, C., Başar, E. (1999) Detection of P300 waves in single trials by the wavelet transform (WT). *Brain Lang* 66(1), (1999 Jan), pp. 108–128, ISSN 0093-934X.

Devos, D., Labyt, E., Derambure, P., Bourriez, J. L., Cassim, F., Reyns, N., Blond, S., Guieu, J. D., Destée, A., Defebvre, L. (2004) Subthalamic nucleus stimulation modulates motor cortex oscillatory activity in Parkinson's disease. *Brain* 127(2), (2004 Feb), pp. 408–419, ISSN 0006-8950.

Donoghue, J. P., Sanes, J. N., Hatsopoulos, N. G., Gaal, G. (1998) Neural discharge and local field potential oscillations in primate motor cortex during voluntary movements. *J Neurophysiol* 79(1), (1998 Jan), pp. 159-173, ISSN 0022-3077.

Durka, P. J., Zygierewicz, J., Klekowicz, H., Ginter, J., Blinowska, K. J. (2004) On the statistical significance of event-related EEG desynchronization and synchronization in the time-frequency plane. *IEEE Trans Biomed Eng* 51(7), (2004 Jul), pp. 1167–1175, ISSN 0018-9294.

Dushanova, J., Philipova, D., Nikolova., G. (2009) Event-related synchronization/ desynchronization during discrimination task conditions in patients with Parkinson's disease. *Cell Mol Neurobiol* 29(6-7), (2009 Sep), pp. 971-980, ISSN 0272-4340.

Dushanova, J., Philipova, D., Nikolova, G. (2010) Beta and gamma frequency-range abnormalities in parkinsonian patients under cognitive sensorimotor task. *Journal of the Neurological Sciences* 293(1-2), (2010 Jun 15), pp. 51–58, ISSN 0022-510X.

Ellfolk, U., Karrasch, M., Laune, M., Pesonen, M., Krause, K. (2006) Event related desynchronisation/ERS during an auditory-verbal memory task in mild Parkinson's disease. *Clin Neurophysiol* 117(8), (2006 Aug), pp. 1737–1745, ISSN 1388-2457.

Engel, A. K., Fries, P., & Singer, W. (2001) Dynamic predictions: oscillations and synchrony in top-down processing. *Nat Rev Neurosci* 2(10), (2001 Oct), pp. 704-716, ISSN 1471-003X.

Evarts, E. V., Teräväinen, H., & Calne, D. B. (1981) Reaction time in Parkinson's disease. *Brain* 104(1), (1981 Mar), pp. 167–186, ISSN 0006-8950.

Farmer, S. F. (1998) Rhythmicity, synchronization and binding in human and primate motor systems. *J Physiol* 509(1), (1998 May 15), pp. 3-14, ISSN 0022-3751.

Fries, P., Reynolds, J. H., Rorie, A. E., Desimone, R. (2001) Modulation of oscillatory neuronal synchronization by selective visual attention. *Science* 291(5508), (2001 Feb 23), pp. 1560-1563, ISSN 0036-8075.

Foffani, G., Bianchi ,A. M., Baselli, G., Priori, A. (2005) Movement-related frequency modulation of beta oscillatory activity in the human subthalamic nucleus, *J Physiol* 568(2), (2005 Oct 15), pp 699–711, ISSN 0022-3751.

Fogelson, N., Williams, D., Tijssen, M., van Bruggen, G., Speelman, H., Brown, P. (2006) Different functional loops between cerebral cortex and the subthalmic area in Parkinson's disease. *Cereb Cortex* 16(1), (2006 Jan), pp. 64-75, ISSN 1047-3211.

Fukuda, M., Edward, C., Eidelberg, D. (2001) Functional brain networks in Parkinson's disease. *Parkinsonism Relat Disord* 8(2), (2001 Oct), pp. 91–94, ISSN 1353-8020.

Gatev, P., & Wichmann, T. (2008) Interactions between Cortical Rhythms and Spiking Activity of Single Basal Ganglia Neurons in the Normal and Parkinsonian State. *Cereb Cortex* 19(6), (2009 Jun), pp. 1330-44, ISSN 1047-3211.

Haig, A. R., Gordon, E., Pascalis, V. D., Meares, R. A., Bahramali, H., Harris, A. (2000) Gamma activity in schizophrenia: evidence of impaired network binding? *Clin Neurophysiol* 111(8), (2000 Aug), pp. 1461–1468, ISSN 1388-2457.

Herrmann, C. S., Munk, M. H. J., Engel, A. K. (2004) Cognitive functions of gamma-band activity: memory match and utilization. *Trends Cogn Sci* 8(8), (2004 Aug), pp. 347-355, ISSN 1364-6613.

Higginson, C. I., King, D. S., Levine, D., Wheelock , V. L., Khamphay, N. O., Sigvardt, K. A. (2003) The relationship between executive function and verbal memory in Parkinson's disease. *Brain Cogn* 52(3), (2003 Aug), pp. 343–352, ISSN 0278-2626.

Hoehn, M. M., Yahr, M. D. (1967) Parkinsonism: onset progression and mortality. *Neurology* 17(5), (1967 May), pp. 427-442, ISSN 0028-3878.

Innis, R. B., Seibyl, J. P., Scanley, B. E., Laruelle, M., Abi-Dargham, A., Wallace, E., Baldwin, R. M., Zea-Ponce, Y., Zoghbi, S., Wang, S. (1993) Single photon emission computed tomographic imaging demonstrates loss of striatal dopamine transporters in Parkinson disease. *Proc Natl Acad Sci USA* 90(24), (1993 Dec 15), pp. 11965–11969, ISSN 0027-8424.

Ivry, R. B., Robertson, L. C. (1998) The two sides of perception. MIT Press, Cambridge, MA, ISBN 0-262-09034-1.

Jensen, O., Tesche, C. D. (2002) Frontal theta activity in humans increases with memory load in a working memory task. *Eur J Neurosci* 15(8), (2002 Apr), pp. 1395–1399, ISSN 0953-816X.

Jensen, O., Gelfand, J., Kounios, J., Lisman, J. E. (2002) Oscillations in the alpha band (9–12 Hz) increase with memory load during retention in a short-term memory task. *Cereb Cortex* 12(8), (2002 Aug), pp. 877– 882, ISSN 1047-3211.

Kaiser, J., Ripper, B., Birbaumer, N., Lutzenberger, W. (2003) Dynamics of gamma-band activity in human magnetoencephalogram during auditory pattern working memory. *NeuroImage* 20(2), (2003 Oct), pp. 816-827, ISSN 1053-8119.

Kaiser, J., Rahm, B., Lutzenberger, W. (2009a) Temporal dynamics of stimulus-specific gamma-band activity components during auditory short-term memory. *Neuroimage* 44(1), (2009 Jan 1), pp. 257-264, ISSN 1053-8119.

Kaiser, J., Lutzenberge, W., Decker, C., Wibral, M., Rahm, B. (2009b) Task- and performance-related modulation of domain-specific auditory short-term memory *Neuroimage* 46(4), (2009 Jul 15), pp. 1127-36, ISSN 1053-8119.

Karakaş ,S., & Başar, E. (1998) Early gamma response is sensory in origin: a conclusion based on cross-comparison of results from multiple experimental paradigms. *Int J Psychophysiol* 31(1), (1998 Dec), pp. 13-31, ISSN 0167-8760.

Karakas, S., Erzengin, O. U., & Başar, E. (2000) A new strategy involving multiple cognitive paradigms demonstrates that ERP components are determined by the superposition of oscillatory signals. *Clin Neurophysiol* 111(10), (2000 Oct), pp. 1719–1732, ISSN 1388-2457.

Karrasch, M., Krause, C. M., Laine, M., Lang, H. A., Lehto, M. (1998) Eventrelated ERD and ERS during an auditory lexical matching task. *Electroencephalogr Clin Neurophysiol* 107(2), (1998 Aug), pp. 112–121, ISSN 0013-4694.

Karrasch, M., Laine, M., Rapinoja, P., Krause, C. M. (2004) Effects of normal ageing on event-related ERD/ERS during a memory task in humans. *Neurosci Lett* 366(1), (2004 Aug 5), pp. 18–23, ISSN 0304-3940.

Klimesch, W. (1997) EEG a-rhythms and memory processes. *Int J Psychophysiol* 26(1-3), (1997 Jun), pp. 319–340, ISSN 0167-8760.

Klimesch, W., Doppelmay, M., Russegger, H., Pachinger, T. (1996) Theta band power in the human scalp EEG and the encoding of new information. *Neuroreport* 7(7), (1996 May 17), pp. 1235–1240, ISSN 0959-4965.

Klimesch, W., Doppelmayr, M., Schwaiger, J., Winkler, T., Gruber W. (2000) Theta oscillations and the ERP old/new effect: independent phenomena? *Clin Neurophysiol* 111(5), (2000 May), pp. 781–793, ISSN 1388-2457.

Klimesch, W., Doppelmayr, M., Stadler, W., Pollhuber, D., Sauseng, P., Rohm, D. (2001) Episodic retrieval is reflected by a process specific increase in human electroencephalographic h activity. *Neurosci Lett* 302(1), (2001 Apr 13), pp. 49–52, ISSN 0304-3940.

Klimesch, W., Schnack, B., Sauseng, P. (2005) The functional significance of h and upper a-oscillations. *Exp Psychol* 52, pp. 99–108, ISSN 1618-3169.

Klimesch, W., Hanslmayr, S., Sauseng, P., Gruber, W., Brozinsky, C. J., Kroll, N. E., Yonelinas, A. P., Doppelmayr, M. (2006) Oscillatory EEG correlates of episodic trace decay. *Cereb Cortex* 16(2), (2006 Feb), pp. 280–290, ISSN 1047-3211.

Konig, R., Heil, P., Budinger, E., Scheich, H. (2005) The auditory cortex: a synthesis of human and animal research. Lawrence Erlbaum Associates, NJ. ISBN 0-8058-4938-6.

Krause, C. M. (2006) Cognition- and memory-related ERD/ERS responses in the auditory stimulus modality. *Prog Brain Res* 159, pp. 197–207, ISSN 0079-6123.

Krause, C. M., Lang, H. A., Laine, M., Helle, S. I., Kuusisto, M. J., Pörn, B. (1994) Event-related desynchronization evoked by auditory stimuli. *Brain Topogr* 7(2), (1994 Winter), pp. 107–112, ISSN 0896-0267.

Krause, C. M., Lang, A. H., Laine, M., Kuusisto, M., Pörn, B. (1996) Eventrelated EEG desynchronization and synchronization during an auditory memory task. *Electroencephalogr Clin Neurophysiol* 98(4), (1996 Apr), pp. 319–326, ISSN 0013-4694.

Krause, C. M., Aström, T., Karrasch, M., Laine, M., Sillanmäki, L. (1999) Cortical activation related to auditory semantic matching of concrete versus abstract words. *Clin Neurophysiol* 110(8), (1999 Aug), pp. 1371– 1377, ISSN 1388-2457.

Knoke, D., Taylor, A. E., Saint-Cyr, J. A. (1998) The differential defects of ceing on recall in Parkinson's Disease and normal subjects. *Brain Cogn* 38(2), (1998 Nov), pp. 261–274, ISSN 0278-2626.

Kühn, A. A., Kupsch, A., Schneider, G. H., Brown, P. (2006) Reduction in subthalamic 8–35 Hz oscillatory activity correlat with clinical improvement in PD. *Eur J Neurosci* 23(7), (2006 Apr), pp. 1956-1960, ISSN 0953-816X.

Leblois, A., Boraud, T., Meissner, W., Bergman, H., Hansel, D. (2006) Competition between feedback loops underlies normal and pathological dynamics in the basal ganglia. *J Neurosci* 26(13), (2006 Mar 29), pp. 3567– 3583, ISSN 0270-6474.

Leblois, A., Meissner, W., Bioulac, B., Gross, C. E., Hansel, D., Boraud, T. (2007) Late emergence of synchronized oscillatory activity in the pallidum during progressive parkinsonism. *Eur J Neurosci* 26(6), (2007 Sep), pp. 1701-1713, ISSN 0953-816X.

Lewis, S. J., Dove, A., Robbins, T. W., Barker, R. A., Owen, A. M. (2003) Cognitive impairments in early Parkinson's disease are accompanied by reductions in activity in frontostriatal neural circuitry. *J Neurosci* 23(15), (2003 Jul 16), pp. 6351–6356, ISSN 0270-6474.

Levy, R., Dostrovsky, J. O., Lang, A. E., Sime, E., Hutchison, W. D., Lozano, A. M. (2001) Effects of apomorphine on subtalamic nucleus and globus pallidus internus neurons in patients with Parkinson's disease. *J Neurophysiology* 86(1), (2001 Jul), pp. 249-260, ISSN 0022-3077.

Levy, R., Hutchison, W. D., Lozano, A. M., Dostrovsky, J. O. (2002) Synchronized neuronal discharge in the basal ganglia of parkinsonian patients is limited to oscillatory activity. *J Neuroscience* 22(7), (2002 Apr 1), pp. 2855-2861, ISSN 0270-6474.

Lopes da Silva, F. H., Vos, J. E., Mooibroek, J., Van Rotterdam, A. (1980) Relative contributions of intracortical and thalamo-cortical processes in the generation of a-rhythms, revealed by partial coherence analysis. *Electroencephalogr Clin Neurophysiol* 50(5-6), (1980 Dec), pp. 449–456, ISSN 0013-4694.

Lutzenberger, W., Ripper, B., Busse, L., Birbaumer, N., Kaiser, J. (2002) Dynamics of gamma-band activity during an audiospatial working memory task in humans. J Neurosci 22(13), (2002 Jul 1), pp. 5630-5638, ISSN 0270-6474.

Magnani, G., Cursi, M., Leocani, L., Volonté, M. A., Locatelli, T., Elia, A., Comi, G. (1998) Event-related ERD to contingent negative variation and self-paced movement paradigms in Parkinson's disease. *Mov Disord* 13(4), (1998 Jul), pp. 653–660, ISSN 0885-3185 .

Marsden, J. F., Limousin-Dowsey, P., Ashby, P., Pollak, P. & Brown, P. (2001) Subthalamic nucleus, sensorimotor cortex and muscle interrelationships on Parkinson's disease. *Brain* 124(2), (2001 Feb), pp. 378-388, ISSN 0006-8950.

Mason, M., & Newton, M. (1990) A rank statistics approach to the consistency of the general bootstrap, *Ann. Statist* 20, pp. 1611–1624, ISSN 0090-5364.

Marceglia, S., Fiorio, M., Foffani, G., Mrakic-Sposta, S., Tiriticco, M., Locatelli, M., Caputo, E., Tinazzi M., Priori, A. (2009) Modulation of beta oscillations in the subthalamic area during action observation in Parkinson's disease. *Neuroscience* 161(4), (2009 Jul 21), pp. 1027-1036, ISSN 0306-4522.

Minamoto, H., Tachibana, H., Sugita, M., Okita, T. (2001) Recognition memory in normal ageing and Parkinson's disease: behavioural and electrophysiological measures. *Brain Res Cogn Brain Res* 11(1), (2001 Mar), pp. 23–32, ISSN 0926-6410.

Oldfield, R. C. (1971) The assessment and analysis of handedness: the Edinburgh inventory. *Neuropsychologia* 9(1), (1971 Mar), pp. 97-113, ISSN 0028-3932.

Owen, A. M. (2004) Cognitive dysfunction in Parkinson's disease: the role of frontostriatal circuitry. *Neuroscientist* 10(6), (2004 Dec), pp. 525–537, ISSN 1073-8584.

Ozgoren, M., Basar-Eroglu, C., Basar, E. (2005) Beta oscillations in face recognition. *Int J Psychophysiol* 55(1), (2005 Jan), pp. 51-59, ISSN 0167-8760.

Pantev, C. (1995) Evoked and induced gamma-band activity of the human *Brain Topogr* 7(4), (1995 Summer), pp. 321-330, ISSN 0896-0267.

Pesonen, M., Björnberg, C. H., Hämäläinen, H., Krause, C. M. (2006) Brain oscillatory 1–30 Hz EEG ERD/ERS responses during the different stages of an auditory memory search task. *Neurosci Lett* 399(1–2), (2006 May 15), pp. 45–50, ISSN 0304-3940.

Philipova, D., Gatchev, G., Vladova, T., Georgiev, D (1997) Event related potentials under auditory discrimination tasks. *Int J Psychophysiol* 27(1), (1997 Jul), pp. 69–78, ISSN 0167-8760.

Piccirilli M., D'Alessandro P., Finali G., Piccinin G. L., Agostini L. (1989) Frontal lobe dysfunction in Parkinson's disease: prognostic value for dementia? *Eur Neurol* 29(2), pp. 71–76, ISSN 0014-3022.

Pillon, B., Ertle, S., Deweer, B., Bonnet, A., Vidailhet, M., Dubois, B. (1997) Memory for spatial location in 'de novo' parkinsonian patients. *Neuropsychologia* 35(3), (1997 Mar), pp. 221–228, ISSN 0028-3932.

Priori, A., Foffani, G., Pesenti, A., Bianchi, A., Chiesa, V., Baselli, G. (2002) Movement-related modulation of neural activity in human basal ganglia and its l-DOPA dependency: recordings from deep brain stimulation electrodes in patients with Parkinson's disease. *Neurol Sci* 23(Suppl. 2), (2002 Sep), pp. S101-S102, ISSN 1590-1874.

Priori ,A., Foffani, G., Pesenti, A., Tamma, F., Bianchi, A. M., Pellegrini, M., Locatelli, M., Moxon, K. A., Villani, R. M. (2004) Rhythm-specific pharmacological modulation of subthalamic activity in Parkinson's disease. *Exp Neurol* 189(2), (2004 Oct), pp. 369-379, ISSN 0014-4886.

Pfurtscheller, G., & Klimesch, W. (1991) Event-related ERD during motor behaviour and visual information processing. *Electroencephalogr Clin Neurophysiol Suppl* 42, pp. 58–65, ISSN 0424-8155.

Pfurtscheller, G., & Lopes da Silva, F. H. (1999) Event-related desynchronization Handbook of Electroencephalography and Clinical Neurophysiulogy. Revised Series, Vol. 6, ISBN 0-444-82999-7.

Rauschecker J. P. (1998) Cortical processing of complex sounds. *Curr Opin Neurobiol* 8(4), (1998 Aug), pp. 516-521, ISSN 0959-4388.

Rivlin-Etzion, M., Marmor, O., Heimer, G., Raz, A., Nini, A., Bergman, H. (2006) Basal ganglia oscillations and pathophysiology of movement disorders. *Curr Opin Neurobiol* 16(6), (2006 Dec), pp. 629-637, ISSN 0959-4388.

Robertson, C., Empson, J. (1999) Slowed cognitive processing and high workload in Parkinson's disease. *J Neurol Sci* 162(1), (1999 Jan 1), pp. 27–33, ISSN 0022-510X.

Rolland, A.-S., Herrero, M.-T., Garcia-Martinez, V., Ruberg, M., Hirsch, E. C., Francois, C. (2007) Metabolic activity of cerebellar and basal ganglia-thalamic neurons is reduced in parkinsonism. *Brain* 130(1), (2007 Jan), pp. 265–275, ISSN 0006-8950.

Sawamoto, N., Honda, M., Hanakawa, T., Fukuyama, H., Shibasaki, H. (2002) Cognitive slowing in Parkinson's disease: a behavioural evaluation independent of motor slowing. *J Neurosci* 22(12), (2002 Jun 15), pp. 5198–5203, ISSN 0270-6474.

Schmiedt, C., Meistrowitz, A., Schwendemann, G., Herrmann, M., Basar-Eroglu, C. (2005) Theta and alpha oscillations reflect differences in memory strategy and visual discrimination performance in patients with Parkinson's disease. *Neurosci Lett* 388(3), (2005 Nov 18), pp. 138–143, ISSN 0304-3940.

Schurmann, M., Basar-Eroglu, E. (1994) Topography of alpha and theta oscillatory responses upon auditory and visual stimuli in humans. *Biol Cybern* 72(2), pp. 161–174, ISSN 0340-1200.

Shohamy, D., Myers, C. E., Grossman, S., Sage, J., Gluck, M. A. (2005) The role of dopamine in cognitive sequence learning: evidence from Parkinson's disease. *Behav Brain Res* 156(2), (2005 Jan 30), pp. 191–199, ISSN 0166-4328.

Silberstein, P., Pogosyan, A., Kühn, A. A., Hotton, G., Tisch, S., Kupsch, A., Dowsey-Limousin, P., Hariz, M. I., Brown, P. (2005) Cortico-cortical coupling in Parkinson's disease and its modulation by therapy. *Brain* 128(6), (2005 Jun), pp. 1277-1291, ISSN 0006-8950.

Stam, C. J., Visser, S. L., Op de Coul, A. A., De Sonneville, L. M., Schellens, R. L., Brunia, C. H., de Smet. J. S., Gielen, G. (1993) Disturbed frontal regulation of attention in Parkinson's disease. *Brain* 116(5), (1993 Oct), pp. 1139–1158, ISSN 0006-8950.

Sutoh T., Yabe H., Sato Y., Hiruma T., Kaneko S. (2000) Event-related desynchronization during an auditory oddball task. *Clin Neurophysiol* 111(5), (2000 May), pp. 858-862, ISSN 1388-2457.

Tachibana, H., Toda, K., Sugita, M. (1992) Actively and passively evoked P3 latency of event-related potentials in Parkinson's disease. *J Neurol Sci* 111(2), (1992 Sep), pp. 134–142, ISSN 0022-510X.

Taylor, A. E., Saint-Cyr, J. A., Lang, A. E. (1990) Memory and learning in early Parkinson's disease: evidence for a ''frontal lobe syndrome''. *Brain Cogn* 13(2), (1990 Jul), pp. 211–232, ISSN 0278-2626.

Tallon-Baudry, C., Bertrand, O., Delpuech, C., Pernier, J. (1996) Stimulus specificity of phase-locked and non-phase-locked 40 Hz visual responses in human. *J Neurosci* 16(13), (1996 Jul 1), pp. 4240-4249, ISSN 0270-6474.

Terman, D., Rubin, J. E., Yew, A. C., Wilson, C. J. (2002) Activity patterns in a model for the subthalamopallidal network of the basal ganglia. *J Neurosci* 22(7), (2002 Apr 1), pp. 2963-2976, ISSN 0270-6474.

Tiitinen, H., Sinkkonen, J., Reinikainen, K., Alho, K., Lavikainen, J., Näätänen, R. (1993) Selective attention enhances the auditory 40-Hz transient response in humans. *Nature* 364(6432), (1993 Jul 1), pp. 59-60, ISSN 0028-0836.

van Spaendonck, K. P., Berger, H. J., Horstink, M. W., Cools, A. R. (1998) Cognitive deficits in Parkinson's disease. *Tijdschr Gerontol Geriatr* 29(4), (1998 Aug), pp. 189-195, ISSN 0167-9228.

Vieregge, P., Verleger, R., Wascher, E., Stüven, F., Kömpf, D. (1994) Auditory selective attention is impaired in Parkinson's disease event-related evidence from EEG potentials. *Brain Res Cogn Brain Res* 2(2), (1994 Sep), pp. 117–129, ISSN 0926-6410.

Visscher, K. M., Kaplan, E., Kahana, M. J., Sekuler, R. (2007) Auditory short-term memory behaves like visual short-term memory. *PLoS Biol* 5(3), (2007 Mar) e56, ISSN 1544-9173.

Wang, H., Wang, Y., Wang, D., Cui L., Tian, S., Zhang, Y. (2002) Cognitive impairment in Parkinson's disease revealed by event-related potential N270. *J Neurol Sci* 194(1), (2002 Feb 15), pp. 49–53, ISSN 0022-510X.

Wang, S. Y., Aziz, T. Z., Stein, J. F. & Liu, X. (2005) Time-frequency analysis of transient neuromuscular events: dynamic changes in activity of the subthalamic nucleus and forearm muscles related to the intermittent resting tremor. *J Neurosci Methods* 145(1-2), (2005 Jun 30), pp. 151-158, ISSN 0165-0270.

Wascher, E., Verleger, R., Vieregge, P., Jaskowski ,P., Koch, S., Kömpf, D. (1997) Responses to cued signals in Parkinson's disease. Distinguishing between disorders of cognition and of activation. *Brain* 120(8), (1997 Aug), pp. 1355–1375, ISSN 0006-8950.

Weinberger, M., Mahant, N., Hutchison, W. D., Lozano, A. M., Moro, E., Hodaie, M., Lang, A E., Dostrovsky, J. O. (2006) Beta oscillatory activity in the subthalamic nucleus and its relation to dopaminergic response in Parkinson's disease. *J Neurophysiol* 96(6), (2006 Dec), pp. 3248-3256, ISSN 0022-3077.

Weinberger, M., Hutchison, W. D., Lozano, A. M., Hodaie, M., Dostrovsky, J. O. (2008) Increased gamma oscillatory activity *J Neurophysiol* 101(2), (2009 Feb), pp. 789-802, ISSN 0022-3077.

Williams, D., Kühn, A., Kupsch, A., Tijssen, M., van Bruggen, G., Speelman, H., Hotton, G., Yarrow, K. & Brown, P. (2003) Behavioural cues are associated with modulations of synchronous oscillations in the human subthalamic nucleus. *Brain* 126(9), (2003 Sep), pp. 1975-1985, ISSN 0006-8950.

Williams, D., Tijssen, M., van Bruggen, G., Bosch, A., Insola, A., Di Lazzaro, V., Mazzone, P., Oliviero, A., Quartarone, A., Speelman, H., & Brown, P (2002) Dopamine dependent changes in the functional connectivity between basal ganglia and cerebral cortex in the human. *Brain* 125(7), (2002 Jul), pp. 1558-1569, ISSN 0006-8950.

In: Horizons in Neuroscience Research. Volume 15
Editors: Andres Costa and Eugenio Villalba

ISBN: 978-1-63321-523-8

Chapter 11

The Impact of Subthalamic Stimulation on Autonomic Dysfunction in Parkinson's Disease

E. Trachani[1, 2], *E. Chroni*[2] *and C. Constantoyannis*[1]
[1]Functional Neurosurgery Unit, Department of Neurosurgery
[2]Department of Neurology, Medical School of Patras, Patras, Greece

Abstract

Parkinson's disease (PD) is a degenerative movement disorder, with rigidity, bradykinesia and tremor being its cardinal symptoms. Even in the early stages of PD, Autonomic Nervous System (ANS) involvement is established with cardiovascular, urogenital, gastrointestinal, sudomotor and thermoregulatory manifestations. Degeneration changes in hypothalamus, intermediolateral nucleus of the thoracic cord, sympathetic ganglia and in sacral parasympathetic nuclei might be responsible for the dysautonomia in PD. There is substantial evidence that Deep Brain Stimulation of the Subthalamic Nucleus (STN-DBS) is particularly effective in improving motor impairment and reducing the medication needs in patients with PD. However, the influence of this modern therapy on the autonomic regulation is not yet clarified. There are several standard tests to estimate the integrity of each autonomic function and therefore the STN-DBS influence upon it, such as: Sympathetic Skin Response, Quantitative Sudomotor Axon Reflex Test, Thermoregulatory Sweat Test, salivation test, plasma catecholamine levels and urodynamic studies; specifically for the cardiovascular dysautonomia the tests are divided into those estimating the sympathetic (blood pressure changes to standing, to head-up Tilt testing and during Valsalva maneuver) and those estimating the parasympathetic part of the regulation (recording of heart rate changes to standing, to head-up Tilt-table test and during Valsalva maneuver, heart rate variation to deep breathing) or both (spectral analysis of RR interval variation). The DBS mechanism of action in the basal ganglia is complex and not completely understood, but since the basal ganglia project to a number of nuclei involved in the control of the ANS, as well as to thalamus, a possible direct effect of STN-DBS on ANS functions might be expected. It is worth mentioning that STN-DBS has been found to increase blood flow in the anterior cingulated cortex and the insular cortex, which are primary centers of autonomic control.

On the other hand, ANS manifestations could subside due to post-DBS medication reduction and general improvement of motor function seen in the majority of patients. Nevertheless, the results of the - up to date- reports on whether STN-DBS could influence the dysautonomia in PD are controversial. Some study groups have reported in direct effect of STN-DBS on autonomic cardiac function, whereas others have found improvement in several cardiovascular variables and in Sympathetic Skin Response, as well as an amelioration of constipation, dyshidrosis and urological dysfunction. In this chapter we present the current status of the investigations on DBS effect on ANS function in humans.

I. Introduction

Basal ganglia have been reported to play a crucial role not only in movement functions but also in the control of autonomic activity [1]. The autonomic impairment in Parkinson's disease (PD) is established even in early stages of the disease and it has a great negative impact on the quality of everyday life [2]. Dysautonomia in PD includes clinical manifestations from several systems such as cardiovascular, gastrointestinal, urogenital and thermoregulatory system. Deep Brain Stimulation of Subthalamic Nucleus (STN-DBS) is a recently developed and widely adopted treatment method for parkinsonian patients. Its effectiveness upon the motor dysfunction is well recognized and discussed in detail; more specifically it has been shown to reduce the motor Unified Parkinson's Disease Rating Scale (UPDRS) by 50-70%, to decrease dyskinesias, motor fluctuations and the postoperative L-dopa daily dose (LEDD) and improves in general the physical aspects of quality of life [3-9]. On the contrary, the potential influence of STN stimulation on autonomic dysfunction in PD is less well established. Since the DBS has been found τo improve the quality of patients' life by inducing reduction of the frequency of non motor symptoms [10], there remains the need for further research to investigate with more objective methods and in larger samples the net effect of STN-DBS on autonomic dysfunction. Here we review the published literature on the frequent autonomic disturbances in PD, their pathophysiology and assessment, the benefit of DBS to autonomic system, the results of our clinical study upon it and discuss the future of this surgical therapy on the dysautonomia of PD.

II. Autonomic Symptoms in PD: Pathophysiology-Clinical and Neurophysiological Evaluation-Treatment

Cardiovascular Dysfunction

The cardinal symptom of cardiovascular dysregulation in PD is Orthostatic Hypotension (OH) with a prevalence of 20-50% [11, 12]. OH is defined as a fall in systolic blood pressure (BP) of 20 mm Hg or more and in diastolic BP of 10 mm Hg or more between lying supine for 15 minutes (min) and then standing for 5 min [13]. It may be symptomatic, causing postural light-headedness, vision disturbances, generalized weakness, a sense of lethargy and even syncope or asymptomatic. Until recently, some researchers considered OH as a possible consequence of dopaminergic treatment. Moreover, an early presentation of OH in the context of an extrapyramidal disease was regarded as a diagnostic sign of, Multiple System Atrophy. However, it has been found in individuals with PD who never took any antidopaminergic medication[14] whereas several reports demonstrated a significant failure in cardiovascular regulation in de novo or/and untreated patients as part of the PD process itself [11, 15]. Even so, no one can argue that the primary manifestations in PD patients are aggravated with most dopaminergic therapies [12]. Lately, radioisotopic studies focusing on the end-organ ANS have shown partial postganglionic sympathetic cardiac denervation in all symptomatic patients but also in approximately 50% of asymptomatic individuals [16]. Cardiac denervation might be associated with overexpression of α-synuclein due to a whole-gene triplication [17] and it is in parallel with increase of α-synuclein in sympathetic nerve fibers in the epicardium and myocardium [18].

Thermoregulatory Dysfunction

Several clinical reports have shown abnormalities in thermoregulation in parkinsonian patients. These abnormalities include hyperhydrosis or hypohydrosis and heat intolerance; the frequency of these symptoms varies from 25 to 64% [19-21]. Hyperhydrosis may be presented as whole body drenching sweats or increased sweating in the head, neck or trunk [22]. This pattern is interpreted as a compensatory mechanism for impaired sweating in the limbs and other body regions [23, 24]. Excessive sweating disturbances are more common during "off" periods or at end-of-dose or peak-of-dose-dyskinesias [25, 26]. In hypohydrosis lower extremities are more often involved and patients complain of less sweating and exhaustion in hot weather [21]. Sweating disturbances might be a sign of dysfunction of both the central nuclei of sympathetic system and the cholinergic postganglionic system [26].

Urological Dysfunction

Urinary dysfunction is common in PD, though less prevalent than the above autonomic symptoms and its frequency ranges from 36% to 90% [23]. The most frequent urodynamic finding is detrusor hyperreflexia (45-93%) often manifesting with nocturia (62%), urinary frequency and urgency (56%) [2]. These irritative urinary symptoms are responsive to anticholinergic drugs since they decrease detrusor overactivity. The involuntary bladder contractions result from the lack of the inhibitory effect that under normal conditions substantia nigra should exert upon the pontine micturation centre and subsequently on the detrusor activity. More specifically, the degeneration of substantia nigra pars compacta has been found to lead to a significant reduction of D1dopamine receptors stimulation in the striatum [27]. Another rather rare urodynamic finding in PD is detrusor areflexia resulting in obstructive symptoms (23-36% of urinary symptoms) such as hesitancy, weak urinary stream and overflow incontinence [23]. Detrusor underactivity is considered as a dopaminergic treatment implication [23] or the consequence of urethral sphincter bradykinesia [28].

Sexual Dysfunction

Although sexual dysfunction is not uncommon in PD and it is one of the most disabling symptoms in patients with early onset disease, it has been poorly investigated [29]. Sexual dysfunction includes impaired and excessive/distorted function [28]. It is presented as decreased libido, arousal and orgasm difficulties in both genders, as well as erectile dysfunction in parkinsonian men and reduced vaginal sensitivity in women [23, 30]. D1 and D2 receptors in hypothalamus, amygdala, mesencephalic tegmentum and insular mesocortex are known to participate in the regulation of libido and erection respectively [31], brain areas affected by PD degeneration [29]. Apart from the central neurological part, other factors i.e., aging, depressive symptoms and antidepressants, hypertension, testosterone deficiency and more hormonal or environmental factors have to be taken in consideration as far as decreased sexual function is concerned [29]. On the other hand, hypersexuality and abnormal sexual behavior may occur as an adverse event of dopaminergic treatment by direct stimulation of the D2 receptors in the medial preoptic area [24] and is often characterized as a behavioral disorder, as part of the dopamine dysregulation syndrome [20] and as a rare sequel of DBS (personal unpublished experience).

Gastrointestinal Dysfunction

Gastrointestinal dysfunction occurs in the majority of patient with PD who have symptoms from the whole gastrointestinal truck. Hypersalivation is the most common and early symptom in 70-80% of parkinsonians [1]. Excess saliva is not due to increased saliva production, but to reduced swallowing and dysphagia leading to drooling [32]. Gastroparesis, disturbed gastric emptying, constipation, pain and sense of incomplete evacuation due to focal dystonia of the external anal sphincter and puborectalis muscles, are also prevalent manifestations [33], which have been recognized and described in detail even in James Parkinson's "Essay on shaking palsy" published in 1817 [34]. Constipation is the most frequent gastrointestinal autonomic symptom and is often aggravated by anticholinergic and dopaminergic medication. Vagal nerve, parasympathetic nucleus of the spinal cord, the enteric nervous system and the intramural ganglia are responsible for the motility of the gastrointestinal tract [35]. Since Braack and his colleagues have demonstrated Lewy bodies and α-synoucleine bodies in the dorsal motor nucleus of the vagal nerve as well as in the intramural autonomic ganglia even at early stages of PD, autonomic dysfunction in the esophagus, the stomach, the small and large intestine is unavoidable [35].

Clinical and Neurophysiological Examination of Autonomic System in PD

Given that autonomic symptoms appear as not specific subjective complaints, an objective confirmation and quantitative assessment is often desirable [36]. There are some widely accepted, validated questionnaires and rating scales to estimate the non motor and the autonomic symptoms in PD such as the Nonmotor Symptoms Questionnaire (NMSQuest)[20, 33, 37]. Moreover, there are several pharmacological or isotopic-based methods to evaluate the function of Autonomic Nervous System (ANS), but in the daily practice only certain tests are used. Standard tests of cardiovagal, sympathetic and sweating function that have been applied in the pre- and post-DBS comparison studies are briefly presented here.

Cardiovascular Autonomic Testing

Time Domain Analysis of Heart Rate Variability (HRV)

The HRV is tested under resting conditions and during deep breathing. The beat to beat variability of heart rate (HR) is mainly controlled by parasympathetic nervous system. RR signal recordings during rest can be easily recorded on standard electromyographic (EMG) equipment at rest and it may be short length (5-min) or longer (24-hour). Several parameters can be calculated from the RR sequence epoch such as: the mean interbeat interval of normal beats (meanRR); the standard deviation of normal-to-normal RR (SDNN); the coefficient of variation (CV) defined as SDNN divided by the mean RR expressed in percent values [38]; the variation of RR calculated by the following formula: (maxRR-minRR)/meanRR, where maxRR was the maximum normal RR interval value and minRR was the minimum normal RR interval value [39].

Deep breathing RR recordings are performed in the supine position and the patient breaths deeply at six respirations per min, allowing 5 seconds (sec) for inspiration and 5 sec for expiration, inducing the maximal HRV in healthy persons [13, 40]. The expiratory – inspiratory ratio (E: I ratio) or the expiratory-inspiratory difference are determined and these values are age and gender dependent.

Frequency Domain Analysis of HRV

Frequency domain analysis offers a quantitative estimation of both sympathetic and parasympathetic influences, reflecting as well, the equilibrium between the 2 parts of ANS [41]. RR intervals are recorded in EMG equipment and several mathematical methods are used in order to calculate the Power Spectrum Density, which includes the following frequency bands: Very Low Frequency band (VLF) ($\leq$ 0.04 Hz); Low Frequency band (LF) (0.04 to 0.15 Hz); High Frequency (HF) band (0.15-0.40 Hz) [42-45]. The HF band is regarded as an index of parasympathetic activity, whereas LF band is considered by some authors as an index of sympathetic activity and by others as a reflection of both sympathetic and parasympathetic function [46-48]. Normalized units of LF and HF can be calculated in order to reduce variations due to individual values in the VLF range, whereas the ratio of LF / HF power serves as a parameter of the sympatho-vagal balance [49] .

Valsalva Manoeuvre

The HR response to this manoeuvre provides a sensitive and reproducible measure of vagal function, whereas BP changes during this test show the integrity of sympathetic adrenergic function [13]. This manoeuvre is based on the respiratory strain and the increase in intrathoracic and intraabdominal pressures which alters cardiovascular functions. The Valsalva ratio (the ratio of the shortest RR interval, during expiration, to the longest RR interval, during the first 20 sec after the expiratory strain) which is derived from this testing, is affected by age and gender [50].

Cardiovascular Response to Standing and Head-Up Tilt Test

HR changes to standing are an index of the integrity of parasympathetic function, whereas BP changes to standing and during head-up tilt test are an index of the integrity of sympathetic function. These are basic and simple tests of autonomic regulation. The BP and HR are recorded at baseline, after 20 min of supine position and then for 1-3 min after postural standing. The 30:15 ratio is the ratio of the longest RR (slowest HR) at 30 sec after standing divided by the shortest RR (fastest HR) at about 15 sec after standing and it is age related [13].

Cardiac Sympathetic Imaging

This technique uses imaging agents that are taken up into sympathetic nerves and are transported in storage vesicles within sympathetic neurons. Visualization of the image agents gives us a picture of cardiac sympathetic nerves. The most widely used agent is I- metaiodobenzylguanidine (^{123}I-MIGB) and can be imaged with single photon emission computed tomography scanning (SPECT). It has been recently used to demonstrate sympathetic cardiac denervation and it is not widely used but increases in availability and practice [36]. Another technique is positron emission tomography (PET) using radiolabeled catecholamines such as 6-[18F] fluorodopamine, which provides superior spatial and temporal resolution [41].

Skin Blood Flow (Skin Vasomotor Reflex)

Skin blood flow (SBF) employed a laser Doppler flowmeter and measures the cytaneous vasoconstrictor response to autonomic manoeuvre such as mental stress, deep breathing or standing. The tests are performed on the toe or finger pads because these areas are supplied mostly by vasoconstrictor sympathetic fibers. The recordings appear as irregular bursts of varying duration often occurring in relationship to the respiratory rhythm [41] and the pathway of this reflex appeared to be multi-synaptic. SBF is mainly used to detect sympathetic denervation in peripheral autonomic neuropathies, since reduced vasoconstrictor response usually suggests damage of vasoconstrictor neurons [51, 52].

Plasma Catecholamines Levels

During upright posture, the vasopressor responses are sympathetic and adrenergic, thus plasma nor-epinephrine levels almost double. Their levels increase poorly when standing in cases of central or peripheral damage of autonomic nervous system [13, 36].

Sudomotor Testing

Quantitative Sudomotor Axon Reflex (QSART)

It measures the axon-reflex mediated sweat response and thus quantifies postganglionic sympathetic sudomotor function. Acetylcholine, which is iontophorised onto the skin, directly stimulates sweat glands and the resulting humidity from a given skin region like the foot, the distal leg, the proximal leg and the forearm is measured by hygrometer. The sweat output depends on age, sex and site of recording and can be absent, decreased or increased. QSART is a sensitive, accurate and quantitative method, but time-consuming and it requires special equipment not available in the daily routine [13, 53].

Sympathetic Skin Response (SSR)

SSR is an easy to perform on most standard EMG equipment, sensitive and semi-quantitative method for testing the integrity of sympathetic system. It measures electrodermal activity of the skin which reflects the sympathetic cholinergic sudomotor function. SSR is a somato-sympathetic reflex with a spinal, a bulbar and a supra-bulbar component [54]. It is recorded from the palms and soles with surface electrodes and the most common type of stimulus is electrical shock delivered to a peripheral nerve. SSR potentials are characterized by latency, amplitude and area or they are reported as "present or absent" [53]. Abnormalities of SSR such as prolonged SSR latency and lower SSR amplitude in parkinsonians compared to normal subjects are reported in published literature [55, 56] and seem to be correlated with disease severity [55].

Gastrointestinal Motility and Manometry Studies

Examination of gastrointestinal motility identifies neurogenic disorders but it does not have the ability to distinguish extrinsic autonomic disorders from intrinsic enteric nervous system disorders. Neurochemical studies such as nor-epinephrine and epinephrine responses to edrophonium can detect sympathetic denervation, whereas parasympathetic denervation might be revealed by measuring plasma pancreatic polypeptide in response to hypoglycemia. Anal sphincter manometry evaluates somatic and not autonomic function. Gastrointestinal autonomic dysregulation is difficult to be diagnosed with the above tests, so tests such as Valsalva ratio and HRV during deep breathing, also sensitive for vagus gastrointestinal parasympathetic regulation, might be more helpful [13].

Evaluation of Genitourinary Autonomic Regulation

Urodynamic studies and cystometry evaluate the overall bladder function and not the component of autonomic regulation. Tests of male erectile function assess both the autonomic and the vascular components. The results of the tests above are indicative for autonomic dysfunction when supported by clinical symptoms and associated by evidence of dysautonomia in other systems. Therefore, in patients with urinary symptoms and a clinical suspicion of autonomic failure, a clinician should perform other tests of autonomic function and look for evidence of other non- genitourinary symptoms [13].

Treatment of Autonomic Symptoms

Autonomic impairment in PD, when diagnosed, should be managed in order to improve the everyday life of the patients. The treatment of OH might be conservative, for instance increase of salt intake, isotonic exercise and avoidance of sudden postural changes ; pharmacological treatment should be given after careful evaluation of the safety profile of the antihypotensive drugs [24, 57]. Constipation can be treated with small and frequent low fat and high fiber meals, adequate hydration, exercise, osmotic laxatives, possibly apomorphine and avoidance of anticholinergic drugs. Domperidone alleviates nausea and vomiting, whereas botulinum toxin might reduce saliva production [58]. Dopaminergic treatment might be helpful in swallowing by acting on the pharyngeal phases of ingestion [24, 57, 59]. Subcutaneous injections of apomorphine have been found to improve perineal muscle control [57] and DBS performed to improve motor performance may also have a positive effect on urinary symptoms as it will be discussed later in detail. Adding pergolide to the treatment and sildenafil citrate often helps erectile function [57]. In general, dopaminergic treatment can be beneficial only for those autonomic symptoms that are related to motor fluctuations [20].

III. The Effect of STN-DBS in PD on Dysautonomia

DBS and Cardiovascular Function

Reviewing the literature, one might observe that the published results on the influence of STN-DBS on autonomic cardiovascular regulation are controversial. There are clinical reports which concluded that DBS had no effect on cardiovascular ANS, comparing patients before and after surgery, or comparing "on" DBS and "off" DBS state [60-64]. Holmberg could not establish any therapeutic or protective effect on sympathetic or parasympathetic function by the reduction of pharmacotherapy or secondary to motor improvement achieved by STN stimulation [62]. The lack of DBS' influence on cardiovascular system could be explained, by the fact that STN-DBS probably acts on fibers and cell bodies close to the implanted electrode [4] and may selectively affect specific autonomic pathways [63], but not others. In agreement with this hypothesis, no change was found in postoperative OH as evaluated by item 42 of Unified Parkinson's Disease Rating Scale (UPDRS) compared to baseline scores [65]. In contrast, Stemper and his colleagues came up with a positive effect of STN-DBS on orthostatic BP regulation, while Kaufmann concluded that switching "on" DBS induced an increase in HR [66, 67]. Another study showed HR and BP increased during high frequency electrical stimulation of the STN [68]. This positive cardiovascular effect of STN stimulation may be due to the projections of the basal ganglia to centers of autonomic control such as hypothalamus, parabrachial nucleus, red nucleus, ventral tegmentum, locus coeruleus and the periaqueductal gray [67, 68]. Experimental stimulation of the mesencephalic locomotor region is known to increase HR in rats [69]. Thus, the DBS spreading electrical stimulation to the pedunculopontine nucleus, which is another projection of basal ganglia and part of the mesencephalic region, might influence cardiovascular drive [68]. Stimulation of STN is also thought to block any pathological neuronal STN hyperactivity [9, 70] and produces improvement in motor function but it also may, in that way, increase HR and BP [67]. Finally, tachycardia and increase of BP have been reported during intraoperative DBS in the postero-dorsal region of STN [71]. The applied mean voltage could possibly affect the median forebrain bundle and the Zona Incerta, which are autonomic structures close to STN. Alternatively, the source of the alleged DBS induced autonomic effects might be the limbic compartment within STN at the end of its medial extremity [72, 73]. One could also hypothesize that the position for the implantation of the electrode is not always the best target as regards to cardiovascular manifestations and thus slightly different locations of the STN electrodes might have different outcomes. Indeed, it has been pointed out that stimulation of the dorsal pole of STN is involved in autonomic control, whereas the ventral pole of STN and the substantia nigra pars reticulate are involved in limbic-related autonomic activity. This came out of a study in which stimulation of the dorsal STN pole under hidden and open conditions induced increase of HR, as well as of HF and LF bands in spectral analysis of HRV [74]. Finally, there is a clinical report, where elevation of BP and 0.05 HZ sinusoidal oscillations in BP occurred after STN stimulation in one patient, who turned out to have the two electrodes projecting to the posteromedial hypothalamus and the lateral hypothalamic area, confirming the important role of these areas in sympathetic activation [75]. In table 1 a review of the published literature referred to the effect of STN-DBS on cardiovascular autonomic regulation is presented.

Table 1. Literature review on long-term effect of STN-DBS on cardiovascular function

Clinical study	Material (N)	Results
Azevedo, 2010	20	HRV, BP variability: no change
Ludwig, 2007	14	HRV, BP variability (deep breathing, Tilt test): no change
Zibetti, 2010	36	OH(as reported by patients): no change
Erola, 2006	14	HRV: no change
Stemper, 2006	14	BP, SBF (Tilt test): improvement
Holmberg, 2005	11	HR, BP variability (deep breathing, Tilt test): no change
Sauleau, 2005	17	HR, BP: increase (intraoperative DBS)
Benedetti, 2004	10	HRV(spectral analysis): increase in HR, LF and HF
Kaufmann , 2002	3	HR: increase, BP: no change
Thornton, 2002	8	HR, BP: increase
Priori, 2001	9	BP, plasma rennin activity (Tilt test): no change

N: number of patients

DBS and Thermoregulatory Function

As long as sudomotor function is concerned, there are some reviews that show beneficial effect of STN-DBS upon it. As a symptom reported by patients in validated questionnaires for autonomic function during the pre- and postoperative period, a significant decrease in the frequency of sweating problems after the implantation has been shown [76, 77]. In a recent study, only 3 out of 11 patients had reported sweating symptoms, which improved in all 3 afterwards and yet one of them showed dramatic alleviation of bilateral sweating even with unilateral STN-DBS [76]. Witzas and colleagues attributed the excellent response of sweating to DBS (85% of patients experienced reduction of hyperhydrosis) to the decrease of motor implications induced by DBS, as the sweating symptoms were associated with the "off" state and dyskinesias [21, 77]. However, this result could have been related as well to the unusually high percent of sweating dysfunction (87, 5%) seen in their patients before surgery. Along the same line, a previous study demonstrated prolonged SSR latency and lower SSR amplitude when the stimulator was turned off [64] and a case report presented cessation of drenching sweats in one patient after STN-DBS surgery, clearly connecting this effect with the stimulation [78]. SSR is an index of mental (emotional) sweating, which is known to be regulated by the limbic system, premotor cortex, basal ganglia and reticular formation [79]. Since these structures are functionally linked to STN [80], its stimulation might affect the activity of them too and thus resulting in normalization of SSR. According to Shangera [78], several mechanisms, could explain the alleviation of sweating by DBS: a. stimulation spreading directly to the hypothalamus, which is the central drive of thermoregulation in humans [26, 64], b. enhanced activity of the limbic region within the STN [74], c. direct activation of autonomic sympathetic fibers located in the dorsal STN in the caudal ventral thalamus. Finally, stimulation spreading to the dopamine neurons in the Zona Incerta might be another justification for the sweating reduction, since excessive sweating is related to low levels of L-dopa [25]. A different aspect has been presented in a clinical study of non motor effects during intraoperative STN stimulation: excessive sweating in the whole body or restricted to one half of the body, of the face and neck or trunk occurred when the dorsal region of STN was stimulated [71] . Profound sweating in face, neck and upper part of chest was also experienced by a patient after STN stimulation, only to be found that the electrodes in this case projected to areas of hypothalamus outside the STN, which, as mentioned above, is the centre of sympathetic sweating regulation [75].

DBS and Urinary Function

There are a few clinical reports on the effect of STN-DBS on bladder complaints and on the urodynamic parameters, which do not clear out the influence of STN stimulation on the urinary function. In an urodynamic evaluation STN-DBS increased significantly bladder capacity and reflex volume, which were attributed to decreased detrusor hyperreflexia after surgery [81]. Likewise, Seif et al. showed a significant increase of the initial desire to void and of the maximal bladder capacity. The authors of this study explained their results either by a direct inhibitory effect of STN stimulation on the micturation reflex pathway at the brainstem level leading to a decrease of detrusor overactivity or by indirect activation of thalamocortical projections such as cingular and prefrontal cortex [82]. These findings are in contrast with the results of a more recent study, where no change was found on the urodynamic parameters 6 months after DBS surgery, although a significant reduction of the irritative symptoms of detrusor hyperreflexia was reported [83]. This study group justified the lack of change in urodynamic measurements with the claim that bladder control may need more than 6 months from the initiation of stimulation to improve just like bladder dysfunction that begins 3 years after the onset of PD. Likewise, Zibetti and his colleagues evaluating the urological dysfunction before and after surgery by collecting data from the patients çlinical charts, reported no alteration in pre- and post DBS scores [65]. There is another clinical report, where the cortical control of urinary bladder in parkinsonian patients with STN-DBS was assessed. Enhanced activity was showed in the anterior cingulated cortex during "on" stimulation state and in the lateral frontal cortex during "off" stimulation state. Therefore the STN-DBS ameliorated bladder function adjusting the cerebral bladder control [84]. The same study group, in a PET study of the effect of STN stimulation on the sensory gaiting of urinary bladder afferents confirmed the increase in bladder capacity after surgery and also explained how STN-DBS modulates the sensory processing of urinary bladder information in the level of visceral afferents [85]. Finally, in an evaluation of patients with the Non-Motor symptom questionnaire before and 1 year after DBS, urinary urgency and nocturia were among the symptoms that at least 5 fewer patients reported them after surgery [10]. Moreover, urinary symptomatology, as reported in an autonomic questionnaire by a small sample of patients, was improved due to DBS [76].

DBS and Sexual Function

As mentioned above, sexual dysfunction in PD is multifactor and has not been studied in detail. To our knowledge there are a few reports evaluating the effect of DBS on sexual impairment in PD. Tyvaert L and his colleagues in a multi-questionnaire evaluation of the DBS influence reported a 47% improvement of sexual dysfunction [86]. Another study yet showed a small effect of STN-DBS on sexual functioning. In particular, there was only a significant improvement on the dissatisfaction subscale referred to male patients under 60 years old alone who showed a significant reduction in the Sexual Functioning Inventory (an adjusted by authors questionnaire) [87].

DBS and Gastrointestinal Function

Little is published about the effect of STN-DBS on the gastrointestinal function. Most data are derived from questionnaires filled in by patients pre and post DBS surgery. Zibetti and his colleagues reported amelioration of swallowing, decrease of salivation, and significant reduction in the frequency of constipation after surgery. This effect is possibly justified by the DBS improvement of bradykinesia at the pharyngeal level, the postoperative decrease in dopaminergic therapy and the reduction of motor fluctuations [65]. The results were similar in two very recent clinical studies in which a moderate postoperative improvement in bowel function was reported by the patients [10, 76].

IV. The Result of DBS Effect on Autonomic Function in the Series of Our Patients

In our study 24 consecutive patients with PD, candidates for STN-DBS surgery were evaluated under two treatment conditions: 3 days before surgery, while receiving their standard antiparkinsonian medication and 6 months after the implantation, "on DBS/ on medication" state. The patients were asked to complete a semi-structural questionnaire based on the NMS questionnaire [20], which included a more detailed part about sweating symptoms previously used [33]. It referred to autonomic symptoms in the past months, such as constipation, urgency, orthostatic dizziness, and sweating disturbances. In each patient BP was measured at supine position, and after 1 and 3 min of standing using an aneroid sphygmomanometer with the cuff placed on the upper non-tremor dominant arm [88].

The neurophysiological examination included I. SSR recordings from both palms and the less affected sole II. RR interval variation with respiration, at rest and during deep breathing III. HR response to Valsalva maneuver IV. HR response to tilt test V. off-line spectral analysis of HRV during 5 minutes of normal breathing at rest. The same neurophysiological examination was performed in 24 healthy controls matched for age and sex.

The most common autonomic symptom was, both before and after surgery constipation with a frequency of 50% before, which reduced to 41.7% afterwards (χ^2, $p = 0.008$). Another common autonomic disturbance in our patients was sweating dysfunction, the prevalence of which decreased by 50 % postoperative (χ^2, $p = 0.005$). The number of patients with orthostatic dizziness did not change significantly after DBS implantation, whereas urgency decreased by 39.90%. Most of our patients reported hyperhydrosis of the face and neck pre and post DBS.

Comparing systolic and diastolic BP measurements from the patients before and after surgery at rest, at the first and at the third minute after sudden change from supine to standing position, no significant difference was found. According to the definition of OH mentioned in the Cardiovascular Dysfunction section, 11 patients (45.8%) had OH preoperative and only 3 out of 24 after the operation.

SSR measurements in patients showed no significant change between pre and post DBS study. SSR's amplitude and area from palms and soles were, in patients pre- and post-operative, significant lower than those of healthy subjects ($p<0.050$). Six out of the 24 patients had abnormal or absent SSR before surgery (25%), whereas 7 after surgery (37.5%) (χ^2, $p=0.114$). Either pre- or post-operatively no correlation between abnormal/ absent SSR and presence of sweating dysfunction was found (χ^2, $p=$ 0.524 and 1.000 respectively). Furthermore, no correlation was found between the difference in SSR measurements and the reduction of motor UPDRS, whereas a correlation existed between the LEDD reduction after DBS and the difference in the measurements of the SSR's latency from the right palm between pre and post DBS ($r=-0.600$, $p=0.005$). The DBS related improvement of SSR reported in our preliminary study [88] was not confirmed in a larger sample size (unpublished data). Thus, we have no evidence to support that STN-DBS has a beneficial effect on SSR parameters.

As far as the effect of subthalamic stimulation on HR is concerned, anecdotal experience from our patients so far, indicates that this neurosurgical treatment for motor improvement in PD does not particularly affect the balance of sympathetic and parasympathetic cardiovascular function.

V. Future Applications of STN-DBS on Autonomic Function in PD

Since autonomic effects have been reported intraoperative [71], it might be useful to monitor in the future the autonomic effects during the operation, so that , not only motor symptoms but autonomic symptoms as well, could play a role for the definition the optimal location of STN electrodes. Having in mind that SPECT imaging of the regional cerebral blood flow is a very useful tool to investigate the neural pathways [89], future prospective systematic observations and measurements with SPECT, PET and other imaging techniques in patients undergoing the implantation may help to identify and better understand the autonomic circuitries affected and the neurophysiological changes elicited by STN-DBS in the living human brain.

References

[1] Pazo, JH; Belforte, JE. Basal ganglia and functions of the autonomic nervous system. *Cell Mol Neurobiol,* 2002, 22, 645-54.

[2] Martinez-Martin, P; Schapira, AHV; Stocchi, F; Sethi, K; Odin, P; MacPhee, G; Brown, RG; Naidu, Y; Clayton, L; Abe, K; Tsuboi, Y; MacMahon, D; Barone, P; Rabey, M; Bonuccelli, U; Forbes, A; Breen, K; Tluk, S; Olanow, CW; Thomas, S; Rye, D; Hand, A; Williams, AJ; Ondo, W; Chaudhuri, KR. Prevalence of nonmotor symptoms in Parkinson's disease in an international setting; study using nonmotor symptoms questionnaire in 545 patients. *Mov Disord,* 2007, 22, 1623-9.

[3] Limousin, P; Martinez-Torres I. Deep brain stimulation for Parkinson's disease. *Neurotherapeutics*, 2008, 5, 309-18.

[4] Volkman, J. Deep brain stimulation for the treatment of Parkinson's disease. *J Clin Neurophysiol,* 2004, 21, 6-17.

[5] The Deep-Brain Stimulation for Parkinson's Disease Study Group. Deep-Brain Stimulation of the Subthalamic Nucleus or the Pars Interna of the Globus Pallidus in Parkinson's Disease. *N Engl J Med,* 2001, 345, 956-63.

[6] Rodriguez-Oroz, MC; Obeso, JA; Lang, AE; Houeto, JL; Pollak, P; Rehncrona, S; et al. Bilateral deep brain stimulation in Parkinson's disease: a multicentre study with 4 years follow-up. *Brain,* 2005, 128, 2240-9.

[7] Drapier, S; Raoul, S; Drapier, D; Leray, E; Lallement, F; Rivier, I; Sauleau, P; Lajat, Y; Edan, G; Vérin, M. Only physical aspects of quality of life are significantly improved by bilateral subthalamic stimulation in Parkinson's disease. *J Neurol,* 2005, 252, 583-8.

[8] Kumar, R; Lozano, AM; Kim, YJ; Hutchison, WD; Sime, E; Halket, E; Lang, AE. Double-blind evaluation of subthalamic nucleus deep brain stimulation in advanced Parkinson's disease. *Neurology,* 1998, 51, 850-5.

[9] Limousin, P; Krack, P; Pollak, P; Benazzouz, A; Ardouin, C; Hoffmann, D; Benabid, AL. Electrical stimulation of the subthalamic nucleus in advanced Parkinson's disease. *N Engl J Med*, 1998, 339, 1105-11.

[10] Nazzaro, JM; Pahwa, R; Lyons, KE. The impact of bilateral subthalamic stimulation on non-motor symptoms of Parkinson's disease. *Parkinsonism Relat Disord,* 2011, doi:10.1016/j.parkreldis.2011.05.009

[11] Oka, H; Mochio, S; Onouchi, K; Morita, M; Yoshioka, M; Inoue, K. Cardiovascular dysautonomia in de novo Parkinson's disease. *J Neurol Sci,* 2006, 241, 59-65.

[12] Goldstein, DS. Dysautonomia in Parkinson's disease: neurocardiological abnormalities. *Lancet Neurol,* 2003, 2, 669-76.

[13] Ravits, JM. Aaem minimonograph #48: Autonomic nervous system testing. *Muscle Nerve,* 1997, 20, 919-37.

[14] Bouhaddi, M; Vuillier, F; Fortrat, JO; Cappelle, S; Henriet, MT; Rumbach, L; Regnard, J. Impaired cardiovascular autonomic control in newly and long-term-treated patients with Parkinson's disease: involvement of L-dopa therapy. *Auton Neurosci,* 2004, 116, 30-8.

[15] Kallio, M; Haapaniemi, T; Turkka, J; Suominenm K; Tolonen, U; Sotaniemi, K; Heikkilä, VP; Myllylä, V. Heart rate variability in patients with untreated Parkinson's disease. *Eur J Neurol,* 2000, 7, 667-72.

[16] Goldstein, DS; Holmes, C; Dendi, R; Bruce, SR; Li, ST. Orthostatic hypotension from sympathetic denervation in Parkinson's disease. *Neurology,* 2002, 58, 1247-55.
[17] Singleton, A; Gwinn-Hardy, K; Sharabi, Y; Li, ST; Holmes, C; Dendi, R; Hardy, J; Singleton, A; Crawley, A; Goldstein, DS. Association between cardiac denervation and parkinsonism caused by alpha-synuclein gene triplication. *Brain,* 2004, 127, 768-72.
[18] Dickson, DW; Fujishiro, H; Orr, C; DelleDonne, A; Josephs, KA; Frigerio, R; Burnett, M; Parisi, JE; Klos, KJ; Ahlskog, JE. Neuropathology of non-motor features of Parkinson disease. *Parkinsonism Relat Disord,* 2009, 15, S1-5.
[19] Cheon, SM; Park, MJ; Kim, WJ; Kim, JW. Non-motor off symptoms in Parkinson's disease. *J Korean Med Sci,* 2009, 24, 311-4.
[20] Chaudhuri, KR; Martinez-Martin, P; Schapira, AHV; Stocchi, F; Sethi, K; Odin, P; et al. International multicenter pilot study of the first comprehensive self-completed nonmotor symptoms questionnaire for Parkinson's disease: The NMSQuest study. *Mov Disord,* 2006, 21, 916-23.
[21] Swinn, L; Schrag, A; Vinswanathan, R; Bloem, BR; Lees, A; Quinn, N. Sweating dysfunction in Parkinson's disease. *Mov Disord,* 2003, 18, 1459-63.
[22] Schestatsky, P; Valls-Sole, J; Ehlers, JA; Rieder, CR; Gomes, I. Hyperhidrosis in Parkinson's disease. *Mov Disord,* 2006, 21, 1744-8.
[23] Pfeiffer, RF. Non-motor parkinsonism. *Parkinsonism Relat Disord,* 2007, 13, S211-20.
[24] Ziemssen, T; Reichmann, H. Non-motor dysfunction in Parkinson's disease *Parkinsonism Relat Disord,* 2007, 13, 323-32.
[25] Park, A; Stacy, M. Non-motor symptoms in Parkinson's disease. *J Neurol,* 2009, 256, S293-8.
[26] Pursiainen, V; Haapaniemi, T; Korpelainen, JT; Sotaniemi, KA; Myllyla, VV. Sweating in Parkinsonian patients with wearing-off. *Mov Disord*, 2007, 22, 828-32.
[27] Sakakibara, R; Shinotoh, H; Uchiyama, T; Yoshiyama, M; Hattori, T; Yamanishi, T. SPECT imaging of the dopamine transporter with [^{123}I]-beta-CIT reveals marked decline of nigrostriatal dopaminergic function in Parkinson's disease with urinary dysfunction. *J Neurol Sci*, 2001, 187, 55-9.
[28] Pfeifer, RF. Gastrointestinal, urological, and sexual dysfunction in Parkinson's disease. *Mov Disord*, 2010, 25, S94-7.
[29] Meco, G; Rubino, A; Caravonam N; Valente, M. Sexual dysfunction in Parkinson's disease. *Parkinsonism Relat Disord*, 2008, 14, 451-6.
[30] Hobson, P; Islam,W; Roberts, S; Adhiyman, V; Meara, J. The risk of bladder and autonomic dysfunction in a community cohort of Parkinson's disease patients and normal controls. *Parkinsonism Relat Disord,* 2003, 10, 67-71.
[31] Sakakibara, R; Uchiyama, T; Yamanishi, T; Kishi, M. Genitourinary dysfunction in Parkinson's disease. *Mov Disord*, 2010, 15, 2-12.
[32] Pfeifer, RF. Gastrointestinal dysfunction in Parkinson's disease *Park Related Disord* , 2011, 17, 10-5.
[33] Siddiqui, MF; Rast, S; Lynn, MJ; Auchus, AP; Pfeiffer, RF. Autonomic dysfunction in Parkinson's disease: a comprehensive symptom survey. *Parkinsonism Relat Disord,* 2002, 8, 277-84.
[34] Parkinson J. *An Essay on shaking palsy.* London: Sherwood, Neely and Jones; 1817.

[35] Braak, H; Del Tredici, K; Rüb, U; de Vos, RA; Jansen Steur, EN; Braak, E. Staging of brain pathology related to sporadic Parkinson's disease. *Neurobiol Aging* , 2003, 24, 197-211.

[36] Weimer, LH. Autonomic testing: common techniques and clinical applications. *Neurologist*, 2010, 16, 215-22.

[37] Visser, M; Marinus, J; Stiggelbout, AM; Van Hilten, JJ. Assessment of autonomic dysfunction in Parkinson's disease: the SCOPA-AUT. *Mov Disord*, 2004, 19, 1306-12.

[38] Toth, V; Hejjel, L; Fogarasi, A; Gyimesi, C; Orsi, G; Szucs, A; Kovacs, N; Komoly, S; Ebner, A; Janszky, J. Periictal heart rate variability analysis suggests long-term postictal autonomic disturbance in epilepsy. *Eur J Neurol*, 2010, 17, 780–7.

[39] Stålberg, EV; Nogues, M. Automatic analysis of heart rate variation: I. Method and reference values in healthy controls. *Muscle Nerve,* 1989, 12, 993-1000.

[40] Hilz, MJ; Dutsch, M. Quantitative studies of autonomic function. *Muscle Nerve*, 2006, 33, 6-20.

[41] Freeman, R. Assessment of cardiovascular autonomic function. *Clin Neurophysiol*, 2006, 117, 716-30.

[42] Akselrod, S; Gordon, D; Ubel, FA; Shannon, DC; Berger, AC; Cohen, RJ. Power spectrum analysis of heart rate fluctuation: a quantitative probe of beat-to-beat cardiovascular control. *Science*, 1981, 213, 220-2.

[43] Bilchick, KC; Berger, RD. Heart rate variability. *J Cardiovasc Electrophysiol,* 2006, 17, 691-4.

[44] Task Force of the European Society Cardiology and the North American Society of Pacing and Electrophysiology. Heart rate variability: standards of measurement, physiological interpretation and clinical use. *Circulation*, 1996, 93, 1043-65.

[45] Tomson, T; Ericson, M; Ihrman, C; Lindblad, LE Heart rate variability in patients with Epilepsy. *Epilepsy Res,* 1998, 30, 77-83.

[46] Mense, L; Reimann, M; Rüdiger, H; Gahn, G; Reichmann, H; Hentschel, H; Ziemssen, T. Autonomic function and cerebral autoregulation in patients undergoing carotid endarterectomy. *Circulation,* 2010, 74, 2139-45.

[47] Pagani, M; Montano, N; Porta, A; Malliani, A; Abboud, MF; Birkett, C; Somers, KV. Relationship between spectral components of cardiovascular variabilities and direct measures of muscle sympathetic nerve activity in humans. *Circulation,* 1997, 95, 1441-8.

[48] Sztajzcl, J. Heart rate variability: a noninvasive electrocardiographic method to measure the autonomic nervous system. *Swiss Med Wkly*, 2004, 134, 514-22.

[49] Badilini, F; Maison-Blanche, P; Coumel, P. Heart rate variability in passive tilt test: comparative evaluation of autoregressive and FFT spectral analyses. *Pacing Clin Electrophysiol,* 1998, 21, 1122-32.

[50] Low, PA; Denq, J-C; Opfer-Gehrking, TL; Dyck, PJ; O'Brien, PC; Slezak, JM. Effect of age and gender on sudomotor and cardiovagal function and blood pressure response to tilt in normal subjects. *Muscle Nerve*, 1997, 20, 1561-8.

[51] Schüller, TB; Hermann, K; Baron, R. Quantitative assessment and correlation of sympathetic, parasympathetic, and afferent small fiber function in peripheral neuropathy. *J Neurol*, 2000, 247, 267-72.

[52] Low, PA. Laboratory evaluation of autonomic failure. In: Low PA, editor. Clinical Autonomic Disorders, Evaluation and Management. Boston: Little, Brown and Company; 1993; 169-195.

[53] Illigens, BMW; Gibbons, CH. Sweat testing to evaluate autonomic function. *Clin Auton Res,* 2008, 19, 79-87.

[54] Vetrugno, R; Liguori, R; Cortelli, P; Montagna, P. Sympathetic skin response: basic mechanisms and clinical applications. *Clin Auton Res*, 2003,13, 256-70.

[55] Haapaniemi, TH; Korpelaien, J; Tolonen, U; Suominen, K; Sotaniemi, KA; Myllylä, VV. Suppressed sympathetic skin response in Parkinson disease. *Clin Auton Res,* 2000, 10, 337-42.

[56] Schestatsky, P; Ehlers, JA; Rieder, CR; Gomes, I. Evaluation of sympathetic skin response in Parkinson's disease. *Parkinsonism Relat Disord,* 2006, 12, 486-91.

[57] Dubow, JS. Autonomic dysfunction in Parkinson's disease. *Dis Mon*, 2007, 53, 265-74.

[58] Truong, DD; Jost, W. Botulinum toxin: clinical use. Parkinsonism. *Relat Disord*, 2006, 12, 331-55.

[59] Jost, WH. Gastrointestinal dysfunction in Parkinson's disease. *J Neurol Sci*, 2010, 289, 69-73.

[60] Azevedo, E; Santos, R; Freitas, J. Deep brain stimulation does not change neurovascular coupling in non-motor visual cortex: An autonomic and visual evoked blood flow velocity response study. *Parkinsonism Relat Disord.* 2010, 16, 600-3.

[61] Erola, T; Haapaniemi, T; Heikkinen, E; Huikuri, H; Myllyä, V. Subthalamic nucleus deep brain stimulation does not alter long-term heart rate variability in Parkinson's disease. *Clin Auton Res,* 2006, 16, 286-8.

[62] Holmberg, B; Corneliusson, O; Elam, M. Bilateral stimulation of nucleus subthalamicus in advanced Parkinson's disease: No effects on, and of, autonomic dysfunction. *Mov Disord,* 2005, 20, 976-81.

[63] Ludwig, J; Remien, P; Guballa, C; Binder, A; Binder, S; Schattschneider, J; et al. Effects of subthalamic nucleus stimulation and levodopa on the autonomic nervous system in Parkinson's disease. *J Neurol Neurosurg Psychiatry*, 2007, 78, 742-5.

[64] Priori, A; Cinnante, C; Genitrini, S; Pesenti, A; Tortora, G; Bencini, C; et al. Non-motor effects of deep brain stimulation of the subthalamic nucleus in Parkinson's disease: preliminary physiological results. *Neurol Sci* , 2001, 22, 85-6.

[65] Zibetti, M; Torre, E; Cinquepalmi, A; Rosso, M; Ducati, A; Bergamasco, B; Lanotte, M; Lopiano, L. Motor and nonmotor symptom follow-up in parkinsonian patients after deep brain stimulation of the subthalamic nucleus. *Eur Neurol*, 2007, 58, 218-23.

[66] Kaufmann, H; Bhattacharaya, K; Voustianiouk, A; Gracies, JM. Stimulation of the subthalamic nucleus increases heart rate in patients with Parkinson disease. *Neurology,* 2002, 59, 1657-8.

[67] Stemper, B; Beric, A; Welsch, G; Haendl, T; Sterio, D; Hilz, MJ. Deep brain stimulation improves orthostatic regulation of patients with Parkinson disease. *Neurology,* 2006, 67, 1781-5.

[68] Thornton, JM; Aziz, T; Schlugman, D; Paterson, DJ. Electrical stimulation of the midbrain increases heart rate and arterial blood pressure in awake humans. *J Physiol,* 2002, 539, 615-21.

[69] Chong, RK; Bedford, T. Heart rate, blood pressure, and running speed responses to mesencephalic locomotor region stimulation in anesthetized rats. *Pflugers Arch* , 1997, 434, 280-4.
[70] Lozano, AM; Dostrovsky, J; Chen, R; Ashby, P. Deep brain stimulation for Parkinson's disease: disrupting the disruption. *Lancet Neurol*, 2002, 1, 225-31.
[71] Sauleau, P; Raoul, S; Lallement, F; Rivier, I; Drapier, S; Lajat, Y; Verin, M. Motor and non motor effects during intraoperative subthalamic stimulation for Parkinson's disease. *J Neurol*, 2005, 252, 457-64.
[72] Groenewegen, HJ; Berendse HW. Connections of the subthalamic nucleus with ventral striatopallidal parts of the basal ganglia in the rat. *J Comp Neurol*, 1990, 294, 607-22.
[73] Parent, A; Hazrati, L. Functional anatomy of the basal ganglia. II. The place of subthalamic nucleus and external pallidum in basal ganglia circuitry. *Brain Res Rev,* 1995, 20, 128-54.
[74] Benedetti, F; Colloca, L; Lanotte, M; Bergamasco, B; Torre, E; Lopiano, L. Autonomic and emotional responses to open and hidden stimulations of the human subthalamic region. *Brain Res Bull,* 2004, 63, 203-11.
[75] Lipp, A; Tank, J; Trottenberg, T; Kupsch, A; Arnold, G; Jordan, J. Sympathetic activation due to deep brain stimulation in the region of the STN. *Neurology,* 2005, 65, 774-5.
[76] Halim, A; Baumgartner, L; Binder, DK. Effect of deep brain stimulation on autonomic dysfunction in patients with Parkinson's disease. *J Clin Neurosci*, 2011, 18, 804-6.
[77] Witjas, T; Kaphan, E; Régis, J; Jouve, E; Chérif, AA; Péragut, J-C; et al. Effects of chronic subthalamic stimulation on nonmotor fluctuations in Parkinson's disease. *Mov Dis,* 2007, 22, 1729-34.
[78] Sanghera, MK; Ward, C; Stewart, RM; Mewes, K; Simpson, RK; Lai, EC. Alleviation of drenching sweats following subthalamic deep brain stimulation in a patient with Parkinson's disease-a case report. *J Neurol Sci*, 2009, 285, 246-9.
[79] Kikkawa, Y; Asahina, M; Suzuki, A; Hattori, T. Cutaneous sympathetic function and cardiovascular function in patients with progressive supranuclear palsy and Parkinson's disease. *Parkinsonism Relat Disord,* 2003, 10, 101-6.
[80] Karimi, M; Golchin, N; Tabbal, SD; Hershey, T; Videen, TO; Wu, J; et al. Subthalamic nucleus stimulation-induced regional blood flow responses correlate with improvement of motor signs in Parkinson disease. *Brain*, 2008, 131, 2710-19.
[81] Finazzi-Agrò, E; Peppe, A; D'Amico, A; Petta, F; Mazzone, P; Stanzione, P; Micali, F; Caltagirone, C. Effects of subthalamic nucleus stimulation on urodynamic findings in patients with Parkinson's disease. *J Urol,* 2003, 169, 1388-91.
[82] Seif, C; Hergoz, J; van der Horst, C; Schrader, B; Volkmann, J; Deuschl, G; Juenemann, KP; Braun, PM. Effect of subthalamic deep brain stimulation on the function of the urinary bladder. *Ann Neurol*, 2004, 55, 118-20.
[83] Winge, K; Nielsen, KK; Stimpel, H; Lokkegaard, A; Jensen, SR; Werdelin, L. Lower urinary tract symptoms and bladder control in advanced Parkinson's disease: effects of deep brain stimulation in the subthalamic nucleus. *Mov Disord*, 2007, 22, 220-5.
[84] Herzog, J; Weiss, PH; Assmus, A; Wefer, B; Seif, C; Braun, PM; Herzog, H; Volkmann, J; Deuschl, G; Fink, GR. Subthalamic stimulation modulates cortical control of urinary bladder in Parkinson's disease. *Brain,* 2006, 129, 3366-75.

[85] Herzog, J; Weiss, PH; Assmus, A; Wefer, B; Seif, C; Braun, PM; Pinsker, MO; Herzog, H; Volkmann, J; Deuschl, G; Fink, GR. Improved sensory gating of urinary bladder afferents in Parkinson's disease following subthalamic stimulation. *Brain,* 2008, 131, 132-45.

[86] Tyvaert, L; Devos, D; Rigot, JM; Blond, S; Reyns, N; Destee, A. Sexual dysfunction and sub-thalamic stimulation in Parkinson's disease. *Park Relat Disord,* 2005,11, 116(PT003-17).

[87] Castelli, L; Perozzo, P; Genesia, ML; Torre, E; Pesare, M; Cinquepalmi, A; Lanotte, M; Bergamasco, B; Lopiano, L. Sexual well being in parkinsonian patients after deep brain stimulation of the subthalamic nucleus. *J Neurol Neurosurg Psychiatry*, 2004, 75, 1260-4.

[88] Trachani, E; Constantoyannis, C; Sirrou, V; Kefalopoulou, Z; Markaki, E; Chroni, E. Effects of subthalamic nucleus deep brain stimulation on sweating function in Parkinson's disease. *Clin Neurol Neurosurg*, 2009,112, 213-7.

[89] Sestini, S; Scotto di Luzio, A; Ammannati, F; De Cristofaro, MT; Passeri, A; Martini, S; Pupi, A. Changes in regional cerebral blood flow caused by deep-brain stimulation of the subthalamic nucleus in Parkinson's disease. *J Nucl Med,* 2002, 43, 725-32.

Index

A

B

F

G

H

I

J

K

L

M

Q

R

S

T

U